# Energy Systems Modeling
# and Policy Analysis

# Energy Systems Modeling and Policy Analysis

B K Bala

Fareast International University, Banani, Dhaka, Bangladesh

**CRC Press**
Taylor & Francis Group
Boca Raton London New York

CRC Press is an imprint of the
Taylor & Francis Group, an **informa** business

First edition published 2022
by CRC Press
6000 Broken Sound Parkway NW, Suite 300, Boca Raton, FL 33487-2742

and by CRC Press
2 Park Square, Milton Park, Abingdon, Oxon, OX14 4RN

CRC Press is an imprint of Taylor & Francis Group, LLC

ISBN: 9781032110998 (hbk)
ISBN: 9781032111018 (pbk)
ISBN: 9781003218401 (ebk)

DOI: 10.1201/9781003218401

Typeset in Times
by Newgen Publishing UK

# Contents

# Foreword

Energy engineering has changed from being largely about centralized, usually fossil-fuel-powered supplies with some attention to energy use, to encompassing a wide range of decentralized renewable energy-driven energy sources. This has meant that previous distinctions between the supply-side and the demand-side are no longer valid. It has also led to new market models and new considerations for policymakers. While these developments have enabled the decarburization of electricity production, they also present challenges in energy systems optimization and simulation.

*Energy Systems Modeling and Policy Analysis* covers the interdisciplinary breadth of energy modeling and policy analysis from a systems perspective. It considers systems at various scales, showing the interconnections between optimal design via simulation and policy analysis using system dynamics and linear programming. Importantly, this text also includes a full discussion of communication and automation technologies essential for the effective operation of smart grids.

As most of the world is rural, this book serves to remedy the fact that rural energy contexts and requirements are often less considered. A full approach to the design of integrated rural energy systems is provided.

All chapters of this book are set in the context of the need to achieve radically reduced environmental impact from energy production and use.

Drawing from Professor Bala's extensive teaching and research experience, *Energy Systems Modeling and Policy Analysis* is an important text for undergraduate and graduate students across a wide range of engineering and associated disciplines. Bridging theory and practice, it will also inform policymakers, system planners and industry professionals.

**Professor Brian Norton DSc MRIA**
*International Energy Research Centre, Ireland*
*Head of Energy Research, Tyndall National Institute, Ireland*
*Professor of Solar Energy Applications, Technological University Dublin, Ireland*
*Research Professor, University College Cork, Ireland*

# Preface

*Energy Systems Modeling and Policy Analysis* has been written primarily for undergraduate and graduate students studying energy engineering, electrical engineering, mechanical engineering, environmental engineering and systems engineering. It is the outcome of several years of teaching and research work carried out by the author.

This book covers a very wide spectrum of energy modeling and policy analysis using a systems approach. The unique feature of this book is that it considers the modeling of energy and the environment with micro- and macro-level applications using simulation and optimization techniques in a single book. Chapter 1 deals with the importance of energy and policy analysis modeling, and Chapters 2 and 3 deal with the fundamentals of modeling and policy analysis using system dynamics and linear programming. Chapter 4 deals with the fundamentals of communication technology, which is essential for communication in SCADA systems and automation in smart grids. Chapters 5 and 6 deal with energy supply, energy demand, the environmental impact of energy production, and the modeling of energy supply, energy demand, CO2 emission, and acid rainfall. Chapter 7 deals with integrated energy systems at micro and macro levels. Chapters 8 and 9 are primarily devoted to the application of simulation techniques for integrated rural energy systems and integrated electric power systems, respectively. Chapter 10 is devoted to the optimization techniques used for the operational planning of electric power systems, SCADA systems and smart grids using linear programming and genetic algorithms. A good number of numerical examples on the simulation and optimization of energy systems have been solved to help understand simulation and optimization in theory and practice. An extensive bibliography will help readers find information on various topics of interest. Practicing energy engineers and energy planners will find this book an excellent reference.

I sincerely express my grateful acknowledgment to Professor Brian Norton, International Energy Research Centre, Ireland, for writing the foreword of this book. I warmly recognize the continuing debt to my teacher Professor Donald R Drew, who introduced me to modeling at the Asian Institute of Technology, Bangkok, Thailand, in 1975, and to my teacher Professor A M Z Huq, at the Bangladesh University of Engineering and Technology, Dhaka, Bangladesh, who inspired me to pursue the field of energy modeling after the publication of his paper on the modeling of integrated energy systems in 1975. I acknowledge the encouragement and assistance that I received from Vice Chancellor Khondker Nasiruddin while at the Bangabandhu Sheikh Mujibur Rahman Science and Technology University, Gopalganj, Bangladesh. I gratefully acknowledge the support I received from Professor Serm Janjai at the Solar Energy Research Laboratory, Siplakorn University, Nakhon Pathom, Thailand. Finally, I owe thanks to Dr. Gholum Kibtia Bhuyan, Bangladesh Rice Research Institute, Joydebpur, Bangladesh, Suman Mia,

Bangladesh Agricultural Research Institute, Joydebpur, Bangladesh and Santo Roy, Master of Science student, Hajee Mohammad Danesh Science and Technology University, Dinajpur, Bangladesh, for their assistance in the graphics and beautiful figure drawings.

**B K Bala**

# Author Biography

**Professor B K Bala** is currently a professor in the Department of Electrical and Electronic Engineering, Fareast International University, Banani, Dhaka 1213, Bangladesh. His previous appointment was as a visiting professor of Modeling and Simulation at the Solar Energy Research Laboratory, Silpakorn University, Nakhon Pathom, Thailand, since 2005. He was Chairman of the Department of Electrical and Electronic Engineering and Dean of the Faculty of Engineering at the Bangabandhu Sheikh Mujibur Rahman Science and Technology University, Gopalganj, Bangladesh, from 2018 to 2020; a professor at the Jessore University of Science and Technology from 2016 to 2018; and a professor in the Department of Farm Power and Machinery at Bangladesh Agricultural University, Mymensingh, Bangladesh, where he was engaged in teaching and research for over 42 years, starting from 1970. He has supervised more than a dozen PhD students in the areas of modeling of integrated energy systems and energy policy, renewable energy, modeling and simulation of solar drying, modeling of climate change impacts, and emerging technologies such as neural networks, exergy, genetic algorithms and computational fluid dynamics.

Professor Bala received a BSc (Eng) degree in Electrical Engineering from the Bangladesh University of Engineering and Technology, Dhaka, Bangladesh, in 1969, and an MEng degree from the Asian Institute of Technology, Bangkok, Thailand, in 1975, where he worked on control theory. He also received a PhD from the University of Newcastle upon Tyne, UK, in 1983, where he worked on the modeling and simulation of the industrial deep-bed drying of malt and provided the foundation and basis for energy conservation and the online control of industrial deep-bed drying of malt. He also worked on the modeling and simulation of solar drying as an EC postdoctoral fellow at the University of Newcastle upon Tyne, UK, from 1992 to 1993. Professor Bala was on a DAAD study visit at the Institute of Agricultural Engineering in the Tropics and Subtropics, University of Hohenheim, Germany, for research on the modeling and simulation of solar drying in 1995. He also served as a research fellow in system dynamics modeling and policy analysis at the Institute of Agricultural and Food Policy Studies, Universiti Putra Malaysia, from 2012 to 2014.

Professor Bala has published 207 scientific papers, mostly in high-impact factor journals, and is the author of 10 textbooks, including *Energy and Environment: Modelling and Simulation* with Nova Science Publishers, USA, in 1998; *Electrical Energy: Transmission and Distribution* (Second Edition) with

Khanna Publishers, India, in 2002; *Renewable Energy* with Agrotech Publishing Academy, India, in 2003; *Drying and Storage of Cereal Grains* with Wiley-Blackwell in 2017; *System Dynamics: Modelling and Simulation* (as the first author) with Springer in 2017; and *Agro-Product Processing Technology: Principles and Practice* with CRC Press Taylor & Francis in 2020.

# 1 Introduction

## 1.1 INTRODUCTION

Energy is needed to meet the subsistence requirement as well as to meet the demand for economic growth and development (Bala, 1997a, 1997b, 1998). Global economic growth for the period 2002–2030 is estimated at 3.2% per year, with China, India and Asian countries expected to lead the peak, and the population worldwide is put at more than 8 billion in 2030, up from 6.2 billion in 2002. To meet these requirements, the International Energy Agency predicts an energy demand rise of 59% between now and 2030. Figure 1.1 shows the projected global electricity consumptions for different regions of the world, which increases from 19895 TWh in 2007 to 42860 TWh in 2050, consistent with economic growth.

Two major supplies of energy for its use as fuels and for the production of electricity are oil and gas. Figure 1.2 shows the production projection of oil and gas. The world oil and gas production reached a peak of 30 billion barrels in 2010, starting from 1.5 billion barrels in 1930. Since oil and gas are non-renewable energy sources and are being depleted, and with production set to reach 12 billion barrels in 2050, this is a clear sign that there is an energy crisis and a serious need for alternative energy resources, such as renewable energies, including nuclear energy.

Per capita consumption of energy is a measure of physical quality of life (Bala, 1998). Though the regional average of energy consumption in Asia has increased

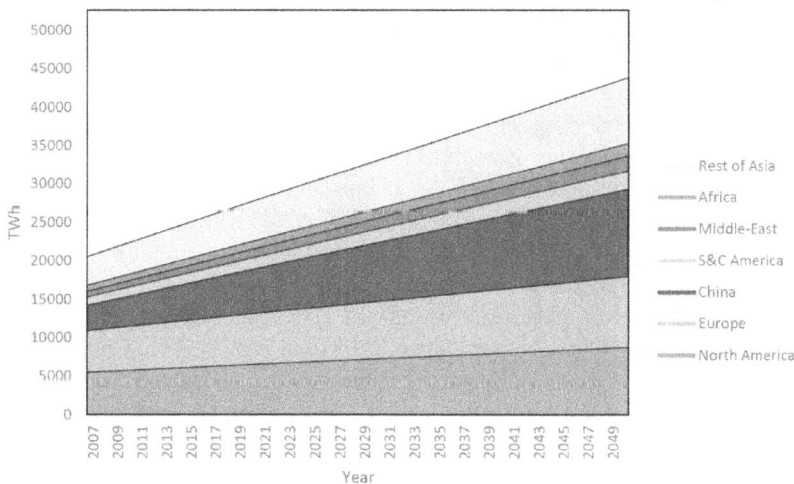

**FIGURE 1.1** World electricity consumption.

DOI: 10.1201/9781003218401-1

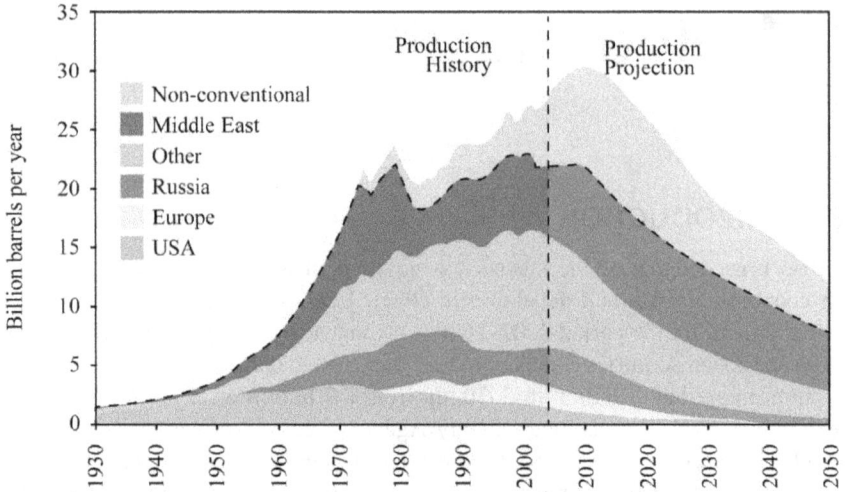

**FIGURE 1.2**   World liquid oil and gas projections.

in recent years, it is far below the world average. Figure 1.3 shows the relationship between physical quality of life and per capita total energy consumption. It may be noted from Figure 1.3 that the lower limit of energy consumption is 3000 kWh/year and reaches saturation at 14000 kWh/year. This means that 14000 kWh/year is the minimum upper limit of physical quality of life in order to lead a decent life.

Energy production and use are major sources of greenhouse gas emissions and may cause serious environmental impacts. These impacts, in turn, can threaten overall social and economic development. At regional and global levels, oil, gas and coal consumption may lead to acid rain, and most likely to global warming. At the local level, continued reliance on traditional biomass fuels for cooking in many developing countries such as Bangladesh can place added stress on farmlands, resulting in decreased relative humidity and environmental degradation (Bala, 2003).

South Asia is home to several of the most polluted cities, including Calcutta, Dhaka, Mumbai, Delhi and Karachi. However, total emissions in these regions account for a small fraction (3%) of global emissions, and these regions contribute only a small amount to global warming and climate change. Carbon emissions of some of the largest contributors are shown in Figure 1.4, and it is evident that China is the largest contributor to global warming. The contribution by Bangladesh towards global warming and climate change is a very small fraction (Warrick, 1996; Bala, 1997a, 1997b, 2006; Bala and Khan, 2003) and is only 5% of China's contribution, but could be seriously affected by climatic change. However, emissions can be controlled through the application of a suitable carbon tax, and high tax levels would result in the substantial penetration of renewable energy technologies, such as solar energy technologies, in developing countries like Bangladesh.

The Third Assessment Report of the United Nation's Intergovernmental Panel on Climate Change confirmed that the earth's climate is changing as a result of

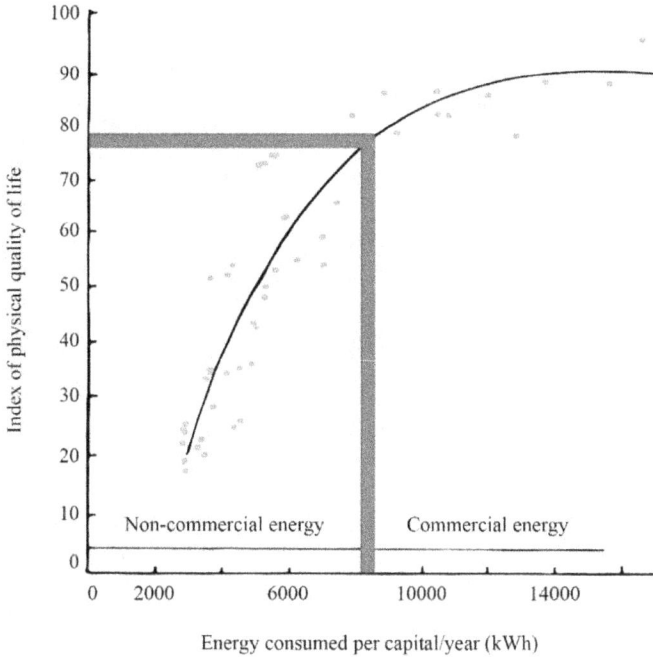

**FIGURE 1.3** Relationship between physical quality of life and per capita total energy consumption.

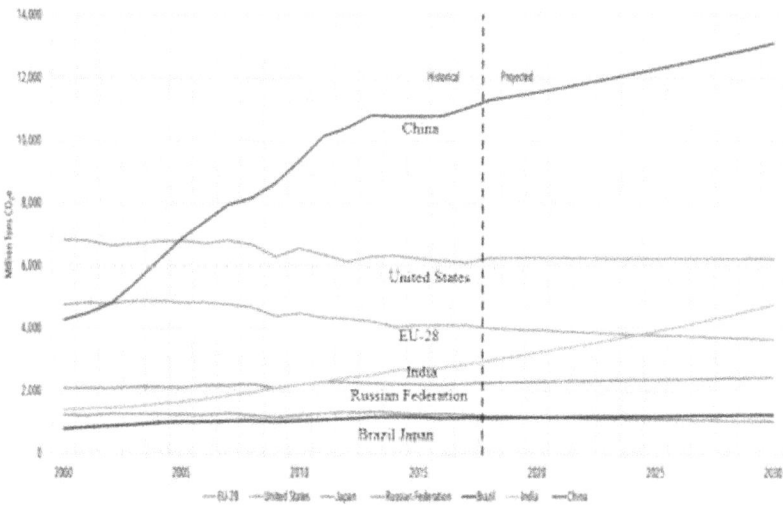

**FIGURE 1.4** Greenhouse gas emissions for major economies, 1990–2030 (Centre for Climate and Energy Solutions, 2021). Data source: Oak Ridge National Laboratory, 2017 and International Energy Agency, 2019)

human activities, particularly from fossil fuel energy use, and that further change is inevitable. Natural ecosystems are already adapting to the change, although some are under threat, and it is evident that human health and habitats will be affected worldwide. Such climate change could also affect the present supplies of renewable energy sources and the performance and reliability of conversion technologies.

The European Commission has strongly supported the generation of green electricity in past decades, but the energy obtained from fossil fuels still prevails throughout the region. Bengochea and Faet (2012) reported that there exists a relationship between green energy, the price of fossil fuels and $CO_2$ emissions, implying that an increase in $CO_2$ leads to an increase in renewable energy supply.

We need energy security for sustainable development. Most of the definitions of energy security focus on energy supplies, particularly supplies of oil (Clawson, 1997). This supply-based focus has, as its cornerstones, reducing vulnerability to foreign threats or pressure, preventing a supply crisis from occurring, and minimizing the economic and military impact of a supply crisis once it has occurred. Energy security usually revolves around the concept of supply security, which means sustainable, reliable and adequate energy supply at a reasonable price. The United Nations Development Programme (UNDP) defines energy security as the availability of energy at all times in various forms, in sufficient quantities and at affordable prices, without an unacceptable or irreversible impact on the economy and the environment.

Energy security and climate change have been at the forefront of energy policy. The International Energy Agency (2007) reported that unless countries change their energy use policy, oil imports, natural gas and coal use, and greenhouse gas emissions threaten to undermine energy security and accelerate climate change. In addition, those who examine specific energy conservation or alternative energy technologies frequently observe a complementarity between the abatement of greenhouse gases and an increase in energy security (Farrell et al., 2006; Tyner, 2007). Although such complementarity can exist for individual technologies, policymakers need to make a trade-off between these two policy objectives and should choose a mix of individual technologies that reduce greenhouse gas emissions and enhance energy security. The policymakers need to model the energy systems and analyze the energy policies to select the optimal policy.

## 1.2  COMPLEXITY AND DYNAMICS OF ENERGY SYSTEMS

We live in a complex world that is always changing and we are confronted with complex technological, environmental, political and socio-economic problems that we need to understand and manage for sustainable development (Bala, 1998; Bala et al., 2017). In a global context, global warming and its impacts on agriculture, energy and the environment are debated seriously, and in reports on economic cycles that cause financial panic at regional and local levels, price fluctuations and energy insecurity in developing countries are just some of the problems of complex and dynamic systems.

In recent years, decision-makers in an increasing number of countries have realized that energy planning should be carried out in an integrated manner (Bala et al., 2014). Traditionally, the planning of oil, gas and electricity has been carried out independently. This approach is good provided the energies are cheap. Fuel price increase and fluctuations, and the sudden energy crisis as shown in Figure 1.5, as well as their contributions to air pollution and global warming, suggest a gradual transition to cheaper and pollution-free environmentally friendly energy sources and integrated energy systems is required. How can we understand the complexity and dynamics of fluctuations in prices and the supply of oil? Can we understand the complexity of integrated energy systems when coordinated planning is adopted in various energy subsectors such as electric power, oil, gas, coal and renewable energy resources? If the answer is yes, how can we do it? We need improved knowledge and analytical capabilities to understand and manage energy price volatility and reduce energy insecurity.

Indeed, we can understand and design management strategies of such complex systems, but we need some structures or guiding principles to understand and manage the complexity and changes of dynamic systems based on a systems approach that considers all systems rather than in isolation. The systems approach is a rational and rather intuitive approach that depends on some formalized methodologies consisting of methods of problem definition, dynamic hypothesis, modeling, policy analysis, etc., and theoretical techniques that are useful for the solving models and sub-models of the problem.

The need to understand and predict the functioning and performance of the individual components of the energy system or the overall system behavior motivates the development of models (Subramanian et al., 2018). Through modeling and

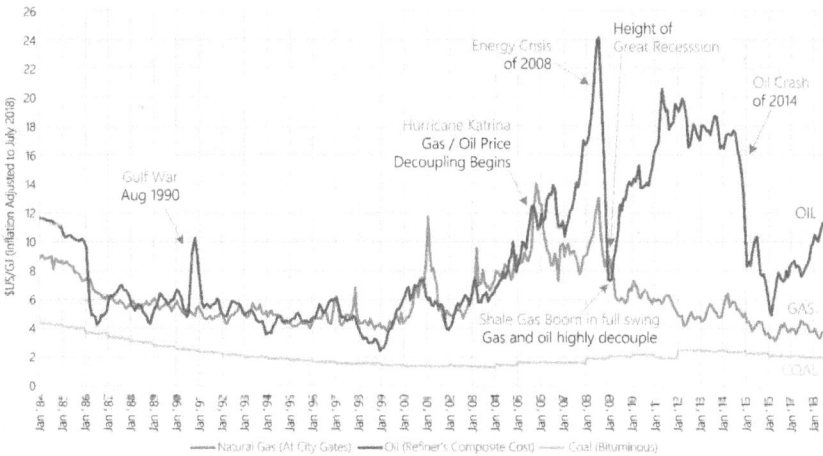

**FIGURE 1.5** Price fluctuations of oil, gas and coal in the international markets (Subramanian et al., 2018).

simulation, we can study the problems and take actions to improve the system. In essence, energy systems must be modeled and simulated to understand the dynamics of energy systems and design management strategies. They must be done before the implementation of management strategies. Forrester's system dynamics provides the methodology: the guiding principles for constructing computer models that simulate complex and dynamic systems such as to understand them and design management strategies (Bala, 1999; Bala et al., 2017; Forrester, 1968). In essence, systems thinking is a formalized methodology consisting of methods of problem definition, dynamic hypothesis, modeling and policy analysis for the understanding and management of complex and dynamic systems (Cavana and Maani, 2000).

As the complexity of our world increases, systems thinking is emerging as a critical factor for success and even for survival. How then can people become skilled systems thinkers for energy modeling and planning? In the world of complex dynamic systems, everyday experience fails because the time horizon and scope of the systems are so vast, we will never experience the majority of the effects of our decisions. When experiments in the real world are impossible for systems such as energy systems, simulation becomes the main way we can learn effectively about the dynamics of complex energy systems. System dynamics is the most appropriate technique for simulating such complex and dynamic systems based on systems thinking for the development of policy scenarios and for learning how to effectively manage the systems.

## 1.3   CONCEPTS OF SYSTEMS AND SYSTEM DYNAMICS

System dynamics is a methodology of constructing computer models based on feedback systems of control theory that was developed by Jay W Forrester at MIT during the 1950s. It can easily handle the non-linearity, time delay and the multi-loop structures of complex and dynamic systems. Forrester's methodology provides a foundation for constructing computer models that do what the human mind cannot – rationally analyze the structure, interactions and modes of behavior of complex social systems, thus providing a framework whereby strategies can be tested and trade-offs can be performed while options are still open. Nowadays, many choices of software are available that have revolutionized system dynamics modeling, such as STELLA and VENSIM. Furthermore, they are icon-operated and allow us to model virtually any process or system.

We discussed some complex and dynamic systems in section 1.2. But what is meant by a system? A system is a combination of components for a common purpose. For example, a transformer is an example of a system that is a combination of windings and a magnetic core for stepping up or stepping down voltages. A system may include people as well as physical parts. A family is a system for living and raising children. A system may include physical, economic, social, biological, technological and political components, and such a system is highly complex. For example, energy systems and electric power consist of physical, technological, environmental, socio-economic and political components and their interactions, and such systems are highly complex. Systems are, therefore, characterized by the elements and structures that influence them to achieve the system's purpose and identity. In modeling and simulation, our main task is to find the essential elements and their functions, and establish the essential influence structure of the system for simulation (Bossel, 2018).

Systems may be classified as (a) open systems and (b) feedback systems. In open systems, the output responds to the input, but the output has no influence on the input. Also, the input is not aware of its own performance. In an open system, past action does not affect future action. For example, a watch is not aware of its inaccuracy and does not correct the time itself. In an open system, the problem is perceived and action is taken, but the result does not influence the action (Figure 1.6). Running an electric fan without an automatic control switch is an example of an open-loop system. When the manual switch is closed, current starts flowing into the electric motor of the fan, and the electric fan continues to run until the manual switch is opened by the user, who is not part of the system.

Feedback systems are closed-loop systems, and the inputs are affected and changed on the basis of the output. A feedback system has a closed-loop structure that brings back the results of the past action to control the future action. In a closed-loop system, the problem is perceived, action is taken and the result influences further action (Figure 1.7). Thus, the distinguishing feature of a closed-loop system is a feedback path of information, decision and action connecting the output to the input.

Feedback systems may be divided into positive feedback and negative feedback systems. Positive feedback systems generate growth and negative feedback systems are goal-seeking. A population growth system is an example of a positive feedback system. A population multiplies to produce more population, which

Information about the problem ──────────▶ Action ──────────▶ Result

**FIGURE 1.6** Open-loop system concept.

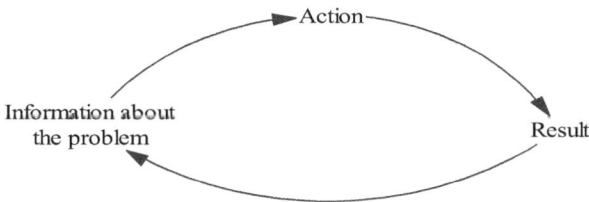

**FIGURE 1.7** Closed-loop system concept.

increases the growth rate at which the population is increased (Figure 1.8). Thus, a positive feedback system generates growth, as shown in Figure 1.8.

To meet the future requirement of electrical energy, power plants are constructed. The construction initiation rate depends on the existing power plant capacity and the forecast for the future need of the power plant, and their difference is negatively related with the existing power plant. The power plant under construction increases with the increase in initiation rate, and the power plant capacity increases with the increase in plants under construction. A balancing loop is formed that causes the power plant system ultimately to reach equilibrium, and thus the system is

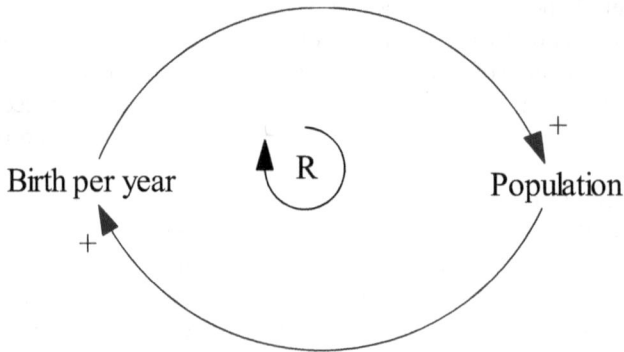

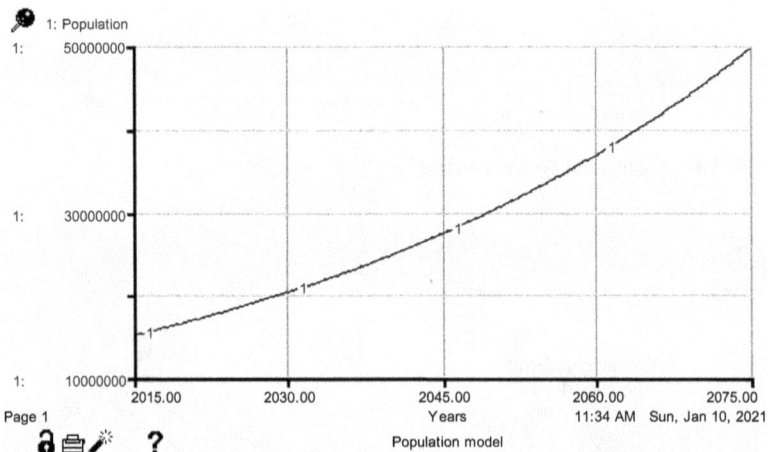

FIGURE 1.8  Population growth system.

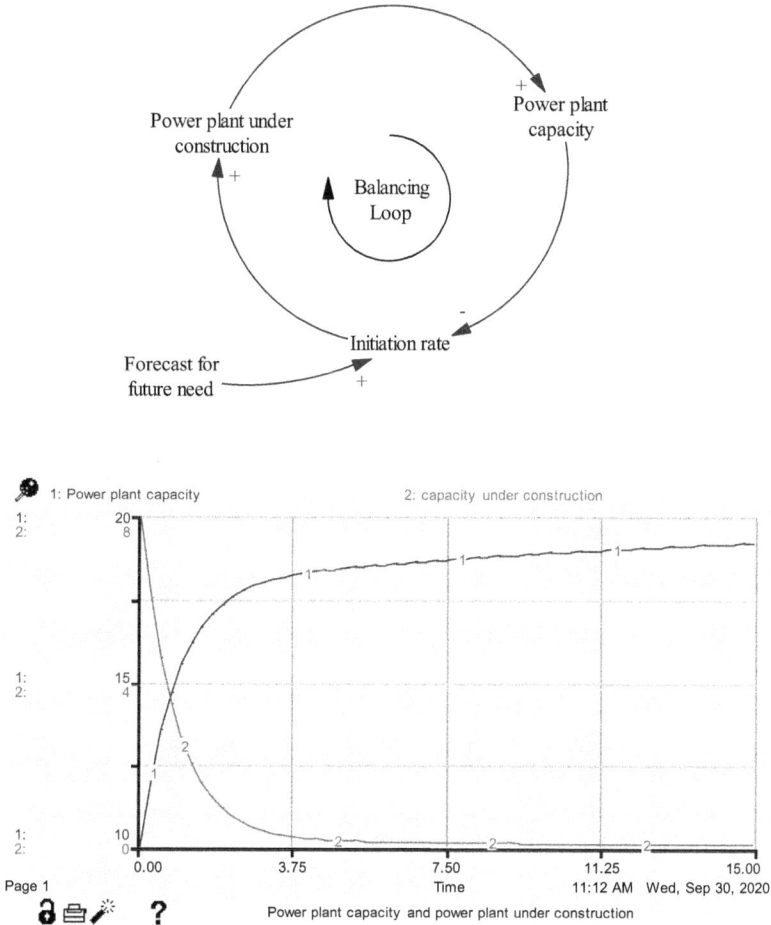

FIGURE 1.9 Power plant capacity and power plant under construction.

goal-seeking. The power plant and its construction system is an example of a negative feedback system (Figure 1.9).

## 1.4 MODES OF BEHAVIOR OF DYNAMIC SYSTEMS

The feedback loop system structure simulates dynamic behavior, and all the dynamics arise from the interactions of two types of feedback loop: positive and negative. Positive feedback loops generate growth, i.e., they are self-reinforcing. The causal loop consisting of population, birth and population in sequence in Figure 1.10 is an example of a positive feedback loop that reinforces the population level. The negative feedback loop in Figure 1.10 consisting of population, death and population is goal-seeking. Figure 1.11 shows the dynamic behavior

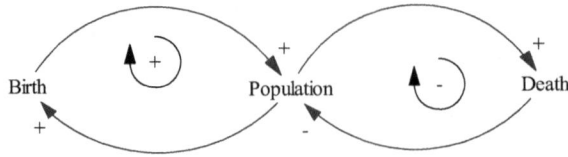

**FIGURE 1.10**   Causal loop of population growth.

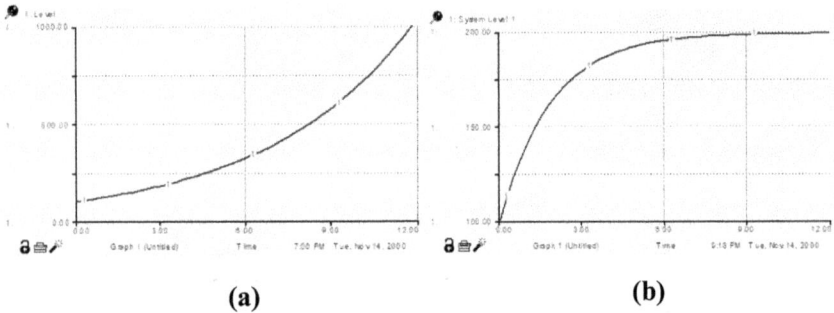

**(a)**                                                    **(b)**

**FIGURE 1.11**   Dynamic behavior of (a) positive feedback and (b) negative feedback systems.

of positive and negative feedback systems. A positive feedback system generates exponential growth, as shown in Figure 1.11(a), while a negative feedback system is goal-seeking, as shown in Figure 1.11(b). The control theory block diagram of a negative feedback system is analogous to the causal loop diagram in system dynamics, and essentially system dynamics is a control theory for social systems (Manetsch and Park, 1982).

Figure 1.12(a) shows a second-order feedback closed-loop system characterized by two stocks, employment and inventory, and Figure 1.12(b) represents the response of a second-order system. The second-order feedback loop system generates oscillation.

Figure 1.13 represents the responses of some complex systems. Figure 1.13(a) shows S-shaped growth, Figure 1.13(b) shows S-shaped growth with overshoot, and Figure 1.13(c) shows overshoot and the collapse of complex systems. Innovation of new ideas and technologies often follows S-shaped patterns (Sterman, 2000), and in S-shaped growth, exponential growth reaches the limit (Figure 1.13(a)). S-shaped growth with overshoot is a pattern of behavior that reaches a limiting effect of a balancing loop with delay (Figure 1.13(b)). The boom and bust of cocoa production in Malaysia is an example of the collapse of a complex system (Bala et al., 2017) and is a reminder that when an industry is prone to exceed and consume environmental carrying capacity, a boom and bust type of development results. The overshoot and collapse in Figure 1.13(c) is essentially a variation on growth with overshoot, but the limit is a floating goal with an extra reinforcing loop. First-order

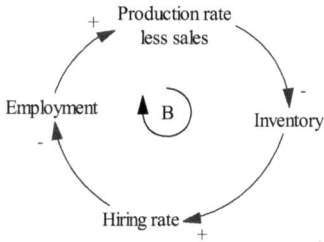

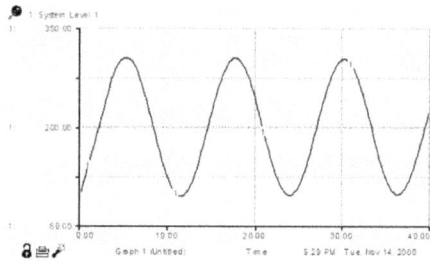

(a)                                              (b)

**FIGURE 1.12**   (a) Causal loop diagram of a second-order system and (b) dynamic behavior of a second-order feedback system.

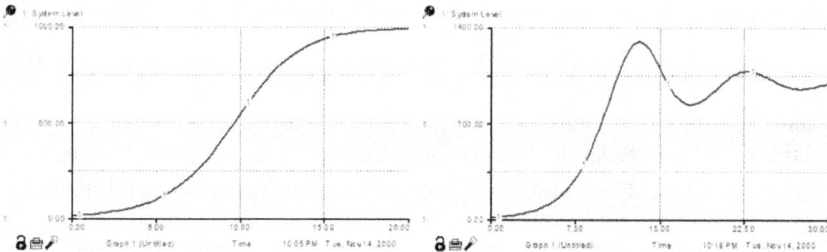

(a)                                              (b)

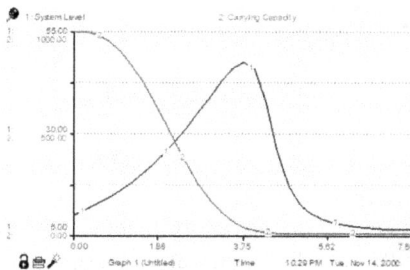

(c)

**FIGURE 1.13**   Dynamic output of complex systems: (a) S-shaped growth, (b) S-shaped growth with overshoot and (c) overshoot and collapse of complex systems.

and higher-order positive feedback systems generate growth. First-order negative feedback systems are goal-seeking, but higher-order negative systems oscillate as they search for a goal. Models in practice may contain thousands of interconnected loops with time delays and non-linearities.

## 1.5   LINEAR PROGRAMMING

Linear programming is the most powerful method of constrained optimization. The standard form of linear programming model consists of two parts, an objective function and constraints. The objective function is an equation that defines the quantity to be optimized. For linear programming, this quantity must be a one-dimensional scalar quantity such as:

$$Z = \sum_i C_i X_i \qquad (1.1)$$

The variables in the objective function $X_i$ are decision variables because we need to seek a decision to optimize the objective. For electrical energy supply, the objective function Z may be the total cost of generation to be minimized, $C_i$ is the unit cost of electricity, and the decision variable $X_i$ is the amount of electricity generated from fuel type i, such as oil, gas or solar.

The constraints are of the form:

$$\sum a_{ij} X_i \gtrless b_j \qquad (1.2)$$

Where $b_j$ represents the upper or lower bounds on a particular feature of a problem, and the $a_{ij}$ parameters define the contribution of each decision variable to that feature. For electricity generation, $b_j$ represents the demands that can be fulfilled using j type of fuel, and $a_{ij}$ represents the contributions of $X_i$.

Because linear programming problems in practice typically deal with a very large number of variables, $X_i$ and the constraints $b_j$, they are almost invariably presented in vector and matrix notation. Thus:

$$\text{Maximise or Minimise } Z = CX \qquad (1.3)$$

$$\text{Subject to } AX \gtrless B \qquad (1.4)$$

Where C, X and B are row and column vectors, and A is the matrix of $a_{ij}$ parameters of the variable in the system of constraints.

In electric power systems, electricity is generated from non-renewable and renewable energy resources. Non-renewable energy sources such as natural gas, oil and coal are limited resources and finite, and their production and use cause air pollution and contribute to global warming, while coal-based electric power generation may also cause acid rainfall. Furthermore, the production of electrical energy is highly capital intensive. Hence, generation cost and pollution must be minimized using optimization techniques such as linear programming. Electrical power systems must not only be modeled and simulated for designing policy options but also optimized for operational planning to minimize cost and pollution.

## 1.6   INTEGRATED ENERGY SYSTEMS AND SYSTEMS APPROACH

Energy systems consist of energy supply, energy demand, price, climate change and environmental quality to facilitate the better supply of energy with minimum impact on climate change and environmental quality. To ensure energy security and reduce contributions to global warming, the transition to distributed systems, i.e., renewable energy, is essential. To meet the electrical energy requirements of tomorrow in a sustainable manner, today's central system should be gradually moved into distributed utility, which is essentially a hybrid system to minimize emissions and maximize reliability. Figure 1.14 shows such a distributed utility.

The global community has started to adopt renewable energy as the primary source of power in order to address the issue of global warming and reduce the production of greenhouse gases due to the burning of fossil fuels. The global production of electricity at present is 32% by oil and petroleum products, 28% by coal, 22% by gas and 5% by nuclear energy, and 19.2% of global energy consumption by users is generated by renewable energy (REN, 2016). The utilization of electrical energy is the key to the growth of any country economically. Particularly, developing nations are very keen to increase renewable-based generation as it is abundant and cheap (Camacho et al., 2011). Huge investments have been made by various countries in developing and implementing renewable energy-based power production. A worldwide investment of US$286 billion has been made for the development of renewable energy.

Renewable energy also plays a vital role in the electrification of rural and remote areas where the transmission of grid power is impossible. The major renewable sources that produce electricity are solar, wind, hydro, tides, geothermal and biomass.

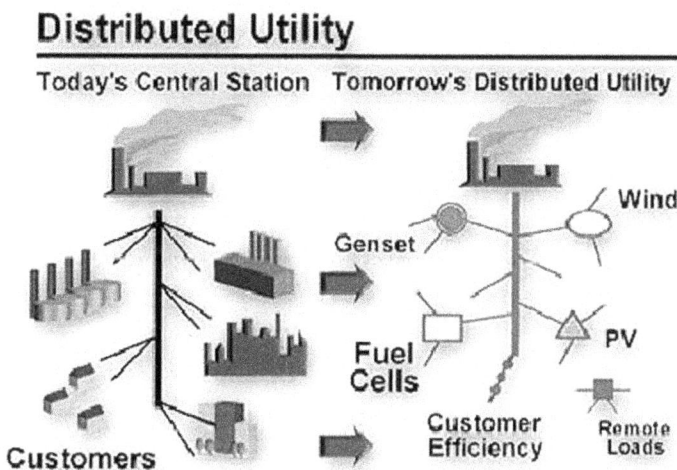

**FIGURE 1.14**   Distributed utility.

Energy systems consist of three main components: supply sector, demand sector and emission sector. The supply sector of a typical developing country is categorized as commercial energy and renewables, mainly biomass fuels and, to a limited extent, solar and wind. Commercial energy comprises natural gas, oil and electricity, while biomass fuels include agricultural wastes, animal wastes and fuelwood. In the demand sector, energy consumers make decisions to utilize gas, oil and electricity based on fuel price and availability, whereas biomass fuel consumption is mainly based on the availability of biomass fuels. In the demand sector, the major energy-consuming sectors are residential, industrial and transportation. To meet the electrical energy demand, the fuels are converted into electrical energy using different conversion technologies. The supply sectors (commercial and biomass) are the fuel supplies and imports. The imports are equal to the shortfalls in domestic supply.

The integrated energy sector comprises the interactions of many components, such as fossil fuels, renewable energy sources such as biomass, solar, wind, etc., and the pricing and management of energy systems, and should be implemented using a systems approach. Integrated energy systems must be simulated to understand the systems clearly before implementation. Since energy systems contain technological, environmental and socio-economic components, they are highly complex. System dynamics, a methodology of the computer simulation of complex socio-economic systems based on a systems approach, is the most appropriate technique for modeling and simulating such complex systems. Figure 1.15 shows the structure of the integrated energy systems of Bangladesh.

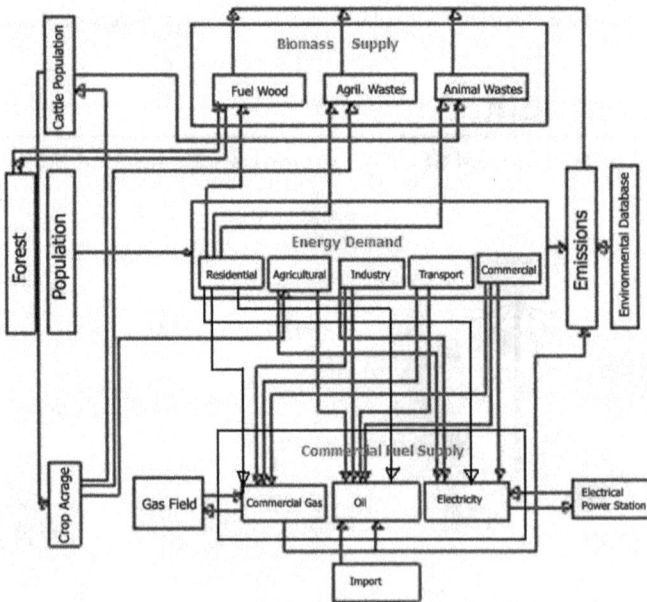

**FIGURE 1.15**    Structure of integrated energy systems of Bangladesh.

Since integrated national energy systems are highly complex, containing technological, environmental, socio-economic and political components, the modeling and simulation of such complex systems for rational policy planning is a formidable challenge. Coordinated integrated energy planning requires detailed analysis of the interrelations between the different sectors and subsectors of the entire national energy system using a systems approach and their potential energy requirements on the one hand, and energy supply capabilities on the other hand, with a gradual transition to renewable energy resources.

Integrated national energy planning is motivated by a flexible and continuously updated energy strategy to promote the best use of available energy resources for socio-economic development and the improvement of quality of life. The specific tasks of integrated national energy planning are the determination of energy needs, selecting an energy mix to meet energy requirements with minimum cost, the conservation of energy, preservation of the environment and so on. Integrated energy system planning provides a systematic analysis of all the factors that influence the evolution of energy systems with a gradual transition to renewable energy for developing an effective energy strategy that supports national sustainable development goals.

## 1.7   ENERGY MODELING AND SYSTEMS APPROACH

Energy planning is required for sustainable development and effective environmental management due to the limitations of fossil fuel resources, the high capital cost of renewable energy development, and various concerns about energy supply and demand. This requirement has motivated a number of studies on the development of energy systems models and their applications in the planning of energy activities at different levels. There has also been an upsurge of interest among researchers and planners in the modeling of energy systems since the oil embargo of the mid-1970s. Today, many energy system models are available that have been developed for the planning of large systems at national or regional levels. These models can be classified into two main categories: simulation and optimization. Simulation models simulate the dynamics of energy systems, i.e., scenarios for management strategies, while optimization models evaluate the competition between various sources.

Energy modeling has multiple purposes: it provides a better understanding of current and future markets, including supply, demand and prices; it facilitates the better design of energy supply systems in the short, medium and long term; it ensures sustainable exploitation of scarce energy resources; and it provides an understanding of the present and future interactions of energy and the rest of the economy, and the potential implications on environmental quality.

Energy modeling is an art, and it provides insights, *not* answers. To provide better understanding, facilitate better design and give greater insights into energy systems, the modeling of energy systems should be conducted. A systems approach is needed for such modeling. A systems approach of energy modeling considers the entire energy system, including all the components of energy systems, rather than a component in isolation.

## 1.8   SYSTEMS THINKING AND MODELING

Indeed, we need some structures or guiding principles to model and simulate complex dynamic systems such as integrated energy systems, and a systems approach that considers entire systems rather than in isolation refers to a set of conceptual and analytical methods used for systems thinking and modeling (Cavana and Maani, 2000). The general methodological approach towards systems thinking and modeling is discussed here. Many contributions have been reported on systems thinking and system dynamics. Systems thinking and modeling essentially consists of a problem statement, a causal loop diagram, a stock–flow diagram, scenario planning and modeling, and implementation and organization learning. The character of systems thinking makes it extremely effective on the most difficult types of problems to solve, which include complex issues, those that depend a great deal on the past or the actions of others, and those stemming from ineffective coordination among those involved. The steps to be followed for simulating a system dynamics model based on systems thinking (Bala et al., 2017) are summarized below:

- Identify the problem and formulate the mental model in the form of a verbal description (problem identification/conceptualization), and develop a dynamic hypothesis to account for problematic behavior in terms of causal loop diagrams and the stock and flow structure of the system.
- Create the basic structure of a causal diagram from the verbal model.
- Augment causal loop diagrams into system dynamics flow diagrams.
- Translate the system dynamics flow diagrams into STELLA or VENSIM, or a set of simultaneous difference equations.
- Estimate the parameters.
- Validate the model, analyze the sensitivity and analyze the policy.
- Application of the model.

## 1.9   USEFULNESS OF MODELS

The validity and usefulness of dynamic models such as integrated energy systems should be judged not against imaginary perfection but in comparison with other mental and descriptive models that are available. The usefulness of a mathematical simulation model should be judged in comparison with the mental image or other models that would be used instead (Forrester, 1968). There is nothing in either physical or social science about which we have perfect knowledge and information. We can never say that a model is a perfect representation of reality. On the other hand, we can say that there is nothing of which we know absolutely nothing. So, models should not be judged on an absolute scale but on a relative scale if the models clarify our knowledge and provide insights into systems.

## 1.10 ENERGY POLICY ANALYSIS

We need energy policy planning and analysis since we need to analyze and test the effectiveness of policy measures as regards compliance with environmental constraints and climate objectives, investment requirements and financial viability (finance), social/public/political commitment and acceptance, economic development and environmental protection, regional approaches and infrastructure sharing, and communication tools (e.g., public, investors, stakeholders and neighbors). Comprehensive energy policy planning and analysis are essential for sustainable (energy) development as a prerequisite for informed decision-making, assessing future energy demand, evaluating options and reviewing the different ways to meet those needs, identifying risks and benefits, exploring "what if" questions, allocating optimal domestic resources, inherently long-lead and lifetimes, and shifts from sequential stop-gap measures to integrated energy systems planning.

We need integrated energy planning since this provides a systematic analysis of all the factors that influence the evolution of energy systems. It facilitates problem-solving and makes it possible to explore linkages, evaluate trade-offs and compare consequences, thereby helping countries develop an effective energy strategy that supports national sustainable development goals.

Energy planning can be defined by three levels of planning: (1) energy policy planning, (2) strategic energy planning and (3) operational energy planning (Hoffman and Wood, 1976). Energy policy planning involves the formulation of goals or objectives; strategic energy planning concentrates on the development of a set of alternative paths to the desired goal; and operational energy planning deals with the determination of the steps necessary to implement the desired strategy.

Energy planning is about choices and dealing with current and future uncertainties. It addresses the energy trilemma: (a) energy security/supply security and reliability, (b) economic competitiveness/affordability and access and (c) environmental considerations/climate change and local and regional pollution.

Energy planning is not about predicting the future. It is about the analysis and evaluation of a set of different possible futures, as well as being a communication tool (informed policy and decision-making). No analysis is perfect. However, it can be used to explore many more "what if" questions and provide new information. Previously plausible assumptions no longer stand the test of time, and energy planning is a never-ending process. Finally, energy planning is required for sustainable development and effective environmental management due to the limitations of fossil fuel resources, the high capital costs of renewable energy development and various concerns about energy supply and demand.

## 1.11 STRUCTURE OF THE BOOK

This book consists of 10 chapters. Chapter 1 presents the importance, concepts, methodology and techniques of energy modeling and policy analysis. Chapter 2 introduces the model concepts and simulation and provides an overview of systems thinking methodology with examples and causal loop and stock–flow diagrams with examples. In Chapter 3, we introduce an optimization method of linear programming with examples and Pareto optimality. Chapter 4 introduces the fundamentals of communication technologies and presents the standards for information exchange and the medium of communication systems. Then smart metering, advanced metering infrastructure, substation automation and cyber security issues, and the requirements of a smart grid are explained, and SCADA systems and the architecture of communications technology for smart grid and multi-agent systems are also presented. Chapter 5 introduces the concepts and projections of energy demand, supply and balance and the concept of price-setting energy, and the modeling of these for electric power systems is illustrated using system dynamics. In Chapter 6, energy production and its impact on environmental pollution, contribution to global warming and acid rainfall are analyzed, and the modeling of these is illustrated using system dynamics. Chapter 7 provides the energy modeling of integrated farming systems, the optimization of rural energy in agriculture using linear programming, the modeling of national energy systems using LEAP and the modeling of a microgrid using a genetic algorithm. Chapters 8 and 9 present the modeling of rural energy systems and national integrated electric systems using system dynamics. Finally, Chapter 10 presents the optimal operational planning of electric power systems using linear programming and MARKAL, and SCADA and smart grid concepts and their potentialities in economic operation. The reliability and automation of modern power systems, and the optimal operational planning of SCADA systems and smart grids using linear programming and genetic algorithms, respectively, are also discussed.

In summary, the modeling of integrated energy systems for different policy options and the search for optimal policy and operational planning is the scope of this book.

## REFERENCES

Bala, B. K. (2006). Computer modelling of energy and of environment for Bangladesh. *International Agricultural Engineering Journal,* 15: 151–160.
Bala, B. K. (2003). *Renewable Energy.* Agrotech Publishing Academy, Udaipur, India.
Bala, B. K., Alam, M. S., & Debnath, N. (2014). Energy perspective of climate change: The case of Bangladesh. *Strategic Planning for Energy and the Environment,* 33(3): 6–22.
Bala, B. K., Fatimah, M. A., & Kushairi, M. N. (2017). *System Dynamics: Modelling and Simulation.* Springer.
Bala, B. K, & Khan, M. F. R. (2003). Computer modelling of energy and environment. In U. Pandel & M. P. Poonia (Eds.), *Energy Technologies for Sustainable Development,* Prime Publishing House, Delhi, India.
Bala, B. K. (1999). *Principles of System Dynamics* (1st ed.). Agrotech Publishing Academy, Udaipur, India.
Bala, B. K. (1998). *Energy and Environment: Modelling and Simulation.* Nova Science Publishers, New York, USA.
Bala, B. K. (1997a). Computer modelling of the rural energy system and of CO2 emissions for Bangladesh. *Energy,* 22: 999–1003.

Bala, B. K. (1997b). *Computer modelling of energy and environment: The case of Bangladesh.* Proceedings of 15th International System Dynamics Conference, Istanbul, Turkey, August 19–22, 1997.

Bengochea, A., & Faet, O. (2012). Renewable energies and CO2 emissions in the European Union. *Energy Sources,* 2012; Part B 7: 121–130.

Bossel, H. (2018). *Modeling and Simulation.* CRC Press/Taylor and Francis, New York, USA.

Camacho, E. F., Samad, T., Garcia-Sanz, M. & Hiskens, I. (2011). Control for Renewable Energy and Smart Grids. In *The Impact of Control Technology.* Control System Society. 2011: 69–88.

Cavana, R. Y., & Maani, K. E. (2000). *A methodological framework for systems thinking and modelling (ST&M) interventions.* 1st International Conference on Systems Thinking in Management. Internet.

Centre for Climate and Energy Solutions. (2021). *Greenhouse Gas Emissions for Major Economies,* 1990–2030. Global Emissions (Data source: Oak Ridge National Laboratory, 2017, and International Energy Agency, 2019).

Clawson, P. (1997). Energy security in a time of plenty. *Strategic Forum Paper No.130.* National Defense University Press, Washington, DC, October 1997. www.ndu.edu/inss/strforum/SF130/forum130.html.

Farrell, A. E., Plevin, R. J., Turner, B. T., Jones, A. D., O'Hare, M., & Kammen, D. M. (2006). Ethanol can contribute to energy and environmental goals. *Science,* 311 (5760): 506–508.

Forrester, J. W. (1968). *Principles of Systems.* Allen Wright Press, Cambridge, Massachusetts, USA.

Hoffman, K. C., & Wood, D.O. (1976). Energy system modelling and forecasting. *Annual Review of Energy,* 1: 423-453.

International Energy Agency. (2007). *World Energy Outlook.* OECD/IEA, Paris, France.

International Energy Agency. (2019). *World Energy Outlook.* OECD/IEA, Paris, France.

Manetsch, T. J., & Park, G. L. (1982). *System Analysis and Simulation with Applications to Economic and Social Systems.* Department of Electrical Engineering and System Science, Michigan State University, USA.

Oak Ridge National Laboratory. (2017). *Carbon Dioxide Information Analysis Center.* Oak Ridge, Tennessee, USA.

REN. (2016). *Renewables 2016 Global Status Report.* REN21 Secretariat, Paris, France, 2016.

Sterman, J. D., (2000). *Business Dynamics: Systems Thinking and Modeling for a Complex World.* McGraw-Hill Higher Education, Boston, USA.

Subramanian, A. S. R., Gundersen, T., & Adams II, T. A. (2018). Modeling and simulation of energy systems: A review. *Processes,* 6(238): 1–45.

Tyner, W. E. (2007). Policy alternatives for the future biofuels industry. *Journal of Agricultural & Food Industrial Organization,* 5(2): 1–13.

Warrick, R. A. (1996). Integrated model systems for national assessments of the effects of climate change: Applications in New Zealand and Bangladesh. *Water, Air, & Soil Pollution,* 92(1–2): 215–227.

# 2  Modeling and Simulation

## 2.1  INTRODUCTION

Modeling has been an integral part of national energy planning since the 1970's energy crisis. Energy systems are highly complex, containing technological, socio-economic, environmental and political components, and the modeling of such complex systems is a formidable challenge. The twin goals of understanding and managing highly complex energy systems are to provide the desired goal of energy systems, and in order to manage an energy system effectively, the energy system under consideration must be modeled and simulated and well understood. System dynamics, a methodology of computer modeling of highly complex systems containing technological, socio-economic, environmental and political components, is the most appropriate technique to analyze and design policy options for national energy systems. System dynamics methodology is based on non-linear, time-lagged feedback concepts of control theory and systems thinking. Modeling using system dynamics can assist in policy planning and decision-making on a rational basis.

Energy modeling started after the energy crisis in 1970. Energy modeling in Bangladesh was initiated by Huq in 1975 (Huq, 1975) and was further developed by Bala (Bala et al., 2014; Bala, 1997a, 1997b; Bala and Satter, 1986), and in India, energy modeling was initiated by Parikh (Parikh, 1997). Nail (1992) reported an integrated model of US energy supply and demand that is used to prepare projections for energy policy analysis in the US Department of Energy's Office of Policy, Planning and Analysis. This model represents one of the real success stories of system dynamics modeling. It was implemented at the Department of Energy in 1978 as an in-house analytical tool and has been used regularly for national policy analysis since that time. Bala et al. (2014) reported an integrated model of Bangladesh's energy supply and demand and its climate change perspective to address policy issues.

In this chapter, systems thinking methodology is elaborated, and the applications of system dynamics in the modeling of energy systems are illustrated to show the potentiality of system dynamics in the modeling of energy systems.

## 2.2  MODELS AND SIMULATION

We can study the dynamic behavior of a physical system by experimentation with the system itself. Sometimes it may be expensive and time-consuming. An

DOI: 10.1201/9781003218401-2

alternative to this method is to construct a number of prototypes of physical models to experiment with. Sometimes it may not even be possible or practical to experiment with the existing system or construct a physical model for experimentation. Consequently, the most inexpensive and less time-consuming method is to use a mathematical or computer model.

A model may be defined as a substitute for any object or system. Everyone uses models in their daily lives. A mental image used in thinking is a model and is not the real system. A written description of a system is a model that presents one aspect of reality. The simulation model is logically complete and describes the dynamic behavior of the system. Models can be broadly classified as (a) physical models and (b) abstract models, as shown in Figure 2.1. A children's model of a car and an airplane are examples of physical models. Mental models, mathematical models and computational models are examples of abstract models.

Models can be classified as shown in Figure 2.1 (Bala, 1999; Bala et al., 2017). A model is dynamic if it portrays time-varying characteristics; otherwise, the model is static. Steady state and transient characteristics are the properties of dynamic systems, whereas in static systems, the characteristics do not undergo a substantial change in time. In dynamic systems, the system states change substantially in time. Computational models are multi-agent systems, genetic algorithms and neural network models, and these modeling approaches are emerging technologies in the field of modeling energy systems.

It is sometimes possible to solve mathematical models by analytical methods. But for complex systems, the solution of the mathematical model of systems by analytical methods is extremely difficult or may be beyond the reach of today's mathematics. For such complex systems, only a step-by-step numerical solution is possible. This process of step-by-step numerical solution is called simulation.

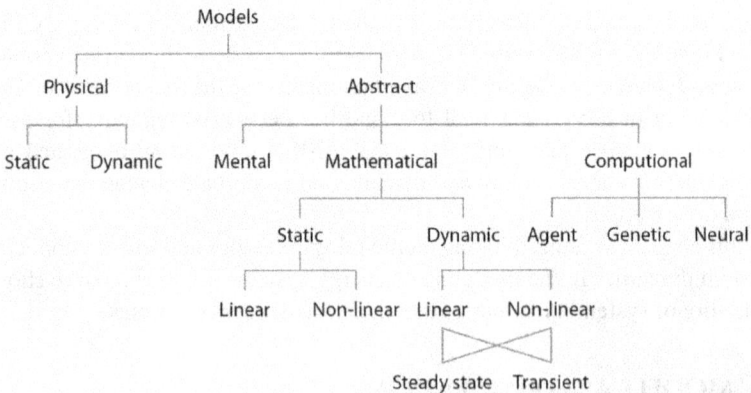

**FIGURE 2.1** Classification of models.

Simulation models are used in place of real systems. A computer simulation is an inexpensive and rapid method of experimenting with a system to give useful information about the dynamics of the real system. Scenarios based on simulated results can provide guidelines for policy planning and the management of complex and dynamic systems such as energy systems.

Forrester's system dynamics methodology provides a foundation for constructing computer models to do what the human mind cannot – "rationally analyze the structure, the interactions and mode of behavior of complex socio-economic, technological and environmental systems." The advantages of a computer model over a mental model are (Forrester, 1968):

(i)   It is precise and rigorous instead of ambiguous and unquantified.
(ii)  It is explicit and can be examined by critics for consistency and error.
(iii) It can contain much more information than any single model.
(iv)  It can proceed from assumption to conclusion in a logical, error-free manner.
(v)   It can easily be altered to represent different assumptions or alternate policies.

## 2.3   SYSTEMS THINKING

Systems thinking is a method of studying the dynamic behavior of a complex system based on a systems approach, i.e., considering the entire system rather than in isolation, and system dynamics is a tool for understanding the complexity and change over time of a dynamic system. When a system is broken into components, the components may give a false impression of the dynamic behavior that is far from the real behavior of the actual system. Thus, systems thinking should consider all the interacting components influencing the dynamics of the complex system, and system dynamics as a methodology based on the feedback concepts of control theory is the most appropriate technique for handling such complex systems to enhance systems thinking and systems learning.

### 2.3.1   SYSTEMS THINKING METHODOLOGY

To enhance systems thinking and systems learning, the system must be modeled and simulated. Basically, there are six important steps in building a system dynamics model. It starts with problem identification and definition, followed by system conceptualization, dynamic hypothesis, model formulation, model testing and evaluation, model use, implementation and dissemination, and design of the learning/strategy/infrastructure. The steps for modeling and simulation of complex systems based on systems thinking are:

(1)  Identify the problem.
(2)  Develop a dynamic hypothesis.

(3) Create causal loop diagrams.
(4) Augment the causal loop diagrams with more information.
(5) Convert the augmented causal loop diagrams into stock–flow diagrams.
(6) Translate the stock–flow diagrams into STELLA or VENSIM programs or equations.

### 2.3.1.1  Problem Identification

The first step in model building is to identify the problem, set its boundary and state the specific objectives. The problem should be clearly identified as it is important for successful modeling to solve the real problem. Systems thinking should be used to identify the problem. Neither the whole system nor a part of it should be considered to draw the boundary of the model; rather, a systems approach should be used of considering the entire system that is endogenously responsible for causing the problem from the feedback structure of the stated entire system. Therefore, the system boundary should encompass the portion of the whole system that includes all the important and relevant variables for addressing the problem and the purpose of policy analysis and design. The scope of the study should be clearly stated in order to identify the causes of the problem for a clear understanding of the problem and the policies for solving the problem in the short and long run.

To identify the problem, a detailed description of the system should be prepared based on available reports and studies, expert opinions, and the past behavior of the system. The problem of system identification is the problem of system operation. The problem statement should clearly describe the major factors influencing the dynamics of the system behavior with facts and figures. Next, it should include the purpose and clearly defined objectives. Discussions with all the stakeholders, including focus groups, should be conducted to justify their opinions on existing problems and their views on the solutions to those problems.

Verbal description is the simplest way to communicate the system with others. The more detail in the description, the easier it becomes to model the system. Major subsystems and their relationships within and between the subsystems of the system as a whole should be clearly described. The model should include only the relevant aspects of the study objectives. The verbal description is, in practice, a qualitative model of the system.

In selecting the variables to be included in the model, all the relevant variables or factors should be included, and unnecessary restrictions must be avoided. The boundary should be such that nothing flows across the boundary except perhaps a disturbance for exciting the system, and the factors needed to address the problem must be included inside the system boundary for the proper comprehensiveness of the model with adequacy. The formulation of a model should start with a boundary that encompasses the smallest number of components within which the dynamic behavior under study is generated.

It may be difficult to comprehend a whole system, especially when it is very large and complex. It is convenient to break such a system up into sectors or blocks. The description of the system should be organized in a series of blocks.

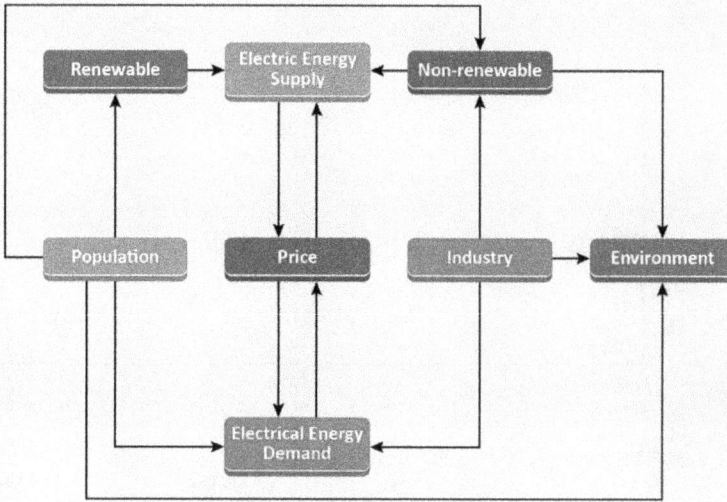

**FIGURE 2.2** Structure of electric energy demand, supply, price and environment model.

The aim in constructing the blocks is to simplify the specification of the interactions within the system boundary. Each block describes part of a system that depends upon a few, preferably one, input variables and results in a few output variables. The system as a whole can be described in terms of the interconnections between the blocks. Correspondingly, the system can be represented in the form of block diagrams. Figure 2.2 shows the structure of an electric energy demand, supply, price and environment model, and it is a typical example of a block diagram. The model is about the study of electric energy demand, electric energy supply, electric energy price and $CO_2$ emission from electric energy production in Bangladesh. The four sectors of the model are electric energy demand (population and industry), electric energy supply (renewable and non-renewable), electric energy price and $CO_2$ emissions (non-renewable, industry and population). The major influences on a sector of other sectors and its influences on other sectors are shown in the diagram.

The system dynamics model of the endogenous structure of feedback loops simulates observed dynamic behavior. This pattern of the change of behavior with time is called reference mode behavior or historical behavior. We need to know the observed reference mode behavior to understand the problem, and hence variables are selected accordingly. Figure 2.3 shows the observed and simulated reference mode behaviors of electricity demand in Bangladesh. The time horizon of reference mode and policy are also important and must be sufficient to cover the problem symptoms and policy issues addressed. The aspects to be addressed in the

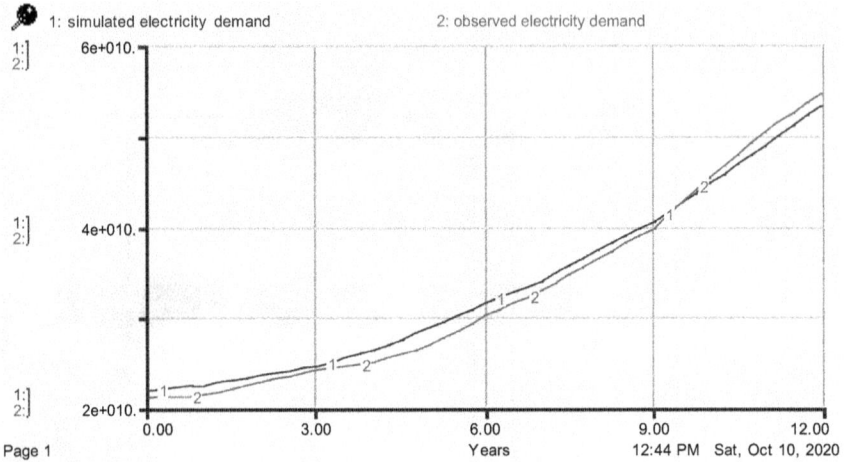

**FIGURE 2.3**    Observed and simulated electricity demand.

development of problem identification are (1) definition of the problem, (2) purpose of the model, (3) systems approach, (4) reference mode and (5) time horizon.

### 2.3.1.2   Dynamic Hypothesis

Once the problem is identified, the next step is to develop a theory called a dynamic hypothesis based on the reference mode behavior over a time horizon. The dynamic hypothesis in terms of a causal loop diagram and stock–flow diagram of the system can explain the dynamics of the problem. The hypothesis is provisional subject to revision and rejection that solely depends on the observed and simulated reference mode behavior over a time horizon (Sterman, 2000).

A dynamic hypothesis is a conceptual model typically consisting of a causal loop diagram, stock–flow diagram, or a combination thereof. The dynamic hypothesis seeks to define the critical feedback loops that drive the system's behavior. When a model based on a feedback concept is simulated, the endogenous structure of the model should generate the reference mode behavior of the system, and thus the endogenous structure causes the changes in the dynamic behavior of the system (Sterman, 2000). For example, energy supply and demand and the price of electric power systems can be represented by a causal loop diagram and stock–flow diagram, and the simulation model based on the causal loop diagram and stock–flow diagram can generate the dynamic behavior of electric power systems. The electric power system in the form of a causal loop diagram and a stock–flow diagram is hypothesized to generate the observed behavior of electric power systems in the reference mode. In essence, the increase of electricity demand with time in the electric power systems results in an increase in population, industries and crop diversification, and in cropping intensity in agriculture for bumper production. In fact, the demand increases with population growth and industrial and agricultural development, and this dynamic results from the endogenous consequences of the feedback structure.

The next step in dynamic hypothesis is how to test it. The hypothesis is tested using both observed and simulated reference mode data. In essence, the goal of dynamic hypothesis is to develop an endogenous explanation of a problematic behavior. Endogenous explanation is that the endogenous structure, i.e., the interactions of the variables inside the system, causes problematic behavior (Sterman, 2000).

The following aspects are to be addressed in the development of a dynamic hypothesis:

(1) Endogenous feedback structure.
(2) Observed and simulated reference mode behavior.
(3) A theory to explain the reference mode behavior.

### 2.3.1.3   Causal Loop Diagram

The system boundary covers the key variables inside the boundary and the variables crossing the boundary. The variables inside the boundary are endogenous variables, and the variables outside the boundary are exogenous variables. The next step in systems thinking is the search for the relationships between the variables and the development of feedback loops. These feedback structures are represented in the form of causal loop diagrams in system dynamics (Sterman, 2000) and in the form of control theory block diagrams in systems analysis (Manetsch and Park, 1982). Figure 2.4 shows the causal loop diagram of a simple electrical energy supply and demand model. In this model, the major variables are power plant installed, power plant under construction, electricity consumption, electricity price, price markup, construction initiation rate, and price markup change rate. Power plant under construction increases with the construction initiation rate, which is decreased with an increase in power plant under construction. This forms the negative feedback loop B1. Also, power plant installed increases with power plant under construction, which increases with construction initiation rate, but the construction initiation rate decreases with the increase in power plant installed. This forms the

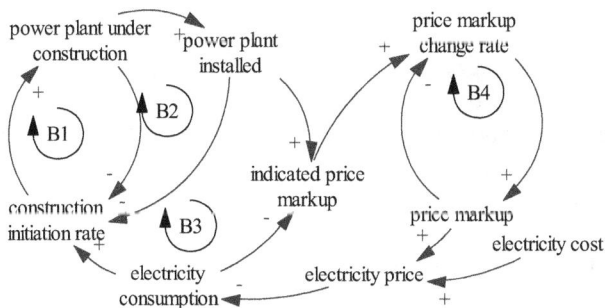

**FIGURE 2.4**   Causal loop diagram of a simple electrical energy supply and demand model.

negative feedback loop B2. Power plant under construction, power plant installed, indicated price markup, price markup change, price markup, electricity price, electricity consumption, construction initiation rate and back to power plant under construction form the negative feedback loop B3, while the price markup and price markup change form the negative feedback loop B4. Thus, the electric supply, demand and price system forms four negative feedback loops. The causal loop diagram represents the feedback loop structure of the system and causes the dynamic behavior of the system. The causal loop diagram represents the feedback structure of systems to capture the hypotheses about the causes of dynamics and the important feedbacks. The causal loop structure generating the reference behavior of the system is hypothesized to be the dynamic hypothesis. The following steps are followed in the development of a causal loop diagram:

(1) Define the problem and the objectives.
(2) Identify the most important elements of the systems.
(3) Identify the secondary important elements of the systems.
(4) Identify the tertiary important elements of the systems.
(5) Define the cause–effect relationships.
(6) Identify the closed loops.
(7) Identify the balancing and reinforcing loops.

Details of the construction of causal loop diagrams are described in Section 2.4.

### 2.3.1.4    Stock–Flow Diagram

The stock–flow diagram is the underlying physical structure of the system in terms of stock and flow. A stock–flow diagram is usually followed by a causal loop diagram. However, a causal loop diagram can follow a stock–flow diagram. Stock represents the state or condition of the system, and flow is changed by decisions based on the condition of the system. It is essentially the physical structure of the system and can be simulated to generate the dynamic behavior of the system. The stock–flow diagram represents integral finite difference equations involving the variables of the feedback loop structure of the system and simulates the dynamic behavior of the system. However, differential equations are formulated in systems analysis based on the control theory block diagram (Manetsch and Park, 1982). The stock–flow diagram or the system of the differential equations representing the feedback structure of systems captures the hypotheses about the causes of dynamics and the important feedbacks. The stock–flow diagram or the system of the differential equations representing the feedback structure of the system generating the reference behavior of the system is hypothesized to be the dynamic hypothesis. Figure 2.5 shows the stock–flow diagram of a solar home model. The four main variables are solar homes installed, solar homes under construction, solar homes construction rate and growth rate of solar homes. Here we have two stock variables. The first, solar homes installed, states the status of the solar homes in operation and is increased by one inflow of solar home construction rate and

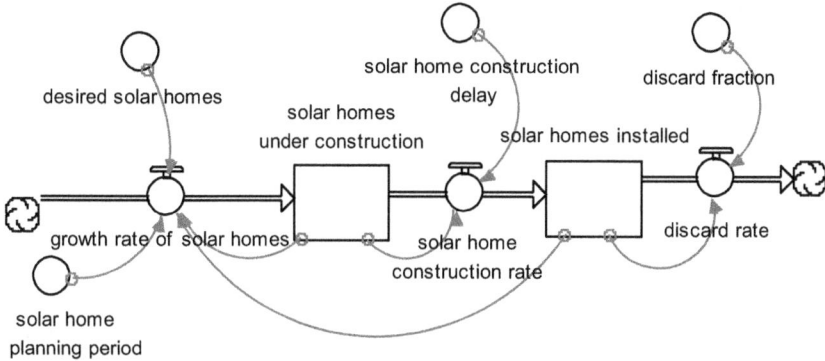

**FIGURE 2.5**   Stock–flow diagram of the solar home model.

decreased by one outflow of solar home discard rate. The second stock variable is the solar homes under construction and is increased by one inflow of growth rate of solar homes and decreased by one outflow of solar home construction rate. Solar homes have the unit of quantity, while the growth rate, construction rate and discard rate have the unit of quantity per unit time. The following steps are followed in the development of a stock–flow diagram:

(1)  Define the problem and the objectives.
(2)  Identify the most important variables of the systems.
(3)  Identify the secondary important variables of the systems.
(4)  Identify the tertiary important variable of the systems.
(5)  Identify the variables representing the stocks, i.e., accumulations.
(6)  Identify the variables representing the flows having a unit of per unit time of the stock.
(7)  Ensure the inflows entering the stock and outflows leaving the stock.

Details of the construction of stock–flow loop diagrams are described in section 2.5.

### 2.3.1.5   Parameter Estimation

Parameter estimation is an important step in system dynamics modeling. Parameter estimation techniques can be classified into three categories: (i) estimation from unaggregated data, (ii) estimation from an equation and data at a level of aggregation of model variables, and (iii) estimation from the knowledge of the entire model structure and data at a level of aggregation of model variables. The details are described in Bala et al. (2017).

### 2.3.1.6   Model Validation, Sensitivity Analysis and Policy Analysis

Tests for building confidence in system dynamics models consist of the validation, sensitivity analysis and policy analysis of system dynamics models. Two important notions of building confidence in system dynamics models are testing and validation. Testing means the comparison of a model with the empirical reality for accepting or rejecting the model, and validation means the process of establishing confidence in the soundness and usefulness of the model. The tests for building confidence in system dynamics models may be broadly classified as:

(i)   tests for structure;
(ii)  tests for behavior; and
(iii) tests for policy implication.

Details of the tests for building confidence in models are described in Bala et al., (2017).

### 2.3.1.7   Application of the Model

Because of their counterintuitive nature, the human mind is not capable of tracing the dynamic behavior of complex systems. System dynamics simulation models can provide a better understanding of and greater insights into such systems. Examining alternative policies for the selection of the best policy for the improved performance of the system is essential for policy planning. A system dynamics simulation model can be used as a computer laboratory for policy analysis and can also be used to assist in the management and control policy design. The optimal management and control policy design of the system with regard to certain criteria and constraints is the ultimate goal of the optimization of the system. A simulation model is essential for the optimization of the system. The applications of system dynamics models in different areas of energy systems are covered in Chapters 8 and 9.

### 2.3.2   CRITICAL ASPECTS OF SYSTEMS THINKING

The following aspects of systems thinking are very important for studying the dynamic behavior of complex systems and need attention to develop models based on systems thinking:

(1) Thinking in terms of cause-and-effect relationships.
(2) Focusing on the feedback linkages among components of a system.
(3) Determining the appropriate boundaries for defining what is to be included within a system.

We are interested in studying and examining the dynamic behavior of systems containing technological, environmental, technological and socio-economic components. Formulating a model with this purpose in mind should start with the question: "Where is the boundary of the dynamic system?" In concept, a feedback system is a closed-loop system and the dynamic behavior arises within the system

### 2.3.3 PARTICIPATORY SYSTEMS THINKING

System dynamics uses simulation models for policy design and policy analysis, and it is based on the feedback concepts of control theory. More specifically, system dynamics uses feedback loops, stock and flow diagrams, and non-linear differential equations. Stakeholders form an important part of system dynamics methodology (Forrester, 1961; Gardiner & Ford, 1980; Vennix, 1996, 1999; Hsiao, 1998; Elias et al., 2000; Maani & Cavana, 2000). Group model building is defined as a model-building process that deeply involves the client group in the process of modeling (Vennix, 1996 & 1999; Andersen & Richardson, 1997; Rouwette et al., 2011).

Participatory modeling includes a broad group of stakeholders in the process of formal decision analysis. It is a process of incorporating stakeholders – often the public – and decision-makers into the modeling process (Voinov and Gaddis, 2008). Non-scientists are engaged in the scientific process, and the stakeholders are involved to a greater or lesser degree in the process. A fully participatory process is one in which participants help identify the problem, describe the system, create an operational computer model of the system, use the model to identify and test policy interventions, and choose one or more solutions based on the model analysis. Involving the stakeholders in the model-building process can build trust among stakeholders (Tàbara and Pahl-Wostl, 2007).

Participatory system dynamics modeling uses a system dynamics perspective in which stakeholders or clients participate to some degree in different stages of the model building process. Participatory system dynamics modeling is more than simply eliciting knowledge from clients about the problem and the system – it involves building shared ownership of the analysis, problem, system description and solutions, or a shared understanding of the trade-offs among different decisions. In other words, it may be termed participatory systems thinking.

## 2.4  CAUSAL LOOP DIAGRAMS

System dynamics methodology is based on the feedback concepts of control theory, and a causal loop is a convenient way of representing the feedback loop structure of systems. A causal loop diagram is used to diagrammatically represent the feedback loops of systems and is a communication tool of feedback structure representing the principal feedback loops of the systems that generate the reference dynamic behavior of the systems.

Causal loop diagrams identify the principal feedback loops of the systems. The causal loop diagrams are used to describe basic causal mechanisms hypothesized to generate the reference mode of behavior of the system over time. A feedback loop contains two or more causally related variables that close back on themselves. The relationship between one variable and the next in the loop can be positive or negative. A positive relationship means that if one variable increases, the other also increases. For example, in Figure 2.6, the arrow from A to B means that an increase in A causes an increase in B. It can also mean that if A decreases, B will

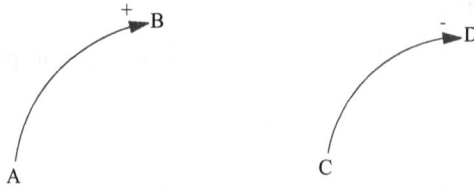

**FIGURE 2.6**   Cause and effect relationships.

also decrease. The arrow starting from A and terminating at B with a (+) sign at its end means the cause–effect relationship is positive. In a negative relationship, the two variables change inversely. For example, in Figure 2.6, the arrow in the direction of C to D means that if C increases, D will decrease. It can also mean that if C decreases, D will also increase. The arrow starting from C and terminating at D with a (-) sign at its end means the cause–effect relationship is negative.

Figure 2.7 shows an example of a causal loop diagram with notation. A simple causal loop diagram of population in Figure 2.7(a) consists of two variables, population growth and population. This figure illustrates a simple positive feedback loop consisting of two cause–effect relationships. In this example, an increase in population will cause an increase in population growth. The cause–effect relationship is positive and is indicated by an arrow with a (+) sign starting at population and terminating at population growth. The cause–effect relationship between population growth and population is also positive. An increase in population growth will cause an increase in population. This is indicated by an arrow with a (+) sign starting at population growth and terminating at population. The loop formed of population to population growth and back to population is a reinforcing loop and is indicated by a (+) sign with an arrow inside the causal loop diagram. Figure 2.7(b) also shows a positive feedback loop of a simple electricity production model consisting of electricity production, revenues from electricity, profitability from electricity, new investment for electricity and installed capacity of power plant. This model has five positive cause–effect relationships. Figure 2.8 shows a negative feedback loop consisting of installed capacity of electric power plant, power plant initiation rate and electric power plant under construction, and has two positive cause–effect relationships and one negative cause–effect relationship. Figure 2.9 illustrates two feedback loops. In loop 1, the number of births increases with the population and the births in turn increases population. These are positive relationships and are represented by arrows with a (+) sign. In loop 1, there are two positive relationships. Hence, it is a positive and re-enforcing loop. In loop 2, the number of deaths increase with population and the population decreases as the deaths increase. In loop 2, the first relationship is positive and is represented by an arrow with a (+) sign, while the second relationship is negative and is represented by an arrow with a (-) sign. One can easily determine if a loop is positive or negative by counting the number of negative relationships in the loop. If there are an

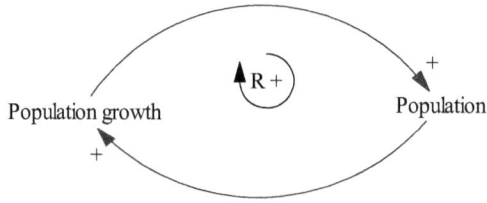

(a)

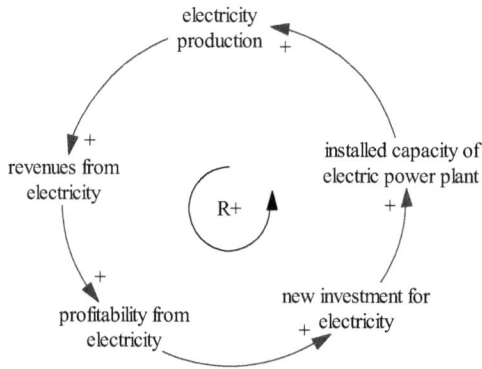

(b)

**FIGURE 2.7**    Positive feedback loop.

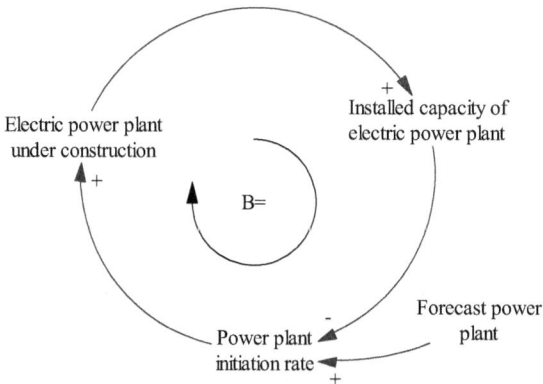

**FIGURE 2.8**    Negative feedback loops.

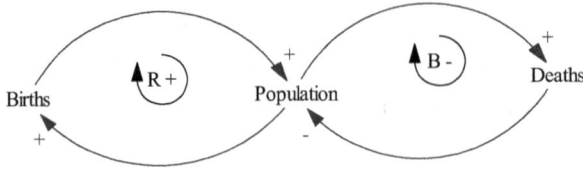

**FIGURE 2.9**  Positive and negative feedback loops.

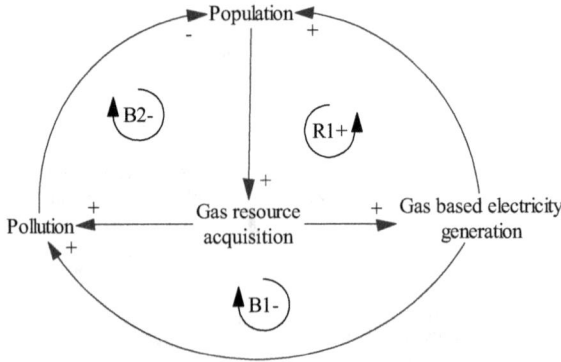

**FIGURE 2.10**  Coupled feedback loops of population, gas-based electricity generation and pollution.

even number of negative relationships in total in a feedback loop, the loop is positive; if there are an odd number of negative relationships, the loop is negative. Positive feedback loops generate growth, i.e., the loop is re-enforcing, and negative feedback loops are goal-seeking. In loop 2, the number of negative relationships is odd, i.e., 1. Hence, this loop is negative and goal-seeking.

Let us now consider a coupled feedback loop of population, gas-based electricity generation and pollution, as shown in Figure 2.10. The loop at the right corner is positive, i.e., re-enforcing, while the loops at the bottom and left corner are negative, i.e., balancing loops. In the positive loop, all the cause–effect relationships are positive, while in the negative feedback loops, the number of (–) relationships is 1, i.e., an odd number.

Let us now consider an interconnected feedback loop of electric supply, demand and price model, as shown in Figure 2.11. There are four re-enforcing loops and two negative, i.e., balancing loops. In the positive loop, all the cause–effect relationships are positive, while in negative feedback loops, the number of (–) relationships is 1, i.e., an odd number.

### 2.4.1  STEPS IN A CAUSAL LOOP DIAGRAM

We must consider the description of the system and the dynamic behavior of the reference modes when constructing the causal loop diagram, and these can aid in

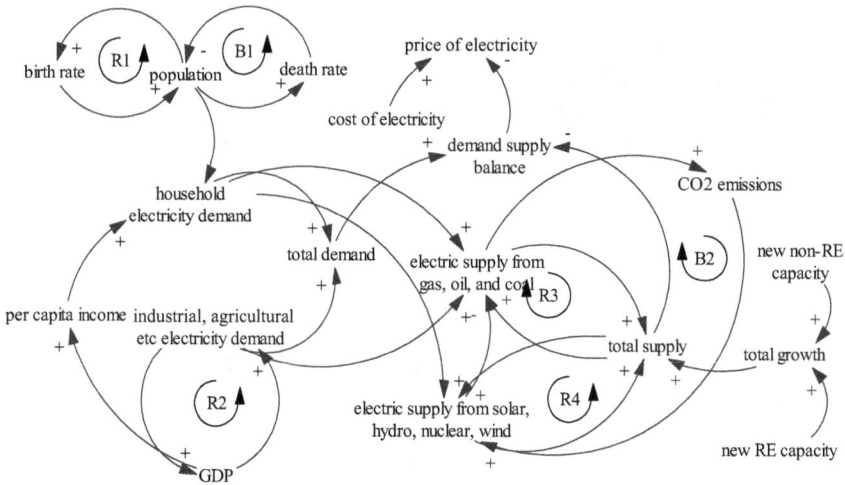

**FIGURE 2.11** Interconnected feedback loops of a simple electric supply, demand and price model.

developing a dynamic hypothesis. The following steps need to be followed in order to develop causal loop diagrams:

(1) **Define the problem and the objectives**

We must first study the system based on the information collected through interviews, focus group discussions, research reports and case studies. We must describe the system and define the problem with the reference mode of the behavior of the system.

(2) **Identify the most important elements of the systems**

We should identify the key variables affecting the behavior of the system, as this should be a good starting point for developing the causal loop diagram. Other variables can be added during the later stages of causal loop development.

(3) **Identify the secondary important elements of the systems**

Secondary variables within the system boundary should be added after careful identification of the most important variables. This would provide an opportunity to consider the secondary variables of the system of importance in the causal loop diagram.

(4) **Identify the tertiary important elements of the systems**

Tertiary variables within the system boundary should be added after careful identification of the secondary variables. However, tertiary variables of little importance can be omitted in the later stages once it is established by simulated studies.

(5) **Define the cause–effect relationships**

Find the cause–effect relationships using arrows with polarity first for the primary variables, then for the secondary and tertiary variables.

(6) **Identify the closed loops**
Trace the closed loops formed by the cause–effect relationships for the variables describing the system.

(7) **Identify the balancing and re-enforcing loops**
Identify the number of negative cause–effect relationships in each of the closed loops. The closed loops with an odd number of negative relationships are negative, i.e., balancing loops, and the others are positive, i.e., re-enforcing loops.

## 2.4.2 EXAMPLES OF CAUSAL LOOP DIAGRAMS

Now we will consider some examples of cause–effect relationships and hence the causal loop diagrams for the problems in the various fields of energy systems such as renewable energy and non-renewable energy. However, importance is placed on electric power systems to provide an understanding of how to develop a qualitative model of electric power systems in the form of causal loop diagrams.

### 2.4.2.1 Population Model

Population has been growing exponentially throughout history. Population increases by a fixed percentage per year and also decreases by a fixed percentage per year because humans have limited lives. Draw the causal loop diagram of a simple population model.

**Solution**
Figure 2.12 shows the causal loop diagrams of a simple population model. The model consists of two fundamental loops. The regenerating loop R generates new birth and adds it to population. The balancing loop B creates the death and depletes it from the population. The birth creates a positive loop since all the

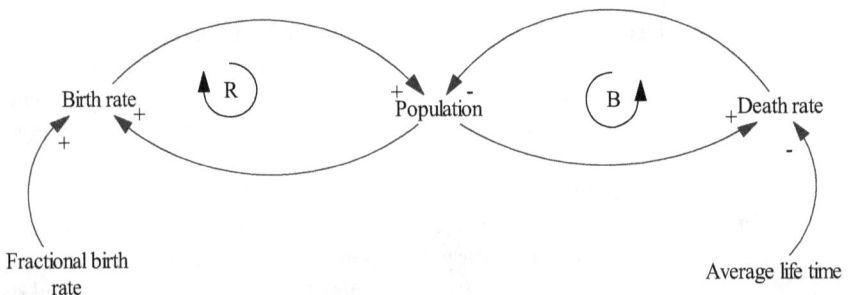

**FIGURE 2.12** Causal loop diagram of a simple population model.

cause–effect relationships within this loop are positive, while there is an odd number of negative relationships within the balancing loop, and hence it is a negative feedback loop.

## 2.4.2.2 Electricity Supply Model

Energy is needed for economic and social development (Bala, 1997a, 1997b, 1998). Per capita consumption of energy is a measure of physical quality of life (Bala, 1998). Per capita consumption of electrical energy is also a measure of physical quality of life. Energy demands are increasing rapidly. Investments for power plants are triggered by the power supply gap, price, and public pressure. Indicated price depends on revenue received, which creates government pressure to increase the price as the indicated price increases, while the public pressure for an increase in price creates pressure to decrease the price. A power plant under construction becomes operational after some time delay and adds to the installed capacity to supply electricity and hence reduces the pressure on new investments. Draw the causal loop diagram of a simple electric supply system.

**Solution**

The causal loop diagram of a simple electricity supply system is shown in Figure 2.13. Investment increases installed capacity, which in turn invites less investment. This forms the negative feedback loop B1. The increase in installed capacity increases power supply, and the increased power supply reduces load-shedding. Load-shedding increases public pressure, which increases investment.

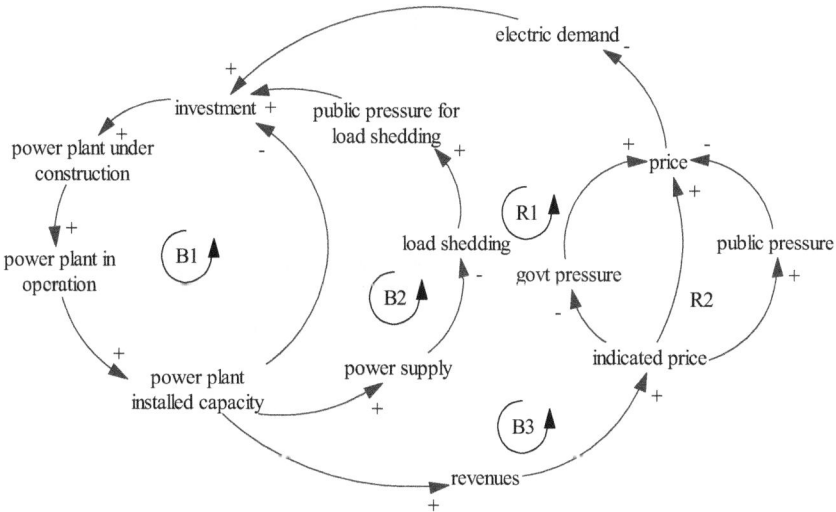

**FIGURE 2.13** Causal loop diagram of an electricity supply system.

Increased investment results in increased installed capacity. This forms the negative feedback loop B2. Revenues, indicated price, price, electric demand, investment and installed capacity form the negative feedback loop B3 while revenues, indicated price, government pressure, public pressure for increase in price, electric demand, investment and installed capacity form two positive feedback loops, and these are indicated as R1 and R2.

### 2.4.2.3  Electricity Supply, Demand and Price Model

Electrical energy demand is increasing globally for economic development and better quality of life. The rapid depletion of fossil fuels, the price upsurge during the 1970's energy crisis and climate change due to energy production and use have created interest in a gradual transition to renewable energy resources to ensure energy security and reduce emissions and contributions to global warming. The simulation of different energy supply strategies, such as the gradual transition to renewable energy resources, including hydro, nuclear and solar, and the target energy is essential for developing energy scenarios for a clear understanding of the different energy supply options and to find the policy to be adopted. Estimates of the projection of energy demand and price are also needed for energy planning, and these are essentially required for optimal operational planning. Draw a causal loop diagram of electricity supply, demand and price.

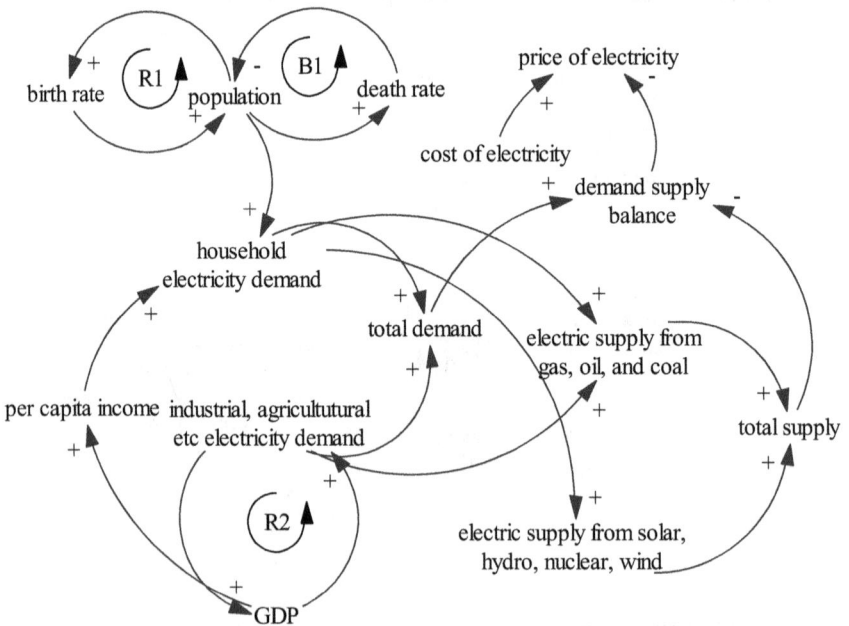

**FIGURE 2.14**  Causal loop diagram of electricity supply, demand and price system.

**Solution**

The causal loop diagram of a simple electricity supply, demand and price system is shown in Figure 2.14. The total demand for electricity increases with the increase in household electricity and industrial and other electricity demand, while the total supply of electricity increases with the increase in supply of non-renewable energy (gas, oil and coal) and renewable energy (solar, hydro, nuclear and wind). Household electricity demand depends on population and GDP. Population and birth rate form a positive feedback loop R1 while population and death rate form a negative feedback loop B1, but the positive feedback loop is dominant. GDP and industrial and other electricity demands form a positive feedback loop R2. The price of electricity is determined by production cost and demand–supply balance. Thus, there are two positive feedback loops and one negative feedback loop simulating the dynamics of electricity supply, demand and price.

### 2.4.2.4   Ethanol Production Model

Fossil fuels such as coal, oil, and gas are the major sources of world energy, accounting for more than 80% of the total energy production. The price of oil has been escalating due to increasing demand, especially from the energy-intensive economies of developing countries, geopolitics, and speculation about the shortage of supplies, among other factors. Ethanol is a biofuel with the potential to promote energy independence. Ethanol is a high-octane fuel that can be used as a gasoline additive and extender. It can be produced by the fermentation of carbohydrates such as sugar, starch and cellulose. It is a relatively clean fuel with wide applications in transportation, and manufacturing would promote sustainable production. Brazil is a pioneer in the production of ethanol from sugarcane. Brazilian production accounted for 38.8 million tons of sugar and 23.9 billion liters of ethanol. This represents a share of 41.8% of the total world production of ethanol, estimated at 49 billion liters. Despite this, the ethanol produced from sugarcane has recently become the subject of intense debate by European authorities that consider the expansion of biofuel production around the world as the reason for a rise in food inflation. To assess the consequences of such expansion in the environment and in food production, a system dynamics model of ethanol production should be developed and simulated to address these policy issues. Draw the causal loop diagram of an ethanol production system.

**Solution**

The causal loop diagram of an ethanol production system is shown in Figure 2.15. Sugar production increases with the increase in sugarcane production, but ethanol production decreases sugar production. Sugarcane price and sugarcane demand form a negative feedback loop B1. Sugarcane demand increases sugar production. Sugarcane production, sugar production, sugar price, revenues from sugar and land for sugarcane forms a negative feedback loop B2. Ethanol price and ethanol demand form the negative loop B3, while diesel price and diesel demand form a negative feedback loop B4. Sugarcane production, sugar production, ethanol production, ethanol price, revenues from ethanol and land for sugarcane production forms a

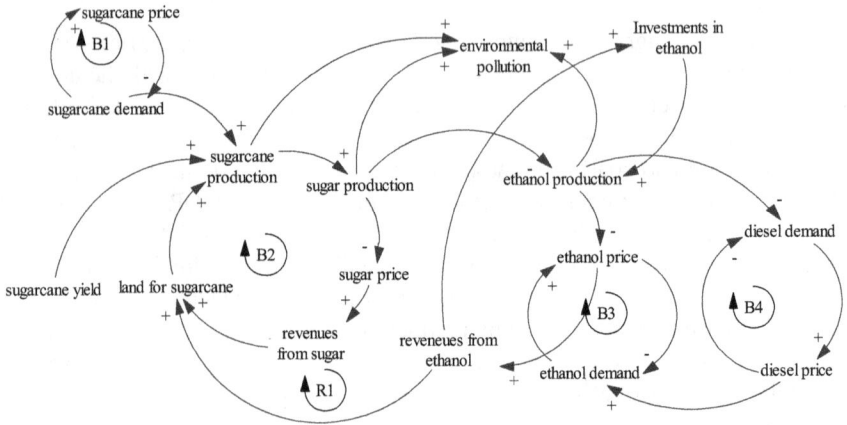

**FIGURE 2.15**   Causal loop diagram of an ethanol production system.

positive feedback loop R1. Thus, there is one positive feedback loop and four negative feedback loops simulating the dynamics of an ethanol production system.

### 2.4.2.5   Fuelwood Supply and Afforestation Model

Accelerating rates of deforestation in tropical and subtropical countries are of great public concern. Land clearing for human settlement, the felling of trees for timber and exploitation of the wood resource for its use as cooking fuels play major roles in deforestation in tropical and subtropical countries. Deforestation causes the loss of carbon sink, which renders these countries susceptible to climatic hazards. To avoid this, allowable cutting should be adopted, and there must be an afforestation program to maintain the recommended level of forest cover. Draw the causal loop diagram of a fuelwood supply and afforestation model.

### Solution

The causal loop diagram of a fuelwood and afforestation system is shown in Figure 2.16. The immature area increases with an increase in planting rate, which depends on an afforestation program and in turn decreases with the increase in immature planting areas. This forms a negative feedback loop B1. The fuelwood forest and felling rate forms a negative feedback loop B2, while the fuelwood forest, felling rate and clear cutting rate form a negative feedback loop B3. Mature area increases with the increase of the immature area, which in turn increases with the increase in planting rate, while the planting rate is decreased with an increase in mature area. This forms a negative feedback loop B4. The fuelwood forest and regrowth rate forms a positive feedback loop R1. Thus, there is one positive feedback loop and four negative feedback loops simulating the dynamics of the fuelwood forest status, clear cutting area, mature and immature areas.

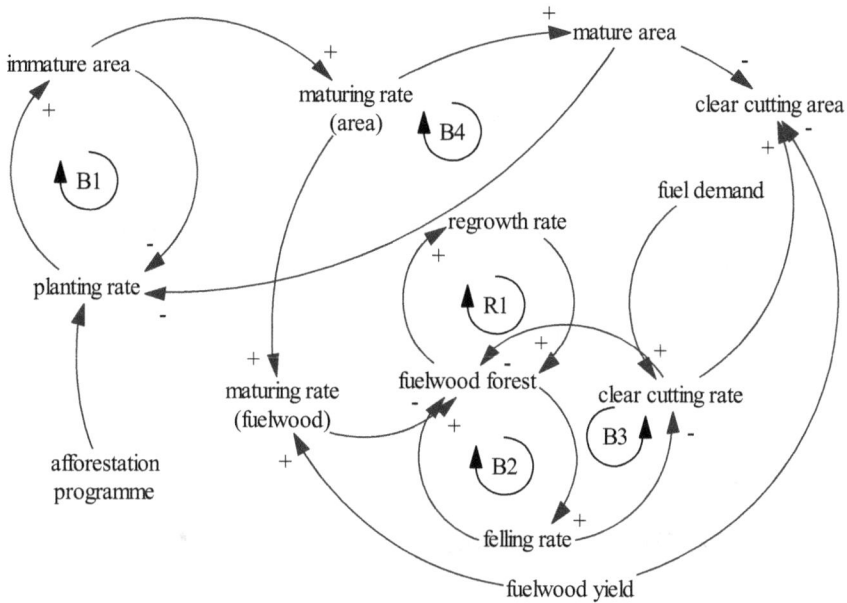

**FIGURE 2.16** Causal loop diagram of fuelwood supply and afforestation system.

### 2.4.2.6 Global Warming Model

Global warming refers to an increase in the average atmospheric temperature resulting from the greenhouse effect and is a prime concern in many developing countries. Earth's surface average temperature is increased by the emissions of greenhouse gases and incoming solar radiation. The emissions are reduced by the carbon sink of the plant kingdom and algae in the sea. Water vapor increases the absorbed radiation. Draw a causal loop diagram of this simple global warming model.

**Solution**

The causal loop diagram of a simple global warming model system is shown in Figure 2.17. Absorbed radiation is increased by solar radiation and $CO_2$ in the atmosphere, and water vapor. The increase in air temperature causes an increase in photosynthesis in the plant kingdom and algae in the water body, which decreases the $CO_2$ in the atmosphere. These create two negative feedback loops B1 and B2. Air temperature increases water vapor, which increases the absorbed radiation. This forms a positive feedback loop R1. Thus, there is one re-enforcing loop and two distinct balancing feedback loops B1 and B2, and these three loops simulate the dynamics of air temperature, i.e., global warming.

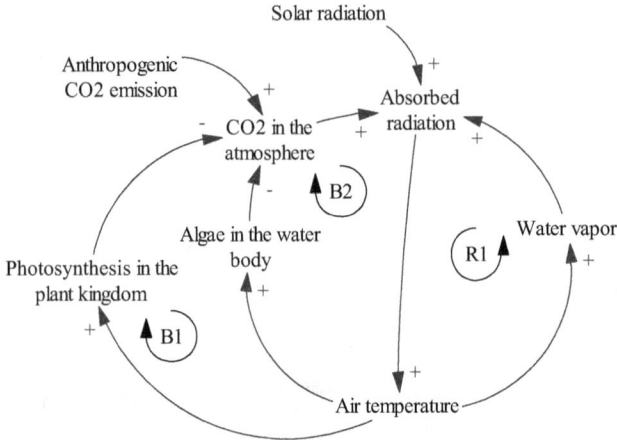

**FIGURE 2.17**   Causal loop diagram of global warming.

## 2.5   STOCK–FLOW DIAGRAMS

One approach to studying a system is by verbal description, and another approach is a causal loop diagram. Causal loops are wonderful for telling the story of the dynamic behavior of complex and dynamic systems to others and represent the verbal and mental model more clearly using cause and effect relationships. But to show the relationship and to accentuate the loop structure of a system, a flow diagram is best since causal loops suffer from their inability to capture the stock and flow of the systems. Essentially, stock and flow represent systems of first-order difference equations central to the system dynamics theory.

A feedback loop is the basic structural element within the system boundary. The boundary separates the system from the surrounding environment. Dynamic behavior is simulated by feedback loops. A feedback loop is a path coupling decision and action, level (or condition), and information with the path returning to the decision point (Figure 2.18). The decision process is one that controls action. It can be the governing process in biological development, the valve and actuator in a chemical plant, or a decision process in electric power plant construction. Whatever the decision process, it is embedded in a feedback loop. The decision is based on available information; the decision controls an action that influences a system level, and new information arises to modify the decision stream. A system may be a single loop or interconnected loops.

Stocks and flows are the two basic building blocks in system dynamics modeling. The equations of a system focus on the composition of each stock and flow. A stock–flow diagram should show how stocks and flows are interconnected to produce feedback loops and how the feedback loops interlink to create the system. The symbols used in STELLA stock–flow diagrams are shown in Figure 2.19. Stock–flow diagrams as regards STELLA stock–flow diagrams are described in the following subsections.

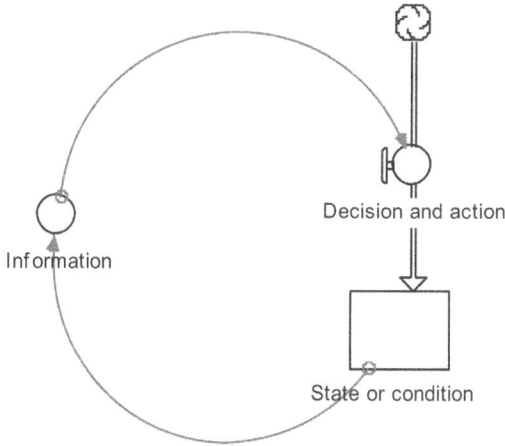

**FIGURE 2.18** Feedback loop.

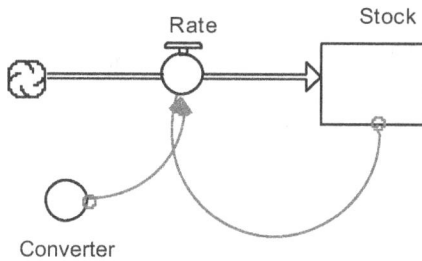

**FIGURE 2.19** Symbols of STELLA flow diagrams.

### 2.5.1 STOCK

The first basic building block in system dynamics modeling is the stock, and this variable describes the condition or state of the system at any particular time. These variables are accumulations in the system. Examples of these variables are population, water in a reservoir, or power plant installed capacity in an electric power system. These are physical accumulations, while examples of non-physical accumulations are public pressure, skill and knowledge.

All stock equations and any special function that represents integration are represented by a rectangle (Figure 2.20). A simple single stock model of population in STELLA is shown in Figure 2.21. The flow into the system is the birth rate and the flow out of the system is the death rate. The accumulation of the population is the stock of the model.

Stocks integrate flows into and out of the stock, and the net flow into the stock is the rate of change of the stock. The stock, which is represented by a first-order integral equation, is:

stock

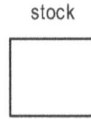

**FIGURE 2.20**   Symbol for stock.

population

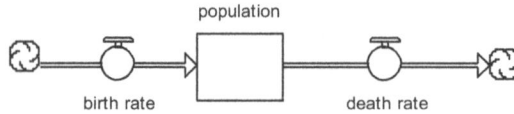

birth rate                                        death rate

**FIGURE 2.21**   Flow diagram of a stock of population.

**Stock**

$$\text{population}(t) = \text{population}(t - dt) + (\text{birth\_rate} - \text{death\_rate}) * dt \qquad (2.1)$$

Equation (2.1) is in the form of a first-order finite difference equation. This simple population model shows that population increases by birth rate and decreases by death rate.

### 2.5.2   Flow

The second building block is the flow, and these variables tell how fast the stocks are changing. Flows are represented by valve symbols (Figure 2.22). Examples of these variables are inflow of birth rate and outflow of death rate to the stock–population, as shown in Figure 2.23. The birth rate depends on the population size and birth fraction, while the death rate depends on the population size and death fraction.

The inflow and outflow equations are:

**Inflow**

$$\text{birth rate} = \text{population*birth fraction} \qquad (2.2)$$

**Outflow**

$$\text{death rate} = \text{population*death fraction} \qquad (2.3)$$

### 2.5.3   Converter

In the STELLA software, the converter serves a utilitarian role. It holds the values for constants, defines external inputs to the model, calculates algebraic

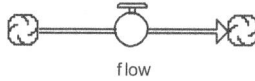

**FIGURE 2.22**   Symbol for flow.

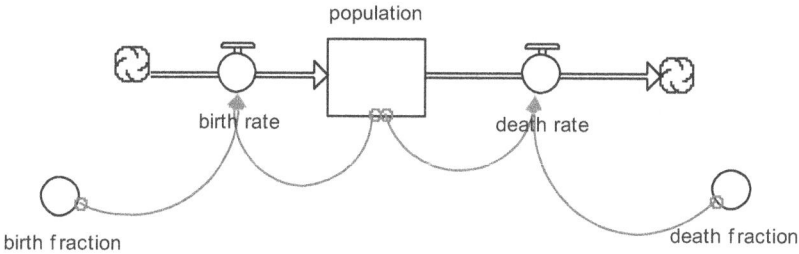

**FIGURE 2.23**   Flow diagram showing flows to stock of population.

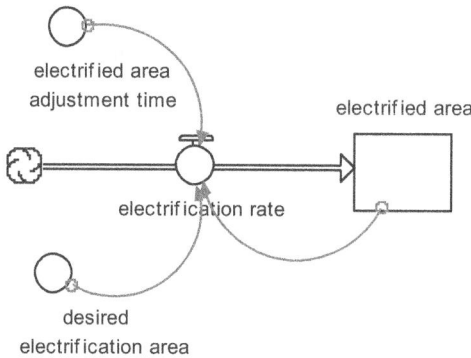

**FIGURE 2.24**   Flow diagram illustrating converter.

relationships, and serves as the repository for graphical functions. In general, it converts inputs into outputs, hence the name converter. In Figure 2.24, the desired electrification area and electrified area adjustment time are converters.

### 2.5.4   DELAYS

Delay is the time between the action and the result (consequence) of this action. A delay is a commonly observed phenomenon in material and information flow, and it takes time for the material to flow and the information to process. A time delay is defined as the time required for the flow of material or information, and more precisely is the time by which the output lags behind the input. Thus, it may be a material delay or an information delay. A material delay causes a delay in

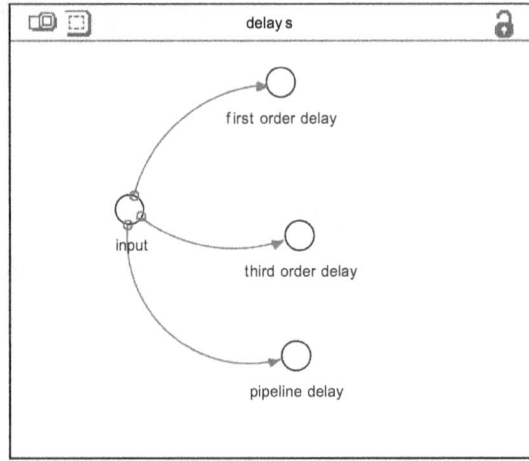

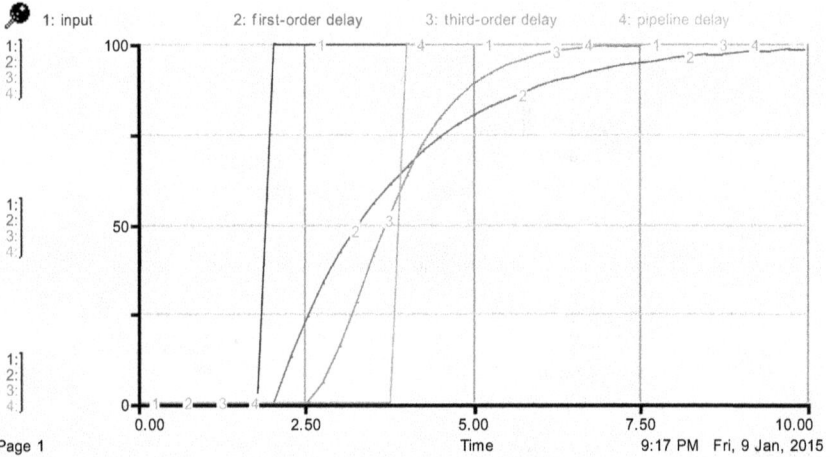

**FIGURE 2.25** STEP response of first, third order and pipeline delay.

the supply of the material, and an information delay is a delay in processing the information.

In a pipeline delay, the delay time is constant, and the order of output is precisely the same as the order of entry. The delay may be first-order, second-order, third-order or even higher order. In a first-order delay, the outflow is always proportional to the stock of the material in transit. A second-order delay consists of two first-order delays in series, with each time delay being half of the original time delay. A third-order delay consists of three first-order delays in series, with each time delay being one-third of the original time delay. Higher-order delays approach the pipeline delay. Figure 2.25 shows the responses of different types of delay functions.

**Delays**
first_order_delay = SMTH1(input,2,0)
input = STEP(100,2)
pipeline_delay = DELAY(input,2,0)
third_order_delay = SMTH3(input,2,0)

### 2.5.4.1   Role of Delay

Delay plays an important role in information and material flow, hence the behavioral performance of the system and the key points that deserve special attention are:

(1) Acknowledge delay as a **factor** in decision-making.
(2) Respect delay as an **element** to understanding success or failure.
(3) Regard delay as a **force** in determining the value of change.

### 2.5.4.2   Choice of Delay Function

The choice of delay function is quite important when inputs of sharp variations such as STEP functions are used. The difference in the response of third- and higher-order delays are not quite large. For practical purposes, the three main possibilities are a first-order delay, a third-order delay and a pure or pipeline delay. When the input has an immediate effect and then a long tail, a first-order delay is recommended. But when the output is the exact copy of the input, a pipeline delay is appropriate. In intermediate cases where the input has a delayed response and a spread of inflow, a third-order delay is recommended. A third-order delay is usually a good compromise well within the accuracy of the commonly available data, which is why it is commonly used in system dynamics modeling.

### 2.5.5   IDENTIFICATION OF STOCK AND FLOW

In control theory, stocks are known as state variables and flows are known as derivatives, while in system dynamics, these are known as levels and rates, respectively. In system dynamics modeling, first, the variables within the system boundary must be identified. This is followed by the identification of which variables are affected or influenced by other variables and the affected variables that represent the state of the system, i.e., the quantities of material or other materials are stocks. The others, having the unit per unit time of the unit of the stock variables, are flows. It should be remembered that stocks are quantity and flows are quantity per unit time. Stocks are only changed by flows, and the contents of stocks and flow networks are conserved. Converters (auxiliaries) are an elaboration of flows and hence are affected by stocks and constants. For a simple population model, the variables are birth rate, death rate and population. The population has a unit of quantity, while the birth rate and death rate have a unit of quantity per unit time. The population level is changed by birth rate and death as inflow and outflow, respectively. Hence, the population is a stock variable, and birth rate and death rate are flow variables. For a simple solar home model, the variables are solar home construction rate, solar home discard rate and solar home. The solar home has a unit of quantity, while the solar home construction rate and solar home discard rate have a unit of quantity per unit time. The solar home level is changed by the solar home construction rate and the solar home discard rate as inflow and outflow, respectively. Hence, the solar home is a stock variable, and the solar home construction rate and solar home discard rate are flow variables.

### 2.5.6  MATHEMATICAL REPRESENTATION OF STOCK AND FLOW

Integration or accumulation creates the dynamic behavior of systems, and integration occurs naturally both in physical and biological systems. A stock equation can be represented by a first-order finite difference equation and expressed as:

$$stock(t) = stock(t_0) + \int_{t_0}^{t} (inflow(t) - outflow(t))dt \qquad (2.4)$$

This integration equation in the differential equation form is:

$$\frac{d(stock)}{dt} = infflow(t) - outflow(t) \qquad (2.5)$$

Equation (2.5) is an initial value problem, and several methods are available to solve it. STELLA provides three methods: (i) the Euler method, (ii) the second-order Runge–Kutta method, and (iii) the fourth-order Runge–Kutta method. For better prediction of the model, the Runge–Kutta fourth-order method is recommended. The numerical solution procedures of these methods are detailed in Bala et al. (2017).

### 2.5.7  SOLUTION INTERVAL

The proper length of the solution interval is related to the shortest delays that are represented in the model. If the solution interval is too long, instability is generated that arises from the computing process, not from any inherent dynamic characteristic of the system itself. If the solution interval is too short, the equation will be evaluated unnecessarily, often using extra computer time. As a practical rule of thumb, the solution interval should be half or less of the shortest first-order delay in the system.

A model should always be tested for sensitivity to step size. The model should initially be run for the full length of the simulation. Some variables should be plotted and printed. Next, the step size should be cut in half, and the simulation should be repeated. This process should be continued until the improvements in results become negligible.

## 2.5.8   FUNCTIONS WITHOUT INTEGRATION

There are several types of special functions having the nature of flows (rates) or converter (auxiliary). These are special computation procedures, table interpolation, test inputs, randomness, and logical choices. These functions can be imposed in any flow channel, and they would not introduce any time delay or periodicity in the system. They only alter the instantaneous amplitude of the signals.

The group of special computation procedures includes the square root, exponential and logarithmic functions. In STELLA, there is no special symbol for each of these functions, and they are represented by converters.

**SQRT(<expression>)**
The SQRT function gives the square root of *expression*. *Expression* can be variable or constant. For meaningful results, *expression* must be greater than or equal to 0.

**Examples**

SQRT(400) returns 20

**EXP(<expression>)**
The EXP function gives e raised to the power of *expression*. *Expression* can be variable or constant. The base of the natural logarithm e is also known as Euler's number. e is equal to 2.7182818... EXP is the inverse of the LOGN function (natural logarithm). To calculate the powers of other bases, use the exponentiation operator (^).

**Examples**

EXP(1) equals 2.7182818 (the value of Euler's constant)
EXP(LOGN(3)) equals 3

**LOGN(<expression>)**
The LOGN function calculates the natural logarithm of *expression*. *Expression* can be variable or constant. Natural logarithms use Euler's constant e, 2.7182818..., as a base. For LOGN to have meaningful results, *expression* must be positive. LOGN is the inverse of the EXP function e raised to the power of *expression*.

**Examples**

LOGN(2.7182818) equals 1
LOGN(EXP(3)) equals 3
LOGN(-10) returns? (undefined for non-positive numbers)

**LOG10(<expression>)**

The LOG10 function gives the base 10 logarithm of *expression*. *Expression* can be variable or constant. For LOG10 to be meaningful, *expression* must evaluate to a positive value. LOG10 is the inverse of the letter E in scientific notation, or base 10 exponentiation (10 raised to the power of *expression*).

**Examples**

LOG10(10) equals 1
LOG10(10E5) equals 5
LOG10(-10) returns? (undefined for non-positive numbers)

**Graphical function**

A second class of functions provides for interpolation in a table. Non-linear relationships appear repeatedly in system dynamics. Table functions locate by linear interpolation, and the values intermediate between the points entered in a table. In STELLA, these functions are represented by converters with a small shadow inside the converters, and the flow diagram symbol of graphical functions is shown in Figure 2.26.

In STELLA, a graphical function is essentially a sketch that relates between some input and output. This can be represented as:

graphical_function_name GRAPH (input_ variable_ name) $(x_1, y_1), (x_2, y_2), (x_3, y_3)$
.........$(x_n, y_n)$

where
   graphical_function_name is the name of the graphical function
   input_variable_name is the input variable for which the corresponding graph-
      ical/table entry is made.

$(x_1, y_1)$ is first graphical point with input variable $x_1$, and output variable $y_1$, $(x_2, y_2)$ is the second graphical point with input variable $x_2$, and output variable $y_2$. Similarly, $(x_n, y_n)$ is the nth graphical point. n is the number of points in the graphical function. Figure 2.27 illustrates the graphical function of ecological effect as a function of insect attack infestation. We need to provide data on the minimum and

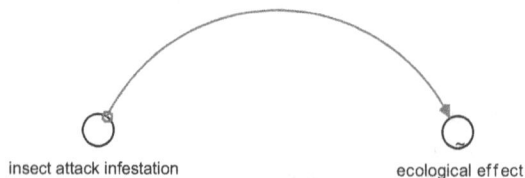

insect attack infestation                                      ecological effect

**FIGURE 2.26**   Graphical function.

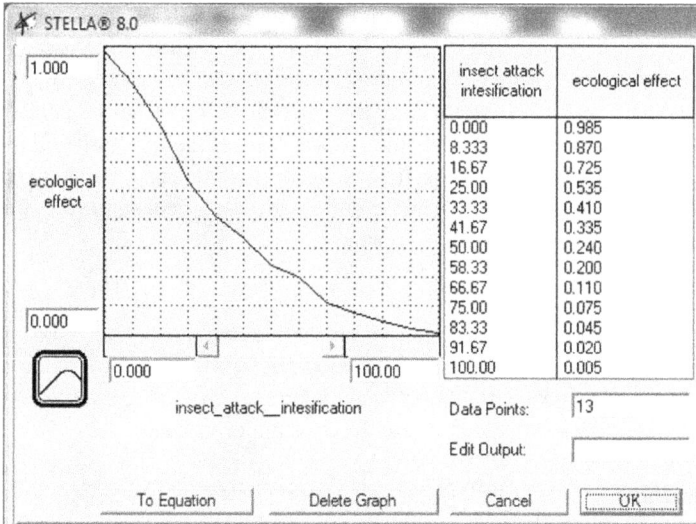

**FIGURE 2.27**   Illustration of graphical function.

maximum values of the X-axis and Y-axis, the number of data points and the x and y values of each point for the graphical function.

For example, indicated industrial electricity demand growth rate is computed from GDP and the relationship between indicated industrial electricity demand growth rate and GDP is expressed graphically as:

indicated_industrial_electricity_demand_growth_rate = GRAPH(gdp)
(1500, 0.0288), (2850, 0.033), (4200, 0.0384), (5550, 0.0432), (6900, 0.048), (8250, 0.0534), (9600, 0.0588), (10950, 0.063), (12300, 0.0672), (13650, 0.0726), (15000, 0.0756)

**Test input functions**
A group of functions consisting of STEP, RAMP, SIN and COS are primarily used as test inputs.

**STEP(<height>,<time>)**
STEP function produces a specified constant height starting at a specified time and is frequently used as a shock excitation to determine the dynamics of the system recovery.

**RAMP(<slope>[,<time>])**
The RAMP function generates a linearly increasing or decreasing input over time with a specified slope (*slope*). Optionally, you may set the time at which the ramp begins. Slope and time can be either variable or constant.

**PULSE(<volume>[,<first pulse>,<interval>])**
The PULSE function generates a pulse input of a specified size (*volume*). In using the PULSE function, you have the option of setting the time at which the PULSE will first fire (*first pulse*), as well as the interval between subsequent PULSEs (*interval*). Each time it fires a pulse, the software pulses the specified volume over a period of one time step (DT). Thus, the instantaneous value taken on by the PULSE function is volume/DT. Volume can be either variable or constant. The first pulse and interval should be specified as constants.

**SIN(<radians>)**
SIN generates a sinusoidal fluctuation having a unit amplitude and specified period. The SIN function gives the sine of *radians*, where *radians* is an angle in radians. To convert measurement between degrees and radians, use the identity: pi (radians) = 180 (degrees). *Radians* can be constant or variable.

**COS(<radians>)**
COS, in a similar fashion, generates a cosine function of unit amplitude and specified period. The COS function gives the cosine of *radians*, where *radians* is an angle in radians. To convert measurement between degrees and radians, use the identity: pi (radians) = 180 (degrees). *Radians* can be constant or variable.

**RANDOM(<min>,<max>[,<seed>])**
The RANDOM function generates a series of uniformly distributed random numbers between *min* and *max*. RANDOM samples a new random number in each iteration of a model run. If you wish to replicate the stream of random numbers, specify *seed* as an integer between 1 and 32767. To replicate a simulation, all variables that use statistical functions must have seeds specified. Each variable using a statistical function should use a separate seed value.

**MAX(<expression>,<expression>,...)**
The MAX function gives the maximum value among the expressions contained within parentheses.

**MIN(<expression>,<expression>,...)**
The MIN function gives the minimum value among the expressions contained within parentheses.

**SWITCH(<Input1>,<Input2>)**
The SWITCH function is equivalent to the following logic:
    If *Input1* > *Input2* then 1 else 0.

**Logical function**
The logical functions (IF, THEN, ELSE, AND, OR, NOT) are used to create expressions and then give values based upon whether the resulting expressions

are TRUE or FALSE. When you have multiple conditions, the expressions to be evaluated must be enclosed in parentheses ().

In STELLA, the logical functions are used to create expressions and give values based upon whether the resulting expressions are TRUE or FALSE. The example of bonus payments logic given below is essentially a logical function:

bonus_payments = IF (TIME = 5) OR (sales > 5000) THEN Bonus ELSE 0

This statement sets Bonus payments to the value of Bonus at simulated time 5, or whenever the value of sales is greater than 5000. When neither condition is met, the statement gives the value 0.

The example of a radiation pollution given below is also essentially a logical function:

radiation_pollution = IF (nuclear_installed > 0) THEN (0.035) ELSE (0)

The radiation pollution is set to a value of 0.035 to indicate the expected impact of radiation pollution if there is a power plant; otherwise, it is 0.

### 2.5.9 FUNCTIONS CONTAINING INTEGRATION

The functions SMOOTH, information delay, and material delay contain integration. These are normally inserted in the flow channel or information channel in a model. Since these functions include stock, they change the time shape of quantities moving between their inputs and outputs. These functions are assemblies of elementary stock and flow equations.

### SMOOTH

SMOOTH produces the first-order exponential average of a physical flow and is represented as shown in Figure 2.28.

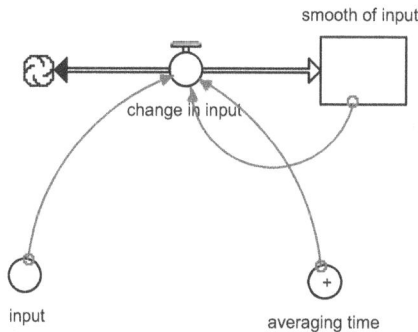

**FIGURE 2.28** Flow diagram of SMOOTH.

## 2.5.10   Information Delay

### 2.5.10.1   First-Order Information Delay

First-order information delay is used in the information channel to produce a first-order exponential delay. It represents the process of gradual, delayed adjustment of recognized information moving towards a value being supplied by a source. It is used to generate a delayed awareness of a changing situation. The flow diagram for a first-order information delay is shown in Figure 2.29 (top).

### 2.5.10.2   Third-Order Information Delay

Third-order information delay is a cascade of three first-order information delays, as shown in Figure 2.29 (bottom). The equations and flow diagram show three sections like that for first-order information delay, except in each section, the delay is one-third of the total.

The delay may be first-order, second-order, third-order and even higher order. Third-order and first-order information delays are placed in the information channel, but third-order information delays produce a different time shape in output. Figure 2.30 shows the step responses of first-order and third-order information delays.

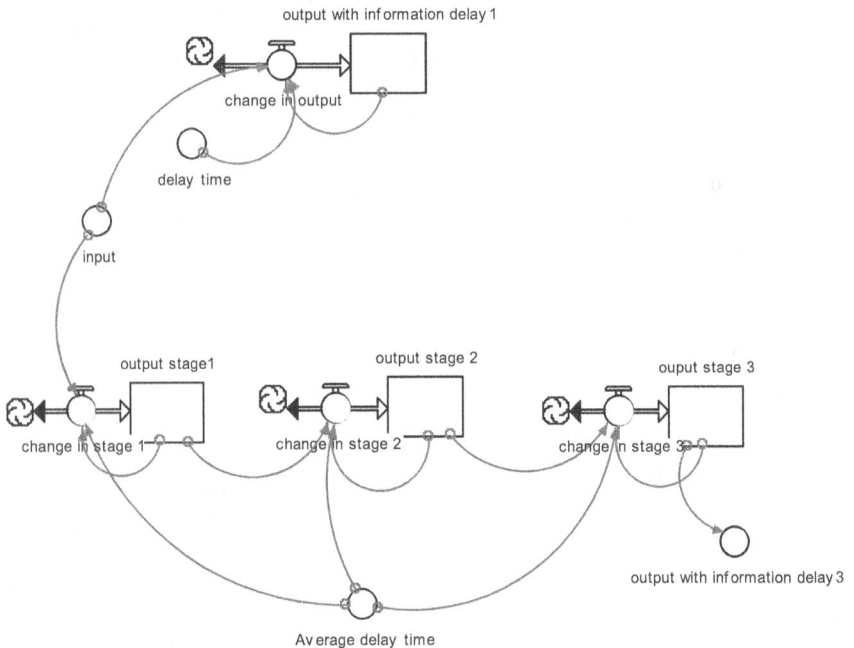

**FIGURE 2.29** Flow diagram of first-order information delay (top) and third-order information delay (bottom).

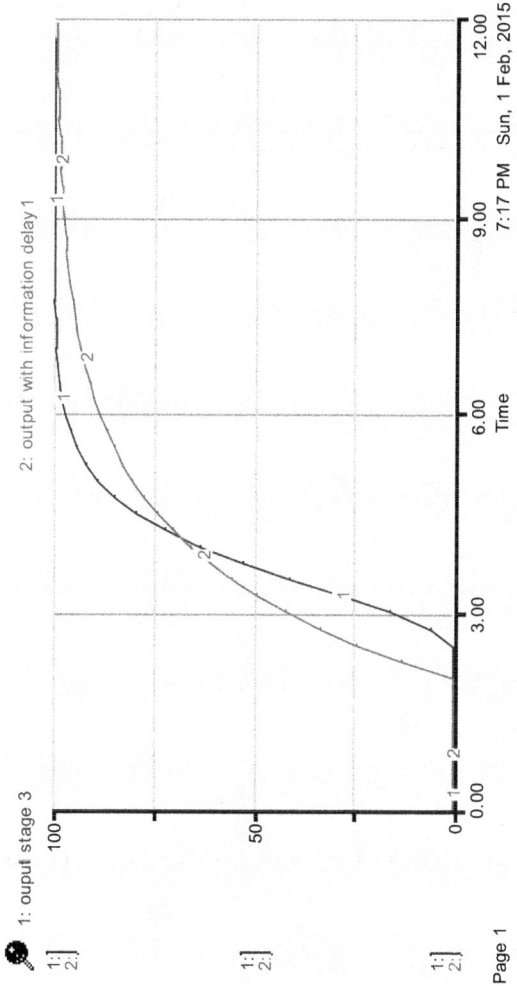

**FIGURE 2.30** STEP response of first and third-order information delay.

### 2.5.11 MATERIAL DELAY

The outflow from a first-order material delay is always proportional to the stock of material in transit. The flow diagram of a first-order material delay is shown in Figure 2.31 (top). A third-order material delay is similar to a third-order information delay and is also the three cascaded first-order exponential loops, as shown in Figure 2.31 (bottom). Therefore, it has the same response. It differs from a third-order information delay being placed in a flow channel and creates a delay in the transmission of quantity from the input to the output. Figure 2.32 shows the step responses of first- and third-order material delays.

### 2.5.12 EXAMPLES OF STOCK–FLOW DIAGRAMS

We have already discussed some examples of causal loop diagrams as qualitative modeling. Here we present some examples of stock–flow diagrams for simulating problems in the various fields of energy systems such as renewable energy and non-renewable energy using STELLA software. Importance has been placed on electric power systems so as to provide an understanding of how to develop a system dynamics simulation model of electric power systems using STELLA.

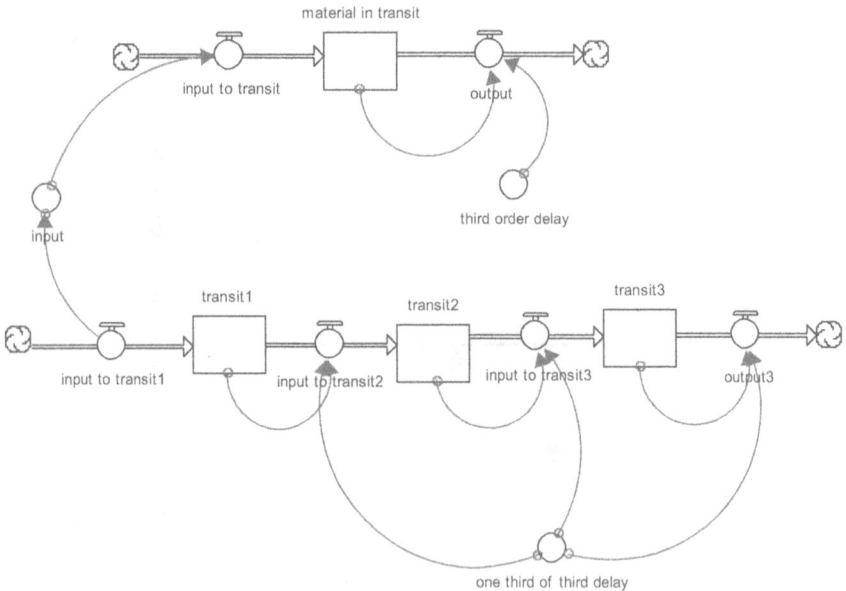

**FIGURE 2.31**  Flow diagram of first-order material delay (top) and third-order material delay (bottom).

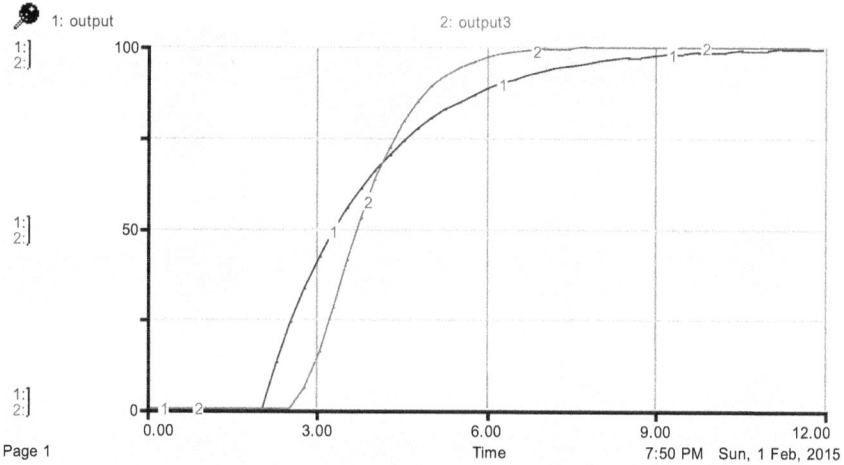

**FIGURE 2.32** STEP response of first and third-order material delay.

## 2.5.12.1 Electricity Demand Model

Electricity is needed for economic growth and a better standard of living. Demand for electrical energy is increasing with the increase in industries, agricultural development and population. Estimates of electrical energy demand are essential for energy planning. The stock–flow diagram of a simple electricity demand model is shown in Figure 2.33.

**Electricity demand**

Electricity demand in GWh is computed from population and per capita electricity demand.

electricity_demand = (population * per_capita_electricity_consumption)/1000000

**Per capita electrical energy consumption**

Per capita electrical energy consumption is expressed as a function of time graphically.

per_capita_electricity_consumption = GRAPH(TIME)
(0.00, 320), (3.00, 358), (6.00, 383), (9.00, 400), (12.0, 423), (15.0, 433), (18.0, 445), (21.0, 455), (24.0, 470), (27.0, 485), (30.0, 495)

**Population**

Population is increased by population growth rate.

population(t) = population(t - dt) + (population_growth_rate) * dt
INIT population = 165000000

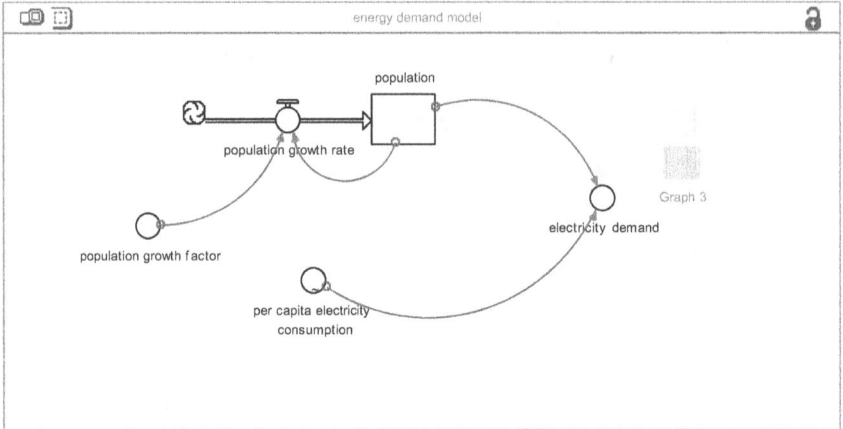

**FIGURE 2.33** Stock–flow diagram of electricity demand model.

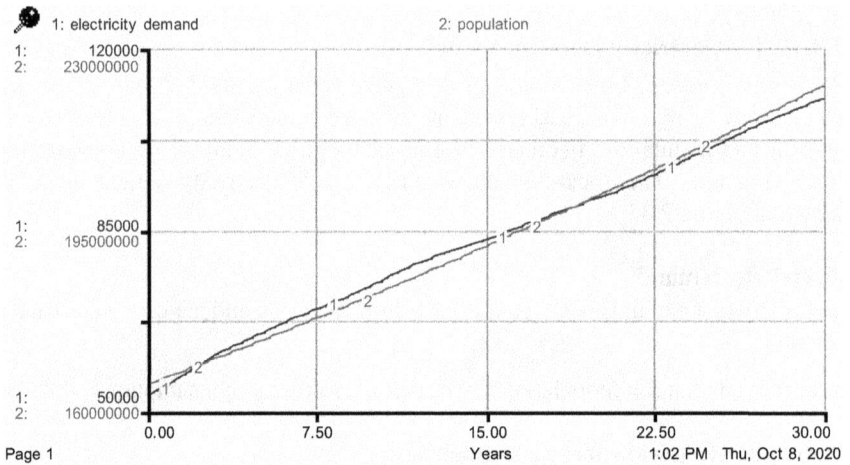

**FIGURE 2.34** Simulated population and electricity demand GWh for a period of 30 years.

**Population growth rate**

Population growth is assumed to depend on population and population growth factor.

population_growth_rate = population * population_growth_factor
population_growth_factor = 0.01

Figure 2.34 shows the simulated population and electricity demand for a period of 30 years and is increasing with time as expected.

## Listing of STELLA program

### Electricity demand model
electricity_demand = (population * per_capita_electricity_consumption)/1000000
population(t) = population(t - dt) + (population_growth_rate) * dt
INIT population = 165000000

### INFLOWS
population_growth_rate = population * population_growth_factor
population_growth_factor = 0.01
per_capita_electricity_consumption = GRAPH(TIME)
(0.00, 320), (3.00, 358), (6.00, 383), (9.00, 400), (12.0, 423), (15.0, 433), (18.0, 445), (21.0, 455), (24.0, 470), (27.0, 485), (30.0, 495)

## 2.5.12.2   Electricity Supply Model
Electricity is produced at power stations from either non-renewable sources or non-renewable and renewable energy sources, and there is some time delay in the construction of power plants. Estimates of energy supply from power stations that are changing with time are essential for energy planning. The stock–flow diagram of a simple electricity supply model is shown in Figure 2.35.

### Capacity under construction
Capacity under construction is increased by construction initiation rate and decreased by construction rate.
capacity_under_construction(t) = capacity_under_construction(t - dt) + (initiation_rate - construction_rate) * dt
INIT capacity_under_construction = 8

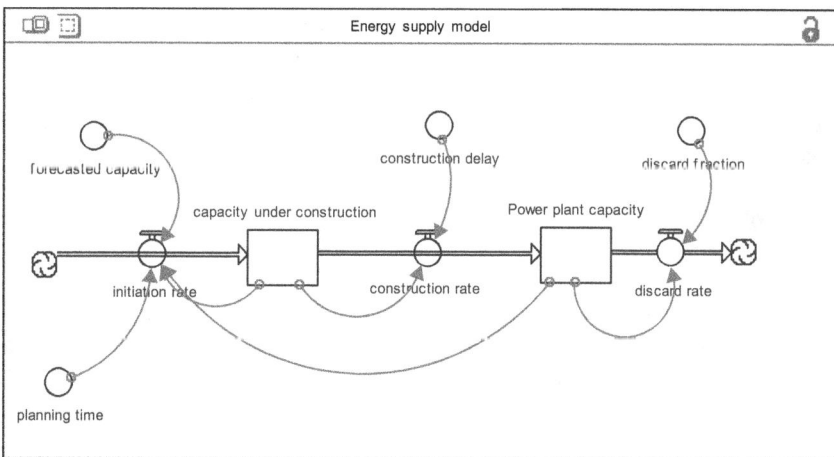

**FIGURE 2.35**   Stock–flow diagram of electricity supply model.

### Construction initiation rate
Construction initiation rate depends on forecasted demand capacity, capacity under construction, installed capacity and planning time.

initiation_rate = (forecasted_capacity-capacity_under_construction-installed_capacity)/planning_time
forecasted_capacity = 20
planning_time = 30

### Construction rate
Construction rate depends on capacity under construction and construction delay.

construction_rate = capacity_under_construction/construction_delay
construction_delay = 1

### Power plant capacity
Power plant capacity is increased by construction rate and decreased by discard rate.

power_plant_capacity(t) = power_plant_capacity(t - dt) + (construction_rate - discard_rate) * dt
INIT power_plant_capacity = 10

### Power plant capacity discard rate
Power plant capacity discard rate depends on power plant capacity and discard fraction.

discard_rate = power_plant_capacity * discard_fraction
construction_delay = 1
discard_fraction = 0.0001

Figure 2.36 shows the simulated installed capacity and capacity under construction for a period of 30 years, and the installed capacity increases rapidly at the initial stage and then slowly approaches the forecasted capacity. The capacity under construction decreases rapidly and then slowly approaches zero since the installed capacity is approaching the forecasted capacity.

### Listing of STELLA program

### Electricity supply model
capacity_under_construction(t) = capacity_under_construction(t - dt) +
(initiation_rate - construction_rate) * dt
INIT capacity_under_construction = 8

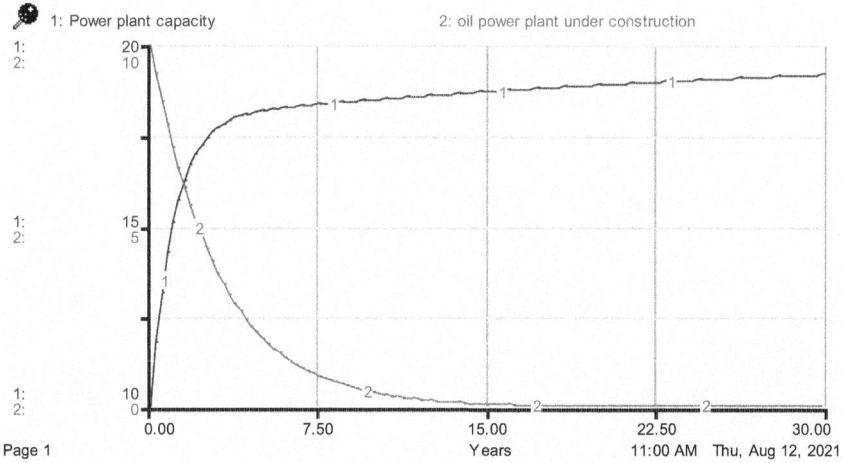

**FIGURE 2.36** Simulated power plant capacity and capacity under construction for a period of 30 years.

**INFLOWS**
initiation_rate = (forecasted_capacity - capacity_under_construction-power_plant_capacity)/planning_time

**OUTFLOWS**
construction_rate = capacity_under_construction/construction_delay
power_plant_capacity(t) = power_plant_capacity(t - dt) + (construction_rate - discard_rate) * dt
INIT power_plant_capacity = 10

**INFLOWS**
construction_rate = capacity_under_construction/construction_delay

**OUTFLOWS**
discard_rate = power plant capacity * discard fraction
construction_delay = 1
discard_fraction = 0.0001
forecasted_capacity = 20
planning_time = 30

### 2.5.12.3   Electricity Supply and Demand, and Price Model
We need electricity for economic development and better quality of life. Demand for electricity is increasing to meet the growing needs of industrial

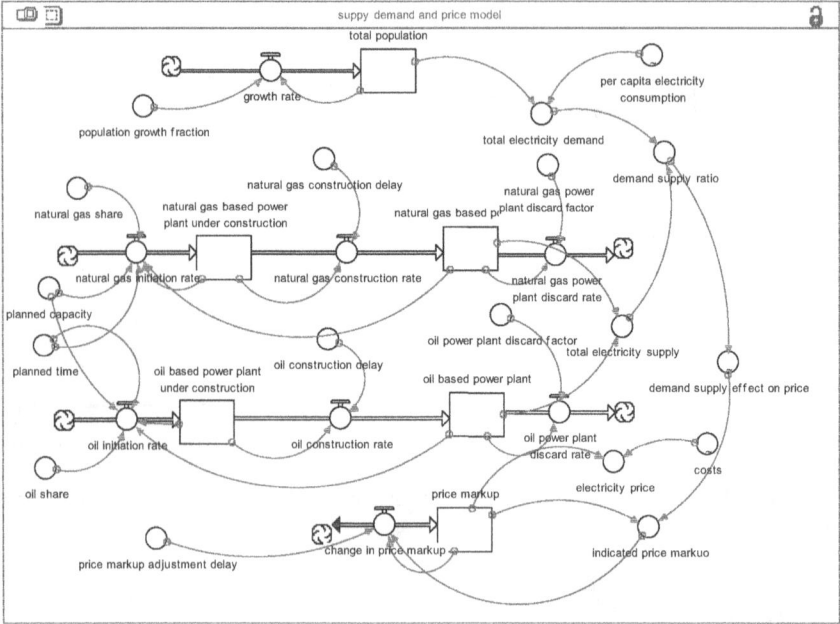

**FIGURE 2.37** Stock–flow diagram of electricity supply and demand, and price model.

and agricultural development and the residential need for a better quality of life. The supply of electricity must match the requirement or the target planned. The supply system consists of power plant installed and power plant under construction. The power plant under construction is associated with construction delay, so the initiation of the construction of power plants should be well planned to ensure the required supply in time with reliability. Again, price of electricity consists of production cost and price markup. The price markup depends on the demand–supply balance, and knowledge of price is essential for the optimal operational planning of power systems. The stock–flow diagram of a simple electricity supply and demand and price model is shown in Figure 2.37.

**Natural gas-based power plant**
Natural gas-based power plant is increased by natural gas construction rate and decreased by natural gas power plant discard rate.

natural_gas_based_power_plant(t) = natural_gas_based_power_plant(t - dt) + (natural_gas_construction_rate - natural_gas_power_plant_discard_rate) * dt
INIT natural_gas_based_power_plant = 7500

**Natural gas-based power plant construction rate**
Natural gas-based power plant construction rate depends on natural gas-based power plant under construction and natural gas-based power plant construction delay.

natural_gas_construction_rate =
natural_gas_based_power_plant_under_construction/natural_gas_construction_
delay
natural_gas_construction_delay = 3

## Natural gas-based power plant discard rate
Natural gas-based power plant discard rate depends on natural gas-based power plant and natural gas power plant discard factor.

natural_gas_power_plant_discard_rate = natural_gas_based_power_plant * natural_gas_power_plant_discard_factor
natural_gas_power_plant_discard_factor = 0.0001

## Natural gas-based power plant under construction
Natural gas-based power plant under construction is increased by natural gas-based power plant initiation rate and decreased by natural gas-based power plant construction rate.

natural_gas_based_power_plant_under_construction(t) = natural_gas_based_power_plant_under_construction(t - dt) + (natural_gas_initiation_rate - natural_gas_construction_rate) * dt
INIT natural_gas_based_power_plant_under_construction = 2500

## Natural gas-based power plant initiation rate
Natural gas-based power plant initiation rate depends on planned capacity, natural gas share, natural gas-based power plant, natural gas-based power plant under construction and planned time.

natural_gas_initiation_rate = (planned_capacity * natural_gas_share - natural_gas_based_power_plant-natural_gas_based_power_plant_under_construction)/planned_time
planned_capacity = 25000
natural_gas_share = 0.75
planned_time = 30

## Oil-based power plant
Oil-based power plant is increased by oil construction rate and decreased by oil power plant discard rate.

oil_based_power_plant(t)=oil_based_power_plant(t-dt)+(oil_construction_rate-oil_power_plant_discard_rate) * dt
INIT oil_based_power_plant = 3500

## Oil-based power plant construction rate
Oil-based power plant construction rate depends on oil-based power plant under construction and oil-based power plant construction delay.

oil_construction_rate = oil_based_power_plant_under_construction/oil_construction_delay

oil_construction_delay = 3

## Oil-based power plant discard rate

Oil-based power plant discard rate depends on oil-based power plant and oil power plant discard factor.

oil_power_plant_discard_rate = oil_based_power_plant * oil_power_plant_discard_factor

oil_power_plant_discard_factor = 0.0001

## Oil-based power plant under construction

Oil-based power plant under construction is increased by oil-based power plant initiation rate and decreased by oil-based power plant construction rate.

oil_based_power_plant_under_construction(t) = oil_based_power_plant_under_construction(t - dt) + (oil_initiation_rate - oil_construction_rate) * dt

INIT oil_based_power_plant_under_construction = 1500

## Oil-based power plant initiation rate

Oil-based power plant initiation rate depends on planned capacity, oil share, oil-based power plant, oil-based power plant under construction and planned time.

oil_initiation_rate = (planned_capacity * oil_share - oil_based_power_plant - oil_based_power_plant_under_construction)/planned_time

oil_share = 0.25

## Total electricity supply

Total electricity supply is the sum of electricity supplied by natural gas-based power plant and oil-based power plant.

total_electricity_supply = natural_gas_based_power_plant + oil_based_power_plant

## Total electricity demand

Total electricity demand is computed from total population and per capita consumption of electricity.

total_electricity_demand = (total_population * per_capita_electricity_consumption)/(0.60 * 8760)

## Total population

Total population is increased by growth rate.

total_population(t) = total_population(t - dt) + (growth_rate) * dt

INIT total_population = 165000000

**Population growth rate**
Population growth is computed from total population and population growth factor.

growth_rate = total_population * population_growth_fraction
population_growth_fraction = 0.01

**Per capita consumption of electricity**
Per capita consumption of electricity is assumed to increase with time and is expressed as a function of time graphically.

per_capita_electricity_consumption = GRAPH(TIME)
(0.00, 0.27), (3.00, 0.315), (6.00, 0.375), (9.00, 0.395), (12.0, 0.435), (15.0, 0.455), (18.0, 0.505), (21.0, 0.54), (24.0, 0.545), (27.0, 0.575), (30.0, 0.585)

**Price markup**
Price markup is increased by change in price markup.

price_mark_up(t) = price_markup(t - dt) + (change_in_pricemark_up) * dt
INIT price_mark_up = 0.10

**Change in price markup**
Change in price markup depends on indicated price markup, price markup and price markup adjustment delay.

change_in_price_markup = (indicated_price_markup-price_markup)/price_markup_adjustment_delay
price_markup_adjustment_delay = 3

**Indicated price markup**
Indicated price markup is computed from price markup and demand–supply effect on price markup.

indicated_price_markup = price_markup * demand_supply_effect_on_price

**Demand–supply effect on price**
Demand–supply effect on price is expressed as a function of demand–supply ratio graphically.

demand_supply_effect_on_price = GRAPH(demand_supply_ratio)
(0.00, 0.5), (0.5, 0.888), (1.00, 1.16), (1.50, 1.50), (2.00, 1.73), (2.50, 1.99), (3.00, 2.14), (3.50, 2.26), (4.00, 2.34), (4.50, 2.41), (5.00, 2.44)

**Demand–supply ratio of electricity**
Demand–supply ratio of electricity is the ratio of total electricity demand to total electricity supply and is expressed as:

demand_supply_ratio = total_electricity_demand/total_electricity_supply

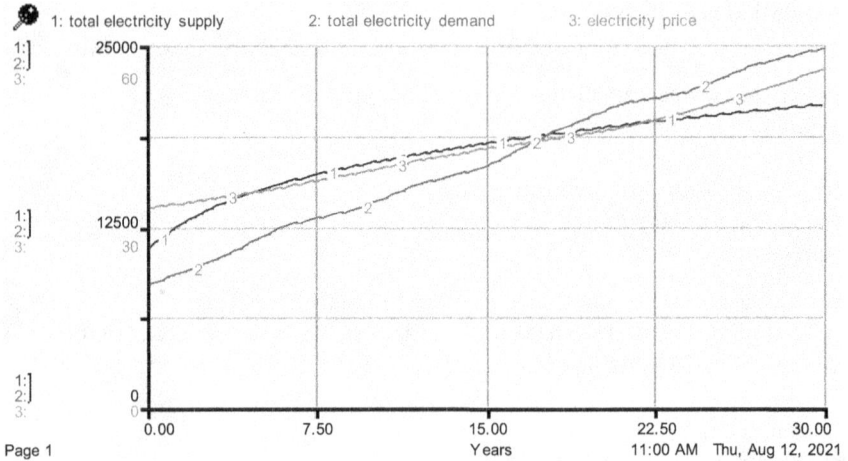

**FIGURE 2.38** Simulated total electricity supply, total electricity demand and electricity price for a period of 30 years.

### Cost of electricity
Cost of electricity is assumed to increase with time and is expressed as a function of time graphically.

costs = GRAPH(TIME)
(0.00, 30.0), (3.00, 31.5), (6.00, 33.0), (9.00, 34.8), (12.0, 36.5), (15.0, 37.8), (18.0, 38.5), (21.0, 39.0), (24.0, 39.8), (27.0, 40.3), (30.0, 40.3)

### Electricity price
Electricity price can not be less than the cost and hence should be either cost or (1 + price markup) × cost, whichever is maximum.

electricity_price = MAX(costs,(1 + price_mark_up) * costs)
Figure 2.38 shows the simulated total electricity supply, total electricity demand and electricity price for a period of 30 years. Total electricity supply and total electricity demand increases with time, but the supply is short in the later period of time since supply is a target supply approaching 2500 MW. Also, price increases with time.

### Listing of STELLA program

### Supply demand and price model
natural_gas_based_power_plant(t) = natural_gas_based_power_plant(t - dt) + (natural_gas_construction_rate - natural_gas_power_plant_discard_rate) * dt
INIT natural_gas_based_power_plant = 7500

**INFLOWS**
natural_gas_construction_rate = natural_gas_based_power_plant_under_construction/natural_gas_construction_delay

**OUTFLOWS**
natural_gas_power_plant_discard_rate = natural_gas_based_power_plant * natural_gas_power_plant_discard_factor
natural_gas_based_power_plant_under_construction(t) = natural_gas_based_power_plant_under_construction(t - dt) + (natural_gas_initiation_rate - natural_gas_construction_rate) * dt
INIT natural_gas_based_power_plant_under_construction = 2500

**INFLOWS**
natural_gas_initiation_rate = (planned_capacity * natural_gas_share - natural_gas_based_power_plant-natural_gas_based_power_plant_under_construction)/planned_time

**OUTFLOWS**
natural_gas_construction_rate = natural_gas_based_power_plant_under_construction/natural_gas_construction_delay
oil_based_power_plant(t) = oil_based_power_plant(t - dt) + (oil_construction_rate - oil_power_plant_discard_rate) * dt
INIT oil_based_power_plant = 3500

**INFLOWS**
oil_construction_rate = oil_based_power_plant_under_construction/oil_construction_delay

**OUTFLOWS**
oil_power_plant_discard_rate = oil_based_power_plant * oil_power_plant_discard_factor
oil_based_power_plant_under_construction(t) = oil_based_power_plant_under_construction(t - dt) + (oil_initiation_rate - oil_construction_rate) * dt
INIT oil_based_power_plant_under_construction = 1500

**INFLOWS**
oil_initiation_rate = (planned_capacity * oil_share - oil_based_power_plant - oil_based_power_plant_under_construction)/planned_time

**OUTFLOWS**
oil_construction_rate = oil_based_power_plant_under_construction/oil_construction_delay
price_markup(t) = price_markup(t - dt) + (change_in_price_markup) * dt
INIT price_markup = 0.10

**INFLOWS**

change_in_price_markup = (indicated_price_markup - price_markup)/price_markup_adjustment_delay

total_population(t) = total_population(t - dt) + (growth_rate) * dt

INIT total_population = 165000000

**INFLOWS**

growth_rate = total_population * population_growth_fraction

demand_supply_ratio = total_electricity_demand/total_electricity_supply

electricity_price = MAX(costs, (1 + price_markup) * costs)

indicated_price_markup = price_markup * demand_supply_effect_on_price

natural_gas_construction_delay = 3

natural_gas_power_plant_discard_factor = 0.0001

natural_gas_share = 0.75

oil_construction_delay = 3

oil_power_plant_discard_factor = 0.0001

oil_share = 0.25

planned_capacity = 25000

planned_time = 30

population_growth_fraction = 0.01

price_markup_adjustment_delay = 3

total_electricity_demand = (total_population * per_capita_electricity_consumption)/(0.60 * 8760)

total_electricity_supply = natural_gas_based_power_plant + oil_based_power_plant

costs = GRAPH(TIME)

(0.00, 30.0), (3.00, 31.5), (6.00, 33.0), (9.00, 34.8), (12.0, 36.5), (15.0, 37.8), (18.0, 38.5), (21.0, 39.0), (24.0, 39.8), (27.0, 40.3), (30.0, 40.3)

demand_supply_effect_on_price = GRAPH(demand_supply_ratio)

(0.00, 0.5), (0.5, 0.888), (1.00, 1.16), (1.50, 1.50), (2.00, 1.73), (2.50, 1.99), (3.00, 2.14), (3.50, 2.26), (4.00, 2.34), (4.50, 2.41), (5.00, 2.44)

per_capita_electricity_consumption = GRAPH(TIME)

(0.00, 0.27), (3.00, 0.315), (6.00, 0.375), (9.00, 0.395), (12.0, 0.435), (15.0, 0.455), (18.0, 0.505), (21.0, 0.54), (24.0, 0.545), (27.0, 0.575), (30.0, 0.585)

### 2.5.12.4    Palm Oil Biodiesel Model

Palm oil biodiesel is an alternative to fossil fuels, and Malaysia and Indonesia have the potential to capture their market share. The rapid increase in the price of crude oil above the price of vegetable oils at the beginning of 2004 prompted palm oil-producing countries to turn vegetable oils into biodiesel for transportation fuel. Palm oil biodiesel production in Malaysia is under its potential level, and the Malaysian government is attempting to popularize palm oil biodiesel in local and international markets, but it has not been successful due to the high price of crude palm oils. Malaysia needs a policy for the development strategies of biodiesel production for the coming years (Sahara et al., 2016). Here we examine whether

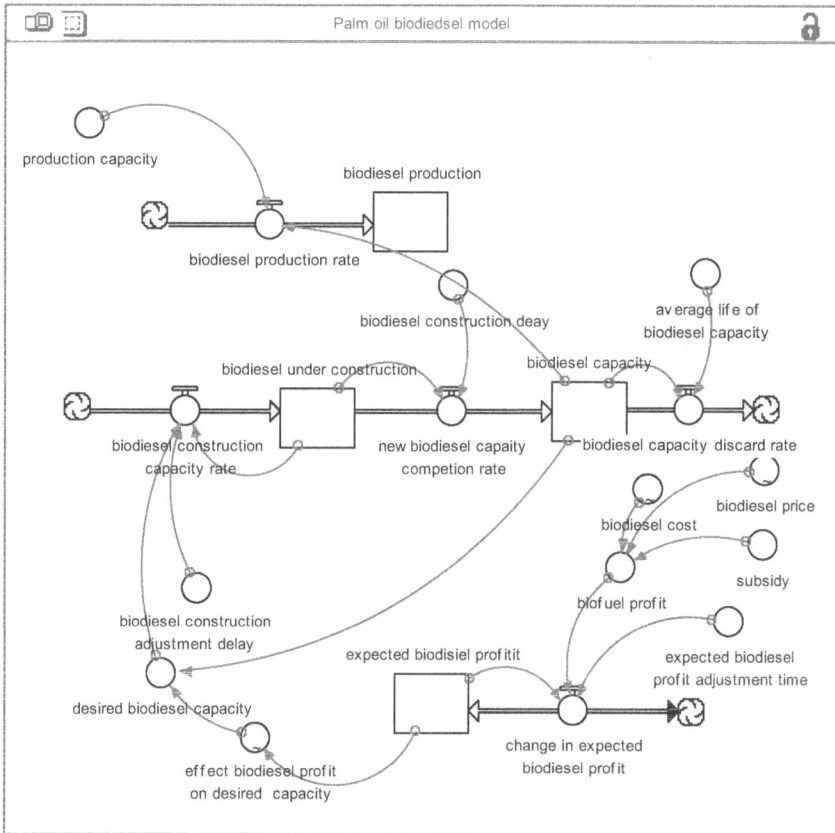

**FIGURE 2.39**   Stock–flow diagram of palm oil biodiesel model.

subsidies can make biodiesel production viable, at least in the local market. The stock–flow diagram of a simple palm oil biodiesel model is shown in Figure 2.39.

**Biodiesel capacity**

Biodiesel capacity is increased by new biodiesel capacity completion rate and decreased by biodiesel capacity discard rate.

biodiesel_capacity(t) = biodiesel_capacity(t - dt) + (new_biodiesel_capacity_completion_rate - biodiesel_capacity_discard_rate) * dt
INIT biodiesel_capacity = 10

**New biodiesel capacity completion rate**

New biodiesel capacity completion rate depends on biodiesel under construction and biodiesel construction delay.

new_biodiesel_capacity_completion_rate   =   biodiesel_under_construction/biodiesel_construction_delay
biodiesel_construction_delay = 5

**Biodiesel capacity discard rate**
Biodiesel capacity discard rate depends on biodiesel capacity and average life of biodiesel capacity.

biodiesel_capacity_discard_rate = biodiesel_capacity/average_life_of_biodiesel_ capacity
average_life_of_biodiesel_capacity = 25

**Biodiesel production**
Biodiesel production is increased by biodiesel production rate.

biodiesel_production(t) = biodiesel_production(t - dt) + (biodiesel_production_ rate) * dt
INIT biodiesel_production = 0

**Biodiesel production rate**
Biodiesel production rate depends on biodiesel capacity and biodiesel production capacity.

biodiesel_production_rate = biodiesel_capacity * production_capacity
production_capacity = 2500000

**Biodiesel under construction**
Biodiesel under construction is increased by biodiesel construction rate and decreased by new biodiesel capacity completion rate.

biodiesel_under_construction(t) = biodiesel_under_construction(t - dt) + (biodiesel_construction_capacity_rate - new_biodiesel_capacity_completion_rate) * dt
INIT biodiesel_under_construction = 5

**Biodiesel construction capacity rate**
Biodiesel construction capacity rate depends on desired biodiesel capacity, biodiesel capacity under construction and biodiesel construction adjustment delay.

biodiesel_construction_capacity_rate = (desired_biodiesel_capacity - biodiesel_ under_construction)/biodiesel_construction_adjustment_delay
biodiesel_construction_adjustment_delay = 2

**Expected biodiesel profit**
Expected biodiesel profit is increased by change in expected profit.

expected_biodiesel_profit(t) = expected_biodiesel_profit(t - dt) + (change_in_ expected_biodiesel_profit) * dt
INIT expected_biodiesel_profit = .2

**Change in expected biodiesel profit**
Change in expected biodiesel profit depends on biofuel profit, expected biodiesel profit and expected biodiesel profit adjustment time.

change_in_expected_biodiesel_profit = (biofuel_profit − expected_biodiesel_profit)/expected_biodiesel_profit_adjustment_time
expected_biodiesel_profit_adjustment_time = 2

## Biofuel profit
Biofuel profit is expressed as a fraction of biodiesel price as:

biofuel_profit = (biodiesel_price - biodiesel_cost + subsidy)/biodiesel_price

## Biodiesel price
Biodiesel price is expressed as a function of time graphically as:

biodiesel_price = GRAPH(TIME)
(0.00, 975), (4.17, 990), (8.33, 1013), (12.5, 1058), (16.7, 1095), (20.8, 1140), (25.0, 1170), (29.2, 1193), (33.3, 1215), (37.5, 1268), (41.7, 1245), (45.8, 1283), (50.0, 1275)

## Biodiesel cost
Biodiesel cost is expressed as a function of time graphically as:

biodiesel_cost = GRAPH(TIME)
(0.00, 1058), (4.17, 1110), (8.33, 1185), (12.5, 1223), (16.7, 1260), (20.8, 1290), (25.0, 1335), (29.2, 1403), (33.3, 1380), (37.5, 1410), (41.7, 1410), (45.8, 1425), (50.0, 1455)

## Biodiesel subsidy
Subsidy is maintained at 270 RM/mt for 10 years and then reduced to 220 RM/mt and is expressed as:

subsidy = 270 + STEP(-50,10)

## Desired biofuel capacity
Desired biofuel capacity is computed from biodiesel capacity and effect of biodiesel profit on desired capacity.

desired_biodiesel_capacity =
biodiesel_capacity * effect_of_biodiesel_profit_on_desired_capacity
expected_biodiesel_profit_adjustment_time = 2
production_capacity = 2500000

## Effect of biodiesel profit on desired capacity
Effect of biodiesel profit on desired capacity is expressed as a function of expected biodiesel profit graphically as:

effect_biodiesel_profit_on_desired_capacity = GRAPH(expected_biodiesel_profit)
(-0.5, 0.435), (-0.37, 0.465), (-0.24, 0.517), (-0.11, 0.622), (0.02, 0.757), (0.15, 0.87), (0.28, 1.02), (0.41, 1.09), (0.54, 1.22), (0.67, 1.31), (0.8, 1.36)

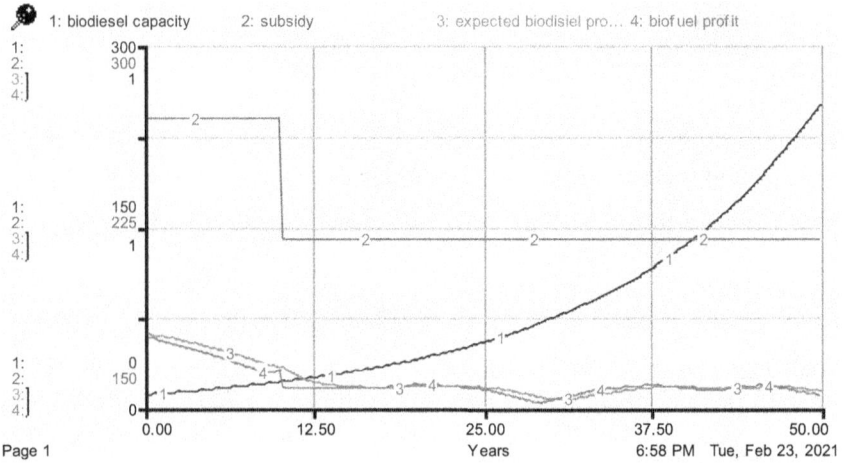

**FIGURE 2.40** Simulated biodiesel production, subsidy, expected biodiesel profit and biofuel profit.

Figure 2.40 shows the simulated biodiesel production capacity, subsidy, expected biodiesel profit and biofuel profit for a period of 30 years. The subsidy is high (RM 270/mt) for an initial period of 10 years, and then it is low (220 RM/mt) for the remaining period. Because of the subsidy, the biofuel profit is always positive, which results in positive expected profit and the expected biodiesel profit is constant. The resulting expected profit causes the growth of the biodiesel industry. As a result, biofuel production capacity increases throughout the simulation period. A factor that does not significantly affect profit and biodiesel capacity construction is the price of biodiesel. The higher the subsidies provided by the government, the higher the profit generated. Arrumaisho and Sunitiyoso (2019) reported that a factor that does not significantly affect profit and production capacity is biodiesel price. From the results of the system dynamics modeling simulation, it can be concluded that the factor that promotes the growth of the biodiesel industry is the use of government subsidy, and the system dynamics model can be used to develop appropriate subsidy policy.

### Listing of STELLA program

#### Palm oil biodiesel model
biodiesel_capacity(t) = biodiesel_capacity(t - dt) + (new_biodiesel_capacity_completion_rate - biodiesel_capacity_discard_rate) * dt
INIT biodiesel_capacity = 10

#### INFLOWS
new_biodiesel_capacity_completion_rate = biodiesel_under_construction/biodiesel_construction_delay

**OUTFLOWS**

biodiesel_capacity_discard_rate = biodiesel_capacity/average_life_of_biodiesel_capacity

biodiesel_production(t) = biodiesel_production(t - dt) + (biodiesel_production_rate) * dt

INIT biodiesel_production = 0

**INFLOWS**

biodiesel_production_rate = biodiesel_capacity * production_capacity

biodiesel_under_construction(t) = biodiesel_under_construction(t - dt) + (biodiesel_construction_capacity_rate - new_biodiesel_capacity_completion_rate) * dt

INIT biodiesel_under_construction = 5

**INFLOWS**

biodiesel_construction_capacity_rate = (desired_biodiesel_capacity - biodiesel_under_construction)/biodiesel_construction_adjustment_delay

**OUTFLOWS**

new_biodiesel_capacity_completion_rate = biodiesel_under_construction/biodiesel_construction_delay

expected_biodiesel_profit(t) = expected_biodiesel_profit(t - dt) + (change_in_expected_biodiesel_profit) * dt

INIT expected_biodiesel_profit = 0.2

**INFLOWS**

change_in_expected_biodiesel_profit = (biofuel_profit - expected_biodiesel_profit)/expected_biodiesel_profit_adjustment_time

average_life_of_biodiesel_capacity = 25

biodiesel_construction_adjustment_delay = 2

biodiesel_construction_delay = 5

biofuel_profit = (biodiesel_price - biodiesel_cost + subsidy)/biodiesel_price

desired_biodiesel_capacity =
biodiesel_capacity * effect_biodiesel_profit_on_desired_capacity

expected_biodiesel_profit_adjustment_time = 2

production_capacity = 2500000

subsidy = 270 + STEP(-50,10)

biodiesel_cost = GRAPH(TIME)

(0.00, 1058), (4.17, 1110), (8.33, 1185), (12.5, 1223), (16.7, 1260), (20.8, 1290), (25.0, 1335), (29.2, 1403), (33.3, 1380), (37.5, 1410), (41.7, 1410), (45.8, 1425), (50.0, 1455)

biodiesel_price = GRAPH(TIME)

(0.00, 975), (4.17, 990), (8.33, 1013), (12.5, 1058), (16.7, 1095), (20.8, 1140), (25.0, 1170), (29.2, 1193), (33.3, 1215), (37.5, 1268), (41.7, 1245), (45.8, 1283), (50.0, 1275)

effect_biodiesel_profit_on_desired_capacity   =   GRAPH(expected_biodiesel_
profit)
(-0.5, 0.435), (-0.37, 0.465), (-0.24, 0.517), (-0.11, 0.622), (0.02, 0.757), (0.15,
0.87), (0.28, 1.02), (0.41, 1.09), (0.54, 1.22), (0.67, 1.31), (0.8, 1.36)

### 2.5.12.5   Emissions from Electricity Production Model

The production and use of electrical energy has an environmental impact in the form
of greenhouse gas emissions that contribute to global warming and climate change.
Emissions from the burning of fuels are determined by multiplying the quantity of
fuel consumed by the emission factors. Emission factors or coefficients describe the
quantity of a pollutant that is released per unit of fuel consumed. Emission factors
are thus numbers that allow energy/environmental planners to estimate quantities
of emissions and other environmental effects or impacts associated with activities
such as tonnes of coal burned or barrels of oils passing through a refinery. Data on
emission factors are available from different documents and databases. There are
several major general sources of emission factors. These are the emission factors
collected by the US Environmental Protection Agency and the Intergovernmental
Panel on Climate Change. A stock–flow diagram of emission from an oil-based
electricity production power plant is shown in Figure 2.41.

**Oil-based power plant under construction**
Oil-based power plant under construction is increased by oil-based power plant
initiation rate and decreased by oil-based power plant construction rate.
oil_power_plant_under_construction(t)   =   oil_power_plant_under_construction
(t - dt) + (oil_power_plant_initiation_rate - oil_power_plant_construction_rate) * dt
INIT oil_power_plant_under_construction = 10

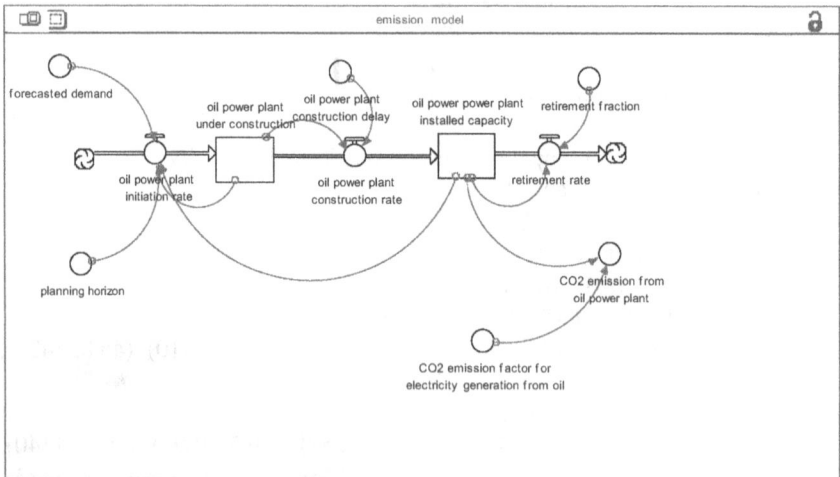

**FIGURE 2.41**   Stock–flow diagram of emission from electricity production model.

### Oil-based power plant initiation rate

Oil-based power plant initiation rate depends on forecasted demand, oil-based power plant under construction, oil-based power plant installed capacity and planning horizon.

oil_power_plant_initiation_rate = (forecasted_demand - oil_power_plant_under_construction - oil_power_plant_installed_capacity)/planning_horizon
forecasted_demand = 30
planning_horizon = 30

### Oil-based power plant construction rate

Oil-based power plant construction rate depends on oil-based power plant under construction and oil-based power plant construction delay.

oil_power_plant_construction_rate = oil_power_plant_under_construction/oil_power_plant_construction_delay
oil_power_plant_construction_delay = 3

### Oil-based power plant installed capacity

Oil-based power plant installed capacity is increased by oil-based power plant construction rate and decreased by retirement rate.

oil_power_power_plant_installed_capacity(t) = oil_power_power_plant_installed_capacity(t - dt) + (oil_power_plant_construction_rate - retirement_rate) * dt
INIT oil_power_power_plant_installed_capacity = 20

### Oil-based power plant retirement rate

Oil-based power plant retirement rate depends on oil-based power plant installed capacity and retirement fraction.

retirement_rate = oil_power_power_plant_installed_capacity * retirement_fraction
retirement_fraction = 0.0001

### $CO_2$ emission from oil-based power plant

$CO_2$ emission from oil-based power plant is computed from oil-based power plant installed capacity and emission factor for electricity generation from oil.

CO2_emission_from_oil_power_plant = oil_power_plant_power_installed_capacity * 8760 * CO2_emission_factor_for_electricity_generation_from_oil
CO2_emission_factor_for_electricity_generation_from_oil = 735

Figure 2.42 shows the simulated oil power plant installed capacity and $CO_2$ emission from oil power plant for a period of 30 years. The emission from the power plant follows the electricity generated.

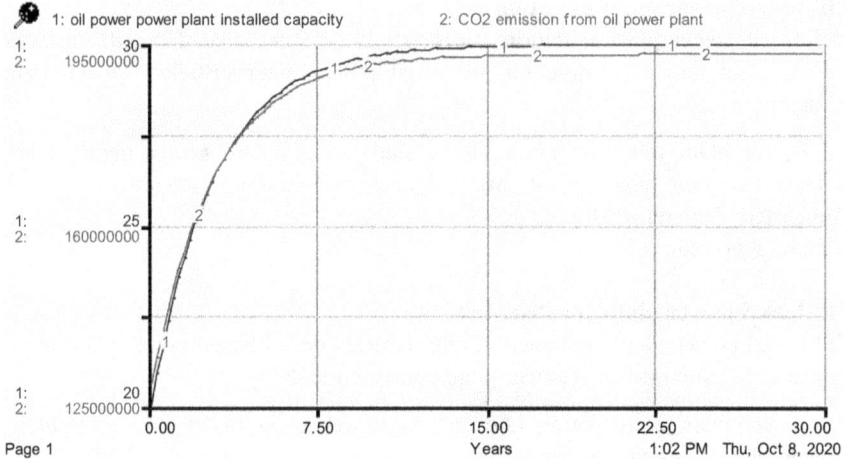

**FIGURE 2.42** Simulated oil power plant installed capacity and $CO_2$ emission from oil power plant.

### Listing of STELLA program

#### Emission from electricity production model

oil_power_plant_under_construction(t) = oil_power_plant_under_construction (t - dt) + (oil_power_plant_initiation_rate - oil_power_plant_construction_rate) * dt
INIT oil_power_plant_under_construction = 10

#### INFLOWS
oil_power_plant_initiation_rate = (forecasted_demand - oil_power_plant_under_construction - oil_power_power_plant_installed_capacity)/planning_horizon

#### OUTFLOWS
oil_power_plant_construction_rate = oil_power_plant_under_construction/oil_power_plant_construction_delay
oil_power_power_plant_installed_capacity(t) = oil_power_power_plant_installed_capacity(t - dt) + (oil_power_plant_construction_rate - retirement_rate) * dt
INIT oil_power_power_plant_installed_capacity = 20

#### INFLOWS
oil_power_plant_construction_rate = oil_power_plant_under_construction/oil_power_plant_construction_delay

#### OUTFLOWS
retirement_rate = oil_power_plant_installed_capacity * retirement_fraction
CO2_emission_factor_for_electricity_generation_from_oil = 735

CO2_emission_from_oil_power_plant = oil_power_power_plant_installed_capacity * 8760 * CO2_emission_factor_for_electricity_generation_from_oil
forecasted_demand = 30
oil_power_plant_construction_delay = 3
planning_horizon = 30
retirement_fraction = 0.0001

### 2.5.12.6  Gradual Transition to Renewable Energy Resources

We need energy for economic and social development, and the energy system consists of energy supply, energy demand, climate change and environmental quality to facilitate the better supply of energy with minimum impact on climate change and environmental quality. To ensure energy security and reduce the contribution to global warming, a transition to renewable energy is essential. A stock–flow diagram of the gradual transition to a renewable energy model is shown in Figure 2.43.

**Grid capacity**

Grid capacity is increased by grid construction rate and decreased by grid discard rate.

grid_capacity(t) = grid_capacity(t - dt) + (grid_construction_rate - grid_discard_rate) * dt
INIT grid_capacity = 7500

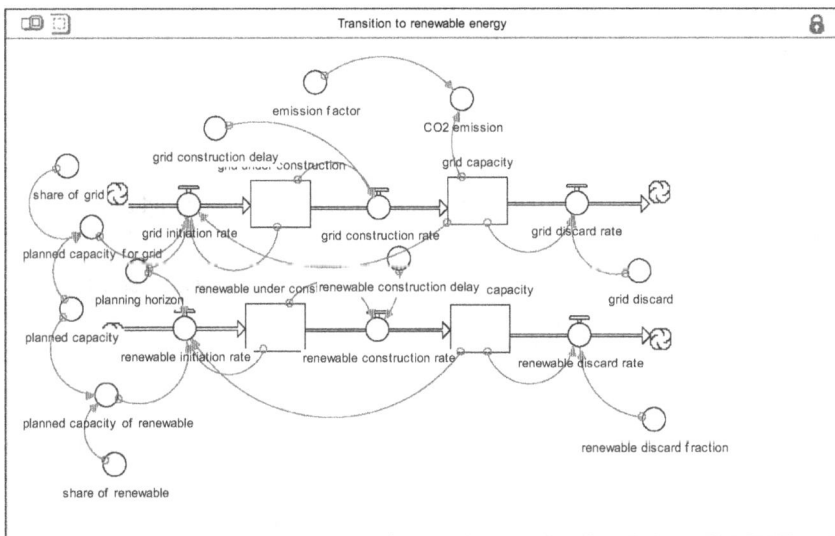

**FIGURE 2.43**  Stock–flow diagram of the gradual transition to a renewable energy model.

### Grid construction rate
Grid construction rate is computed from grid under construction and grid construction delay as:

grid_construction_rate = grid_under_construction/grid_construction_delay
grid_construction_delay = 3.5

### Grid discard rate
Grid discard rate depends on grid capacity and grid discard fraction, and is computed as:

grid_discard_rate = grid_capacity * grid_discard
grid_discard = 0.001

### Grid under construction
Grid under construction is increased by grid-based power plant initiation rate and decreased by grid-based power plant construction rate.

grid_under_construction(t) = grid_under_construction(t – dt) + (grid_initiation_rate – grid_construction_rate) * dt
INIT grid_under_construction = 2500

### Grid initiation rate
Grid initiation rate depends on forecasted demand, grid under construction, grid capacity and planning horizon.

Grid_initiation_rate = (planned_capacity_for_grid-grid_under_construction - grid_capacity)/planning_horizon
planned_capacity_for_grid = planned_capacity * share_of_grid
planned_capacity = 25000
share_of_grid = 0.7
planning_horizon = 20

### Renewable capacity
Renewable capacity is increased by renewable construction rate and decreased by renewable discard rate.

renewable_capacity(t) = renewable_capacity(t - dt) + (renewable_construction_rate - renewable_discard_rate) * dt
INIT renewable_capacity = 2500

### Renewable construction rate
Renewable construction rate depends on renewable under construction and renewable construction delay and is computed as:

renewable_construction_rate = renewable_under_construction/renewable_construction_delay
renewable_construction_delay = 2.5

## Renewable discard rate

Renewable discard rate is computed from renewable capacity and renewable discard fraction as:

renewable_discard_rate = renewable_capacity * renewable_discard_fraction

## Renewable under construction

Renewable under construction is increased by renewable-based power plant initiation rate and decreased by renewable-based power plant construction rate.

renewable_under_construction(t) = renewable_under_construction(t-dt) + (renewable_initiation_rate - renewable_construction_rate) * dt
INIT renewable_under_construction = 500

## Renewable initiation rate

Renewable initiation rate depends on forecasted demand, renewable under construction, renewable capacity and planning horizon.

renewable_initiation_rate = (planned_capacity_of_renewable-renewable_under_construction - renewable_capacity)/planning_horizon
planned_capacity_of_renewable = planned_capacity * share_of_renewable
planned_capacity = 25000
share_of_renewable = 0.3
planning_horizon = 20

## $CO_2$ emission

$CO_2$ emission is computed from grid capacity and emission factor.

CO2_emission = grid_capacity * emission_factor
emission_factor = 500

Figure 2.44 shows the simulated grid capacity, renewable capacity and $CO_2$ emission from a grid power plant for a period of 20 years. The emission from the grid power plant follows the electricity generated by the grid.

### Listing of STELLA program

### Transition to renewable energy
grid_capacity(t) = grid_capacity(t - dt) + (grid_construction_rate - grid_discard_rate) * dt
INIT grid_capacity = 7500

### INFLOWS
grid_construction_rate = grid_under_construction/grid_construction_delay

### OUTFLOWS
grid_discard_rate = grid_capacity * grid_discard

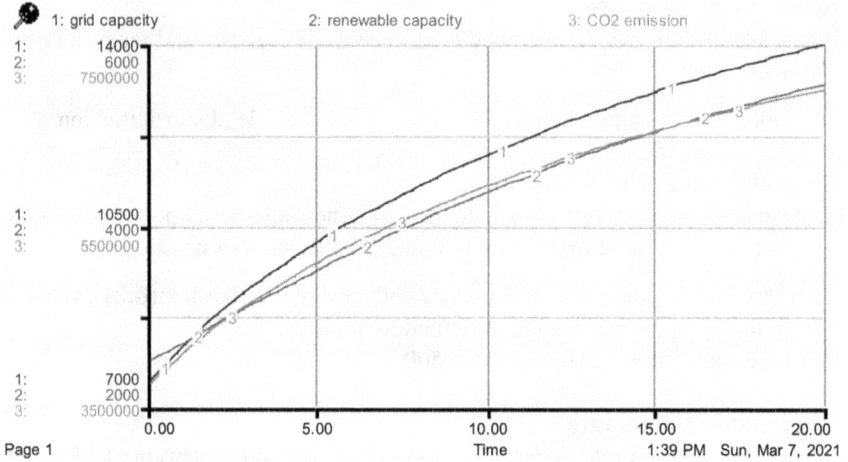

**FIGURE 2.44** Simulated grid capacity, renewable capacity and $CO_2$ emission.

grid_under_construction(t) = grid_under_construction(t - dt) + (grid_initiation_
rate - grid_construction_rate) * dt
INIT grid_under_construction = 2500

**INFLOWS**
grid_initiation_rate = (planned_capacity_for_grid - grid_under_construction -
grid_capacity)/planning_horizon

**OUTFLOWS**
grid_construction_rate = grid_under_construction/grid_construction_delay
renewable_capacity(t) = renewable_capacity(t - dt) + (renewable_construction_
rate - renewable_discard_rate) * dt
INIT renewable_capacity = 2500

**INFLOWS**
renewable_construction_rate = renewable_under_construction/renewable_con-
struction_delay

**OUTFLOWS**
renewable_discard_rate = renewable_capacity * renewable_discard_fraction
renewable_under_construction(t) = renewable_under_construction(t - dt) +
(renewable_initiation_rate - renewable_construction_rate) * dt
INIT renewable_under_construction = 500

**INFLOWS**
renewable_initiation_rate = (planned_capacity_of_renewable - renewable_under_
construction - renewable_capacity)/planning_horizon

**OUTFLOWS**
renewable_construction_rate = renewable_under_construction/renewable_construction_delay
grid_construction_delay = 3.5
grid_discard = 0.001
planned_capacity = 25000
planned_capacity_for_grid = planned_capacity * share_of_grid
planned_capacity_of_renewable = planned_capacity * share_of_renewable
planning_horizon = 20
renewable_construction_delay = 2.5
renewable_discard_fraction = 0.001
share_of_grid = 0.7
share_of_renewable = 0.3
$CO_2$_emission = grid_capacity * emission_factor
emission_factor = 500

### 2.5.12.7   Pollution Model

Pollution is an action that causes environmental degradation and is one of the causes of degradation. Environmental degradation due to pollution caused by human actions can be viewed in terms of population and pollution that is normal due to human action. A stock–flow diagram of the pollution model is shown in Figure 2.45.

**Pollution level**
Pollution level increases by human activities and decreases by absorption.

pollution(t) = pollution(t - dt) + (generation_rate - absorption_rate) * dt
INIT pollution = 10000000

**Pollution generation rate**
Pollution generation is taken as population multiplied by pollution normal.

generation_rate = population * pollution_normal
pollution_normal = 0.25

**Pollution absorption rate**
Pollution absorption depends on the existence of pollution to be absorbed. It depends on the natural processes that determine the amount absorbed in a specified time. Pollution absorption is determined by the amount of pollution divided by absorption time. The pollution absorption time depends on the pollution ratio and is represented as a graphical function. Pollution ratio is defined as the ratio of pollution level to pollution standard.

absorption_rate = pollution/pollution_absorption_time

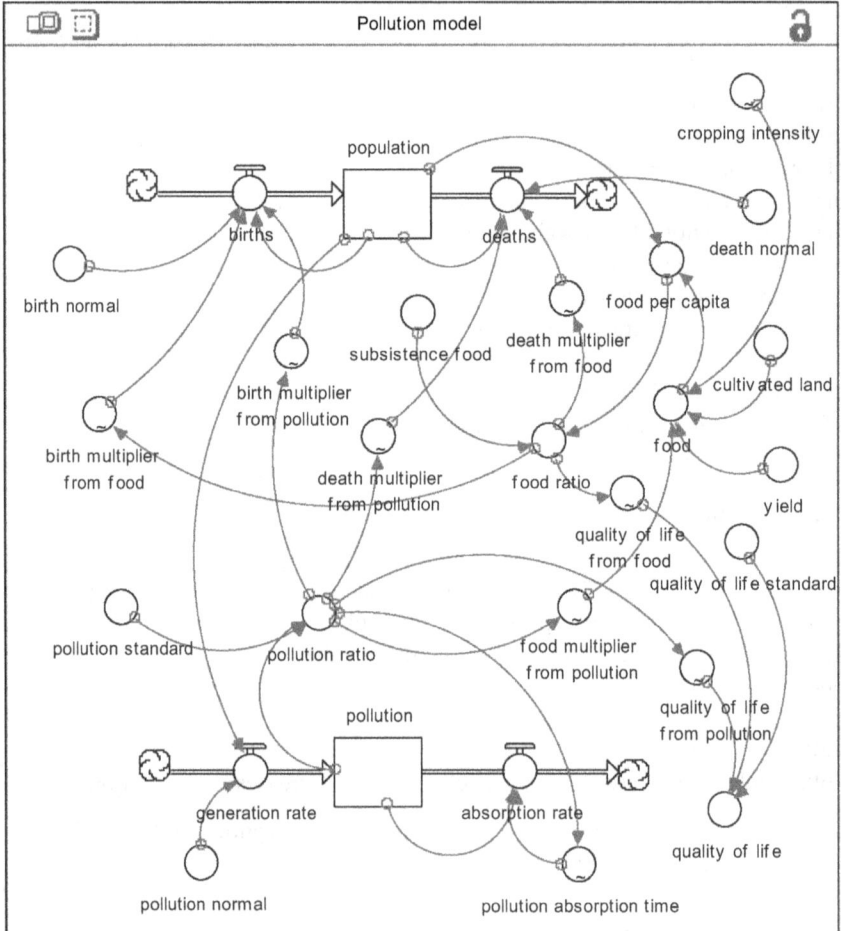

**FIGURE 2.45**   Stock–flow diagram of pollution model.

**Pollution absorption time**
Pollution absorption time depends on pollution ratio and is expressed as a function of pollution ratio.

pollution_absorption_time = GRAPH(pollution_ratio)
(0.00, 0.6), (10.0, 2.50), (20.0, 5.00), (30.0, 8.00), (40.0, 11.5), (50.0, 15.5), (60.0, 20.0)
pollution_ratio = pollution/pollution_standard
pollution_standard = 15000000

**Population**
Population is increased by birth rate and decreased by death rate.

population(t) = population(t - dt) + (births - deaths) * dt
INIT population = 100000000

**Actual birth rate**

Actual birth rate is computed from population level, birth rate normal, birth rate from food multiplier and birth rate from pollution multiplier.

births = population * birth_normal * birth_multiplier_from_food * birth_multiplier_from_pollution
birth_normal = 0.04

**Birth multiplier from food**

Birth multiplier from food is non-linearly related to food ratio and is presented graphically. Food ratio is defined as the ratio of food per capita to food per capita standard.

birth_multiplier_from_food = GRAPH(food_ratio)
(0.00, 0.00), (1.00, 1.00), (2.00, 1.60), (3.00, 1.90), (4.00, 2.00)
food_ratio = food_per_capita/subsistence_food
subsistence_food = 180

**Birth multiplier from pollution**

Birth multiplier from pollution is non-linearly related to pollution ratio and is presented graphically.

birth_multiplier_from_pollution = GRAPH(pollution_ratio)
(0.00, 1.02), (10.0, 0.9), (20.0, 0.7), (30.0, 0.4), (40.0, 0.25), (50.0, 0.15), (60.0, 0.1)
food_ratio = food_per_capita/subsistence_food
subsistence_food = 180

**Death rate**

Death rate depends upon population level, death rate normal, death rate from food multiplier and death rate from pollution multiplier.

deaths = population * death_normal * death_multiplier_from_food * death_multiplier_from_pollution
death_normal = .028

**Death multiplier from food**

Food can be a powerful indicator of population. The non-linear relationship between death rate multiplier from food and food ratio is represented graphically.

death_multiplier_from_food = GRAPH(food_ratio)
(0.00, 30.0), (0.25, 3.00), (0.5, 2.00), (0.75, 1.40), (1.00, 1.00), (1.25, 0.7), (1.50, 0.6), (1.75, 0.5), (2.00, 0.5)

**Death multiplier from pollution**

Death multiplier from pollution is non-linearly related to pollution, and this relationship is expressed graphically.

death_multiplier_from_pollution = GRAPH(pollution_ratio)
(0.00, 0.92), (10.0, 1.30), (20.0, 3.30), (30.0, 4.80), (40.0, 6.80), (50.0, 9.20), (60.0, 9.50)
Food per capita is defined as the ratio of food available to population.
food_per_capita = food/population

## Food availability
Food availability depends on cultivated land, yield, cropping intensity and food from pollution multiplier.

food = cultivated_land * yield * cropping_intensity * food_multiplier_from_pollution
cultivated_land = 7400000
yield = 2200

## Food multiplier from pollution
Food multiplier from pollution is non-linearly related to pollution and is expressed graphically.

food_multiplier_from_pollution = GRAPH(pollution_ratio)
(0.00, 1.02), (10.0, 0.9), (20.0, 0.65), (30.0, 0.35), (40.0, 0.2), (50.0, 0.1), (60.0, 0.05)

## Cropping intensity
Cropping intensity changes with time and is expressed graphically as a function of time.

cropping_intensity = GRAPH(TIME)
(0.00, 1.75), (10.0, 1.80), (20.0, 1.90), (30.0, 2.00), (40.0, 2.30), (50.0, 2.50)

## Quality of life from food
Food has the most powerful influence on quality of life. The quality of life multiplier from food depends on food and is expressed graphically.

quality_of_life_from_food = GRAPH(food_ratio)
(0.00, 0.00), (1.00, 1.00), (2.00, 1.80), (3.00, 2.40), (4.00, 2.70)

## Quality of life from pollution
Pollution affects quality of life and the quality of life multiplier from pollution depends on pollution and is expressed graphically.

quality_of_life_from_pollution = GRAPH(pollution_ratio)
(0.00, 1.05), (10.0, 0.85), (20.0, 0.6), (30.0, 0.3), (40.0, 0.15), (50.0, 0.05), (60.0, 0.02)

## Quality of life

Quality of life is used here as the measure of the performance of the system. It is computed as a multiplication of quality of life standard multiplied by multipliers derived from food and pollution.

quality_of_life = quality_of_life_standard * quality_of_life_from_food * quality_of_life_from_pollution

quality_of_life_standard = 1

Simulated population, food per capita, pollution and quality of life are shown in Figure 2.46. Simulated results show that pollution increases as the population increases, which is the observed phenomenon. Food per capita decreases with time because food product is fixed. As a result, quality of life decreases with time.

## Listing of STELLA program

## Pollution model

pollution(t) = pollution(t - dt) + (generation_rate - absorption_rate) * dt

INIT pollution = 10000000

## INFLOWS

generation_rate = population * pollution_normal

## OUTFLOWS

absorption_rate = pollution/pollution_absorption_time

population(t) = population(t - dt) + (births - deaths) * dt

INIT population = 100000000

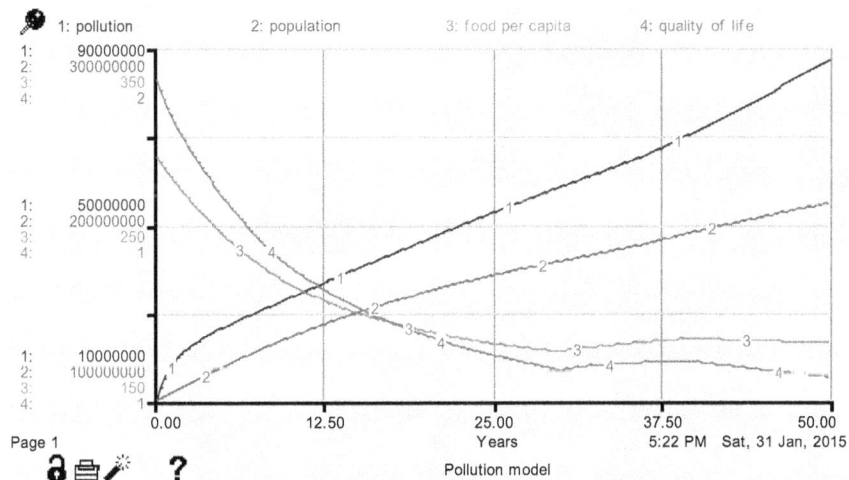

FIGURE 2.46    Simulated population, food per capita, pollution and quality of life.

**INFLOWS**
births = population * birth_normal * birth_multiplier_from_food * birth_multi-
plier_from_pollution

**OUTFLOWS**
deaths = population * death_normal * death_multiplier_from_food * death_multi-
plier_from_pollution
birth_normal = 0.04
cultivated_land = 7400000
death_normal = .028
food = cultivated_land * yield * cropping_intensity * food_multiplier_from_
pollution
food_per_capita = food/population
food_ratio = food_per_capita/subsistence_food
pollution_normal = 0.25
pollution_ratio = pollution/pollution_standard
pollution_standard = 15000000
quality_of_life = quality_of_life_standard * quality_of_life_from_food * quality_
of_life_from_pollution
quality_of_life_standard = 1
subsistence_food = 180
yield = 2200
birth_multiplier_from_food = GRAPH(food_ratio)
(0.00, 0.00), (1.00, 1.00), (2.00, 1.60), (3.00, 1.90), (4.00, 2.00)
birth_multiplier_from_pollution = GRAPH(pollution_ratio)
(0.00, 1.02), (10.0, 0.9), (20.0, 0.7), (30.0, 0.4), (40.0, 0.25), (50.0, 0.15), (60.0, 0.1)
cropping_intensity = GRAPH(TIME)
(0.00, 1.75), (10.0, 1.80), (20.0, 1.90), (30.0, 2.00), (40.0, 2.30), (50.0, 2.50)
death_multiplier_from_food = GRAPH(food_ratio)
(0.00, 30.0), (0.25, 3.00), (0.5, 2.00), (0.75, 1.40), (1.00, 1.00), (1.25, 0.7), (1.50,
0.6), (1.75, 0.5), (2.00, 0.5)
death_multiplier_from_pollution = GRAPH(pollution_ratio)
(0.00, 0.92), (10.0, 1.30), (20.0, 3.30), (30.0, 4.80), (40.0, 6.80), (50.0, 9.20), (60.0,
9.50)
food_multiplier_from_pollution = GRAPH(pollution_ratio)
(0.00, 1.02), (10.0, 0.9), (20.0, 0.65), (30.0, 0.35), (40.0, 0.2), (50.0, 0.1), (60.0,
0.05)
pollution_absorption_time = GRAPH(pollution_ratio)
(0.00, 0.6), (10.0, 2.50), (20.0, 5.00), (30.0, 8.00), (40.0, 11.5), (50.0, 15.5), (60.0,
20.0)
quality_of_life_from_food = GRAPH(food_ratio)
(0.00, 0.00), (1.00, 1.00), (2.00, 1.80), (3.00, 2.40), (4.00, 2.70)
quality_of_life_from_pollution = GRAPH(pollution_ratio)

(0.00, 1.05), (10.0, 0.85), (20.0, 0.6), (30.0, 0.3), (40.0, 0.15), (50.0, 0.05), (60.0, 0.02)

## Exercises

**2.1** Energy sectors supply energy to producers and consumers where it is transformed and consumed. Energy sectors consist of non-renewable energy resources and renewable energy resources. Electricity is produced from both non-renewable energy resources and renewable energy resources, and electricity demands are mainly for residential, industrial, and agricultural purposes. Electricity production depends on its demand and expected price, while the price is determined by its demand–supply balance. Draw the causal loop diagram.

**2.2** Total electrical energy demand depends on fast-growing population, industrial and economic growth. Renewable and non-renewable energy resources play a role in the national level development, and the supply–demand gap is met by the import of electricity. $CO_2$ is one of the dominant greenhouse gases contributing to global warming. To ensure energy security and reduce contributions to global warming and climate change, there should be a gradual transition to renewable energy resources, which requires a policy for energy mix. Draw the causal loop diagram.

**2.3** The causal loop diagram of the electrical energy production capacity of both non-renewable and renewable energy is shown in Figure 2.47. Draw the stock–flow diagram.

**2.4** The causal loop diagram of electrical energy generation based on capacity factor and profitability is shown in Figure 2.48. Draw the stock–flow diagram.

**2.5** The causal loop diagram of electrical energy supply, demand and price is shown in Figure 2.49. Draw the stock–flow diagram.

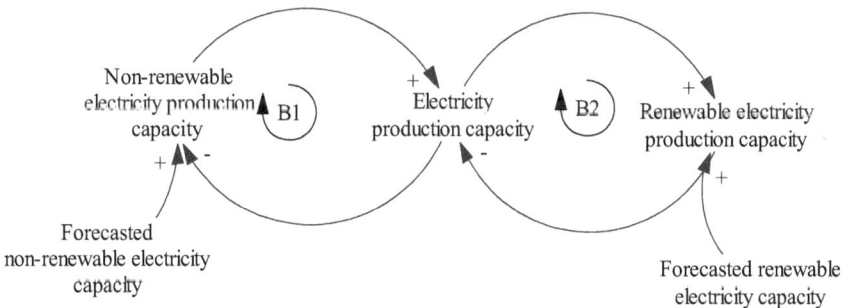

**FIGURE 2.47** Causal loop diagram of electrical energy production capacity of both non-renewable and renewable energy.

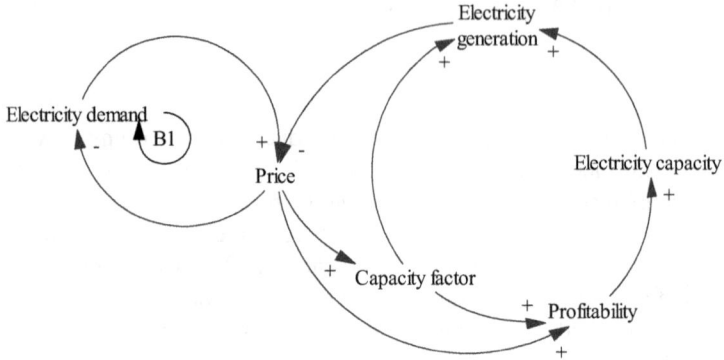

**FIGURE 2.48** Causal loop diagram of electrical energy generation based on capacity factor and profitability.

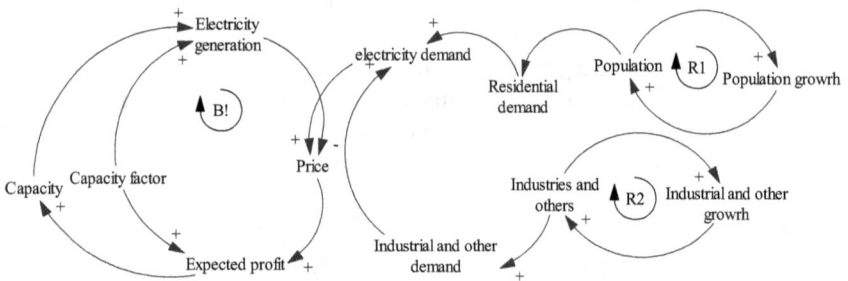

**FIGURE 2.49** Causal loop diagram of electrical energy supply, demand and price.

## REFERENCES

Andersen, D. F., & Richardson, G. P. (1997). Scripts for group model building. *System Dynamics Review,* 13: 107–129.

Arrumaisho, U. S., & Sunitiyoso, Y. (2019). A system dynamics model for biodiesel industry in Indonesia. *The Asian Journal of Technology Management,* 12(2): 149–162.

Bala, B. K. (1997a). Computer modelling of the rural energy system and of CO2 emissions for Bangladesh. *Energy,* 22: 999–1003.

Bala, B. K. (1997b). Computer modelling of energy and environment: The case of Bangladesh. *Proceedings of 15th International System Dynamics Conference, Istanbul, Turkey,* August 19–22, 1997.

Bala, B. K. (1998). *Energy and Environment: Modelling and Simulation.* Nova Science Publishers, USA.

Bala, B. K. (1999). *Principles of System Dynamics* (1st ed.). Agrotech Publishing Academy, Udaipur, India.

Bala, B. K., Alam, M. S., & Debnath, N. (2014). Energy perspective of climate change: The case of Bangladesh. Strategic Planning for Energy and the *Environment,* 33(3): 6–22.

Bala, B. K., Fatimah, M. A., & Kushairi, M. N. (2017). *System Dynamics: Modelling and Simulation*. Springer.

Bala, B. K., & Satter, M. A. (1986a) *Modelling of rural energy systems*. Presented at Second National Symposium on Agricultural Research, BARC, Dhaka, February 12, 1986.

Bala, B. K, & Satter, M. A. (1986b). *Modelling of rural energy systems for food production in developing countries*. Energia and Agricoltura 2 Conferenza Internationale. Sirmione/Brescia (Italia). 3, p. 306, 1986.

Elias, A. A., Cavana, R. Y., & Jackson, L.S. (2000). *Linking stakeholder literature and system dynamics: Opportunities for research*. In 1st International Conference on Systems Thinking in Management. pp. 174–179.

Forrester, J. W. (1961). *Industrial Dynamics*. MIT Press, Cambridge, Massachusetts, USA.

Forrester, J. W. (1968). *Principles of Systems*. MIT Press, Cambridge, Massachusetts, USA.

Gardiner, P. C., & Ford, A. (1980). Which policy run is best, and who says so. *TIMS Studies in Management Science,* 14, pp. 241–257.

Hsiao, N. (1998). *Conflict analysis of public policy stakeholders combining judgment analysis and system dynamics modeling*. In Proceeding of the 16th International Conference of the System Dynamics Society.

Huq, A. Z. M. (1975). *Energy modelling for agriculture units in Bangladesh*. Paper presented at the National Seminar on Integrated Rural Development. Dhaka, 1975.

Maani, K. E., & Cavana, R.Y. (2000). *Systems Thinking and Modelling: Understanding Change and Complexity*. Prentice Hall, New Zealand.

Manetsch, T. J., & Park, G. L. (1982). *Systems Analysis and Simulation with Applications to Economic and Social Systems*. Department of Electrical Engineering and System Science, Michigan State University, USA.

Nail, R. F. (1992). A system dynamics model for national energy policy planning. *System Dynamics Review*. 8: 1–19.

Parikh, J. (1997). *Energy Models for 2000 and Beyond*. Tata McGraw Hill, New Delhi, India.

Rouwette, E. A. J. A., Korzilius, H., Vennix, J. A. M. & Jacobs, E. (2011). Modeling as persuasion: The impact of group model building on attitudes and behavior. *System Dynamics Review*, 27(1), pp.1–21.

Sahara, M., Fatimah, M. A., & Abdulla, I. (2016). Future prospects and policy implications for biodiesel production in Malaysia: A system dynamics approach. *Institutions and Economics*. 8(4): 42–57.

Sterman, J. D. (2000). *Business Dynamics: Systems Thinking and Modeling for a Complex World*. McGraw-Hill Higher Education, Boston.

Tàbara, D., & Pahl-Wostl, C. (2007). Sustainability learning in natural resource use and management. *Ecology and Society*, 12, 3. Available online: www.ecologyandsociety. org/vol12/iss2/art12/ (accessed on April 17, 2010).

Vennix, J. (1996). *Group Model Building*. Irwin McGraw-Hill. New York.

Vennix, J. A. M. (1999). Group model-building: Tackling messy problems. *System Dynamics Review,* 15(4): 379–402.

Voinov, A., & Gaddis, E. J. B. (2008). Lessons for successful participatory watershed modeling: A perspective from modeling practitioners. *Ecological Modelling*, 216: 197–207.

# 3 Optimization Methods

## 3.1 INTRODUCTION

In energy planning, we need to find which courses of action are feasible and then apply a dollar yardstick to see which one would be the most worthwhile. Thus, we need to optimize the use of resources available, i.e., we need to maximize the profits or minimize the costs subject to the restrictions on the availability of the limited resources. The problems arising from allocating scarce resources to competing activities are candidates for linear programming procedures. The studies of energy systems present many opportunities for linear programming applications (Parikh, 1998; Bala, 1998; Muunasinghe and Meier, 2008; Hooman, 2019), and linear programming can be profitably employed in all three stages of energy systems: planning, analysis and control.

In this chapter, the fundamentals of the optimization techniques of linear programming and integer programming are discussed, and their applications are addressed to show the potentiality of these techniques for the optimal operational planning of electric power systems.

## 3.2 LINEAR PROGRAMMING

Energy engineers and managers always work with budget constraints to generate, transmit and distribute electrical energy with environmental restrictions. Linear programming deals with thousands of variables and constraints, yet it is the most powerful method because computers can be used to exploit the computation capability for this type of constrained problem and standard software is also now available to solve this type of constrained optimization problem.

Linear programming uses mathematical models to describe the constrained optimization model. Here, linear means that all the mathematical functions used to solve the constrained optimization problem are linear, and programming means planning. Thus, linear programming means planning activities to obtain an optimal solution to the problems from the feasible alternatives.

The standard form of linear programming consists of two parts: (1) objective function and (2) constraints. Objective function defines the quantity to be optimized, and constraints defines the restrictions on achieving the objectives. For linear programming, the objective function may be stated as maximize or minimize the objective function (de Neufville, 1990):

$$Z = \sum C_i X_i \tag{3.1}$$

The variables $X_i$ in the objective function in Equation (3.1) are known as decision variables and we must find their value in order to optimize the objective function. The constraints in the optimization of the objective function in the standard form are stated as (de Neufville, 1990):

$$\sum a_{ij}X_i \gtreqless b_j \qquad (3.2)$$

where $a_{ij}$ parameters define the contributions of the decision variables to the constraints and $b_j$ are the upper or lower bounds of a particular constraint.

Both maximization and minimization problems can be handled using linear programming, and the typical optimization problems are either maximizing a profit or minimizing the cost and pollution levels subject to the restrictions of the budget. Linear programming problems in practice are dealt with a large number of variables and constraints, and hence these are presented in vector notation as (de Neufville, 1990):

Maximize or minimize

$$Z = CX \qquad (3.3)$$

subject to

$$AX \lesseqgtr B \qquad (3.4)$$

Where C is the row vector, X and B are column vectors, and A is the matrix of $a_{ij}$ parameters of the variables in the equations of the constraints.

### 3.2.1 Example of a Linear Programming Problem

A power company supplies electricity in an isolated remote village from a solar PV- diesel hybrid system. The power company's objective is to maximize the profit from the operation of this microgrid. The profit from solar-operated electricity is Tk. 3.00 per kWh and the profit from diesel-operated electricity is Tk. 5.00 per kWh. The production rates of solar and diesel electricity are $x_1$ kW and $x_2$ kW, respectively. Hence, the objective function is to maximize the profit from the sale of electricity per kWh and can be stated as:

Maximize

$$Z = 3x_1 + 5x_2 \qquad (3.5)$$

The power company wants to search and evaluate three supply options: (1) the solar plant has a maximum capacity of 80 kW, (2) the diesel plant has a maximum capacity of 120 kW and (3) the hybrid of these two has a maximum capacity of 160 kW. Hence, the constraints can be stated as:

$$x_1 \leq 80 \qquad (3.6)$$

$$x_2 \leq 120 \tag{3.7}$$

$$x_1 + x_2 \leq 160 \tag{3.8}$$

$$x_1 \geq 0, \quad x_2 \geq 0$$

The central assumptions of linear programming problems are that the objective function and the restrictions are linear, and the decision variables are continuous and nonnegative.

The problem presented above has two decision variables, $x_1$ and $x_2$, and this two-dimensional problem can be solved graphically with $x_1$ and $x_2$ as the axes. The first step is to find the region of the decision variables ($x_1$ and $x_2$) permitted by no negativity and the constraints. This can be accomplished by drawing lines for $x_1 = 80$, $x_2 = 120$, and $x_1 + x_2 = 160$, and $x_1 \geq 0$ and $x_2 \geq$. 0. The resulting region of the permitted values of the decision variables ($x_1$ and $x_2$) is shown in Figure 3.1.

The next step would be to find the values of the decision variables that maximize the objective function. The feasible values are the corner points of the region of permitted values of the decision variables. The line $Z = 3x_1 + 5x_2$ through the corner point (40, 120) gives the maximum value of the objective function ($Z = 720$) and is the optimal solution.

## 3.2.2 SIMPLEX METHOD

We should now discuss the simplex method, the general procedure for solving linear programming problems (Hillier and Lieberman, 1974; Rao, 1984; Beightler

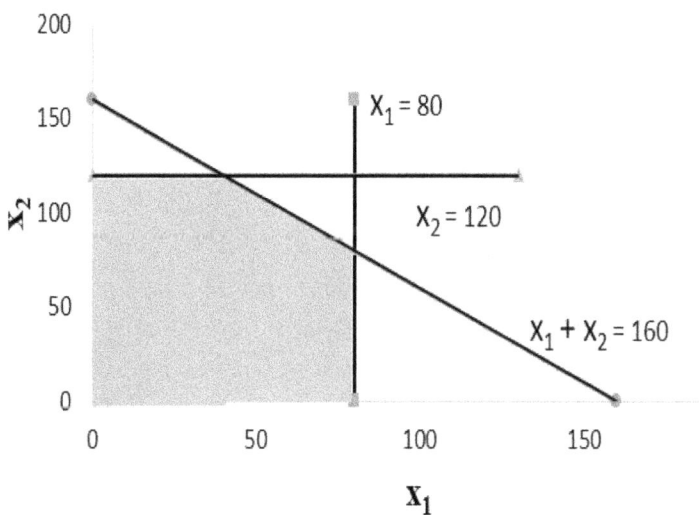

**FIGURE 3.1** Two-dimensional presentation of the linear programming problem.

et al., 1979; Taha, 2017). This is a remarkably efficient method that is routinely used for huge problems using computers. Computers are always used except for small problems, and the software is widely available. Nevertheless, it is important to learn the procedure for how the simplex method works.

The first step is to convert the functional inequality constraints into equality constraints, and this is done by introducing slack variables. Consider the example presented earlier.

$$x_1 \leq 80 \tag{3.9}$$

Introducing slack variable $(x_3)$, Equation (3.9) becomes:

$$x_3 = 80 - x_1 \tag{3.10}$$

Equation (3.10) gives the slack between the two sides of the inequality, and therefore we can write:

$$x_1 + x_3 = 80 \tag{3.11}$$

The original constraint $x_1 \leq 80$ holds for $x_1 \geq 0$. Hence, $x_1 \leq 80$ is entirely equivalent to:

$$x_1 + x_3 = 80 \tag{3.12}$$

$$x_1 \geq 0, \quad x_3 \geq 0$$

Introducing slack variables for the other constraints, the original linear programming model of the example can be replaced by the following equivalent model:
Maximize

$$Z = 3x_1 + 5x_3 \tag{3.13}$$

subject to:

| | | | | | |
|---|---|---|---|---|---|
| (1) | $x_1$ | $+ \ x_3$ | | | $= 80$ |
| (2) | $x_2$ | | $x_4$ | | $= 120$ |
| (3) | $x_1 + x_2$ | | | $+ x_5$ | $= 160$ |

For any linear programming model in standard form, the equivalent constraint model will be in the following form:

| | | | |
|---|---|---|---|
| (1) | $a_{11}x_1 + a_{12}x_2 +$ | $+ a_{1n-1}x_{n-1} + a_{1n}x_n + x_{n+1}$ | $= b_1$ |
| (2) | $a_{21}x_1 + a_{22}x_2 +$ | $+ a_{2n-1}x_{n-1} + a_{2n}x_n \qquad + x_{n+2}$ | $= b_2$ |
| | | | |
| (m) | $a_{m1}x_1 + a_{m2}x_2 +$ | $+ x_{n+m}$ | $= b_m$ |

where $x_{n+1}, x_{n+2}, x_{n+m}$ are the slack variables.

**TABLE 3.1**
**Standard Tabular Form of the Simplex Method**

| Equation number | Coefficient of Z | Coefficient of $x_1$ $x_2$ $x_3$ | | | $x_n$ | $x_{n+1}$ | $x_{n+2}$ | $x_{n+m}$ | Right side of equation |
|---|---|---|---|---|---|---|---|---|---|
| 0 | 1 | $-c_1$ | $-c_2$ | | $-c_n$ | 0 | 0 | 0 | 0 |
| 1 | 0 | $a_{11}$ | $a_{12}$ | | $a_{1n}$ | 1 | 0 | 0 | $b_1$ |
| 2 | 0 | $a_{21}$ | $a_{22}$ | | $a_{2n}$ | 0 | 1 | 0 | $b_2$ |
| m | 0 | $a_{m1}$ | $a_{m2}$ | | $a_{mn}$ | 0 | 0 | 1 | $b_m$ |

**TABLE 3.2**
**Initial Simplex Table of the Example**

| Basic variable | Eq. No. | Coefficient of Z | Coefficient of $x_1$ | $x_2$ | $x_3$ | $x_4$ | $x_5$ | Right side |
|---|---|---|---|---|---|---|---|---|
| Z | 0 | 1 | -3 | -5 | 0 | 0 | 0 | 0 |
| $x_3$ | 1 | 0 | 1 | 0 | 1 | 0 | 0 | 80 |
| $x_4$ | 2 | 0 | 0 | 1 | 0 | 1 | 0 | 120 |
| $x_5$ | 3 | 0 | 1 | 1 | 0 | 0 | 1 | 160 |

The algebraic form is the best for learning the simplex method. The standard tabular form of the simplex method is recommended for solving the problem by hand (Hillier and Liebermann, 1974; Taha, 2017). The standard tabular form for the standard algebraic form of the simplex model is shown in Table 3.1.

Now, introducing the slack variables and selecting the original variable to be the initial non-basic variables and the slack variables as the initial basic variables, the initial simplex table of the example is shown in Table 3.2.

At the initial iteration, the entering basic variable is determined by selecting the variable with the largest negative coefficient of Z and the leaving variable is determined by picking the coefficient in the column (> 0) of the entering variable and dividing the corresponding right-hand coefficient, and finally selecting the one with the minimum ratio. The tableaux of the example at this point is shown in Table 3.3.

The entering variable from the table is $x_2$, and the leaving variable is $x_4$.

The new basic variable by constructing the table is shown in Table 3.4. The first three columns are unchanged except that the leaving basic variable in the first column is replaced by the entering basic variable. To change the coefficient of the new basic variable by +1, the pivot row is changed by dividing the entire row by the pivot number, as shown in Table 3.4.

To eliminate the new basic variable from the other equations, every row except the pivot row is changed for the new table by using the following relationship:

**TABLE 3.3**
**Determination of Entering Variable for the Example**

| Basic variable | Eq. No. | Coefficient of Z | Coefficient of | | | | | Right side | Ratio |
|---|---|---|---|---|---|---|---|---|---|
| | | | $x_1$ | $x_2$ | $x_3$ | $x_4$ | $x_5$ | | |
| Z | 0 | 1 | −3 | −5 | 0 | 0 | 0 | 0 | |
| $x_3$ | 1 | 0 | 1 | 0 | 1 | 0 | 0 | 80 | |
| $x_4$ | 2 | 0 | 0 | 1 | 0 | 1 | 0 | 120 | 120/1 = 120 |
| $x_5$ | 3 | 0 | 1 | 1 | 0 | 0 | 1 | 160 | 160/1 = 160 |

**TABLE 3.4**
**Determination of Leaving Variable for the Example**

| Iteration | Basic variable | Eq. No. | Coefficient of Z | Coefficient of | | | | | Right side |
|---|---|---|---|---|---|---|---|---|---|
| | | | | $x_1$ | $x_2$ | $x_3$ | $x_4$ | $x_5$ | |
| 0 | Z | 0 | 1 | −3 | −5 | 0 | 0 | 0 | 0 |
| | $x_3$ | 1 | 0 | 1 | 0 | 1 | 0 | 0 | 80 |
| | $x_4$ | 2 | 0 | 0 | 1 | 0 | 1 | 0 | 120 |
| | $x_5$ | 3 | 0 | 1 | 1 | 0 | 0 | 1 | 160 |
| 1 | Z | 0 | 1 | | | | | | |
| | $x_3$ | 1 | 0 | | | | | | |
| | $x_2$ | 2 | 0 | 0 | 1 | 0 | 1 | 1 | 120 |
| | $x_5$ | 3 | 0 | | | | | | |

New row = old row − (pivot column coefficient) × new pivot row

where pivot column coefficient is the number in this row that is in the pivot column, and these are illustrated below:

$$\text{Row } 0$$
$$-(-5) \quad \frac{\begin{bmatrix} -3 & -5 & 0 & 0 & 0. & 0 \end{bmatrix}}{\begin{bmatrix} 0 & 1 & 0 & 1 & 0, & 120 \end{bmatrix}}$$
$$\text{New row} = \begin{bmatrix} -3 & 0 & 0 & 5 & 0, & 600 \end{bmatrix}$$

Row 1 is unchanged since the pivot column coefficient is zero.

$$\text{Row } 3$$
$$-(1) \quad \frac{\begin{bmatrix} 1 & 1 & 0 & 0 & 1, & 160 \end{bmatrix}}{\begin{bmatrix} 0 & 1 & 0 & 1 & 0, & 120 \end{bmatrix}}$$
$$\text{New row} = \begin{bmatrix} 1 & 0 & 0 & -1 & 1, & 40 \end{bmatrix}$$

This gives the new Table 3.5 with iteration 1.

**TABLE 3.5**
**Determination of Entering and Leaving Variable for the Example**

| Iteration | Basic variable | Eq. No. | Coefficient of Z | Coefficient of | | | | | Right side |
|---|---|---|---|---|---|---|---|---|---|
| | | | | $x_1$ | $x_2$ | $x_3$ | $x_4$ | $x_5$ | |
| 0 | Z | 0 | 1 | −3 | −5 | 0 | 0 | 0 | 0 |
| | $x_3$ | 1 | 0 | 1 | 0 | 1 | 0 | 0 | 80 |
| | $x_4$ | 2 | 0 | 0 | 1 | 0 | 1 | 0 | 120 |
| | $x_5$ | 3 | 0 | 1 | 1 | 0 | 0 | 1 | 160 |
| 1 | Z | 0 | 1 | −3 | 0 | 0 | 5 | 0 | 600 |
| | $x_3$ | 1 | 0 | 1 | 0 | 1 | 0 | 0 | 80 |
| | $x_2$ | 2 | 0 | 0 | 1 | 0 | 1 | 0 | 120 |
| | $x_5$ | 3 | 0 | 1 | 0 | 0 | −1 | 1 | 40 |

**TABLE 3.6**
**Determination of Second Entering and Leaving Variables for the Example**

| Iteration | Basic variable | Eq. No. | Coefficient of Z | Coefficient of | | | | | Right side |
|---|---|---|---|---|---|---|---|---|---|
| | | | | $x_1$ | $x_2$ | $x_3$ | $x_4$ | $x_5$ | |
| 1 | Z | 0 | 1 | −3 | 0 | 0 | 5 | 0 | 600 |
| | $x_3$ | 1 | 0 | 1 | 0 | 1 | 0 | 0 | 80 |
| | $x_2$ | 2 | 0 | 0 | 1 | 0 | 1 | 0 | 120 |
| | $x_5$ | 3 | 0 | 1 | 0 | 0 | −1 | 1 | 40 |

Each basic variable always equals the right side of the equation. Hence, the new basic feasible solution is (0, 120, 80, 0, 40) and Z = 600. Since the new Eqn (0) still has a negative coefficient (-3), the solution is not optimal. So, a new iterative step is started to find the new solution. Following the instructions stated earlier, we find $x_1$ as the entering basic variable and $x_5$ as the leaving basic variable, as shown in Table 3.6. Row 3 is the pivot row.

$$\text{New row} = \begin{bmatrix} 1 & 0 & 0 & -1 & 1, & 40 \end{bmatrix}$$

$$
\begin{aligned}
\text{Row 0} \qquad & \begin{bmatrix} -3 & 0 & 0 & 5 & 0. & 600 \end{bmatrix} \\
& \underline{-(-3)\begin{bmatrix} 1 & 0 & 0 & -1 & 1, & 40 \end{bmatrix}} \\
& \text{New row} = \begin{bmatrix} 0 & 0 & 0 & 2 & 3, & 720 \end{bmatrix} \\
\text{Row 1} \qquad & \begin{bmatrix} 1 & 0 & 1 & 0 & 0, & 80 \end{bmatrix} \\
& \underline{-(1)\begin{bmatrix} 1 & 0 & 0 & -1 & 1, & 40 \end{bmatrix}} \\
& \text{New row} = \begin{bmatrix} 0 & 0 & 1 & 1 & -1, & 40 \end{bmatrix}
\end{aligned}
$$

Row 2 is unchanged since the pivot column coefficient is zero.

**TABLE 3.7**
**Final Simplex Method Solution Table**

| Iteration | Basic variable | Eq. No. | Coefficient of Z | Coefficient of $x_1$ | $x_2$ | $x_3$ | $x_4$ | $x_5$ | Right side |
|---|---|---|---|---|---|---|---|---|---|
| 0 | Z | 0 | 1 | −3 | −5 | 0 | 0 | 0 | 0 |
| | $x_3$ | 1 | 0 | 1 | 0 | 1 | 0 | 0 | 80 |
| | $x_4$ | 2 | 0 | 0 | 1 | 0 | 1 | 0 | 120 |
| | $x_5$ | 3 | 0 | 1 | 1 | 0 | 0 | 1 | 160 |
| 1 | Z | 0 | 1 | −3 | 0 | 0 | 5 | 0 | 600 |
| | $x_3$ | 1 | 0 | 1 | 0 | 1 | 0 | 0 | 80 |
| | $x_2$ | 2 | 0 | 0 | 1 | 0 | 1 | 0 | 120 |
| | $x_5$ | 3 | 0 | 1 | 0 | 0 | −1 | 1 | 40 |
| 2 | Z | 0 | 1 | 0 | 0 | 0 | 2 | 3 | 720 |
| | $x_3$ | 1 | 0 | 0 | 0 | 1 | 1 | −1 | 40 |
| | $x_2$ | 2 | 0 | 0 | 1 | 0 | 1 | 0 | 120 |
| | $x_1$ | 3 | 0 | 1 | 0 | 0 | −1 | 1 | 40 |

The final simplex method solution for this example is shown in Table 3.7

The optimal solution is $x_1 = 40$, $x_2 = 120$ and $Z = 720$ since none of the coefficient in Eqn (0) is negative.

## 3.3  INTEGER PROGRAMMING

In many practical problems, decision variables are integers. For example, solar panels, nuclear power plants, transformers and electric generators. This may make it difficult to handle such restrictions mathematically. However, such linear programming problems can be solved by being subjected to additional restrictions that the decision variables must have integer values. When all the decision variables are considered to take integer values, it is called integer programming. When some of the decision variables are considered to take integer values, it is called mixed-integer programming. When all the variables are considered to take values 0 or 1, it is called 0-1 integer programming. Among the techniques available for solving integer programming problems, the branch and bound algorithm by Land and Doig is quite popular and is discussed here.

### 3.3.1  BRANCH AND BOUND ALGORITHM

The branch and bound algorithm is particularly useful for solving large classes of mathematical problems, and the general procedure has found extensive applications in integer and mixed-integer problem solutions.

Consider the following mixed integer programming problem:
Maximize

$$Z = f(x) = \sum_{j}^{N} c_j x_j \qquad (3.14)$$

subject to

$$\sum_{j}^{N} a_{ij} x_j \le b_i \qquad i = 1,2,.......,M \qquad (3.15)$$

$$U_j \ge x_j \ge L_j \qquad j = 1,2,........N \qquad (3.16)$$

$$x_k \ge 0 \qquad k = 1,2,........\bar{N} \qquad (3.17)$$

$$x_l \text{ integer} \qquad l = \bar{N}+1, \bar{N}+2..............N$$

The underlying concept of the branch and bound algorithm is that for any value of $x_j$, it states that:

$$[x_j]+1 \ge x_j \ge [x_j] \qquad (3.18)$$

where $[x_j]$ is the largest integer less than (or equal to) the value $x_j$.
   The branch and bound algorithm is summarized in the following steps:

**Step 1 Linear programming solution**
Solve the problem as an ordinary linear programming problem. If the decision variables satisfy the requirement, the solution is optimal. Otherwise, initialize the lower bound of the optimal solution and move on to step 2.
**Step 2 Branching**
Select the variable that is the integer restricted from the linear programming solution and divide it into two new linear programming problems considering:

$$[x_j]+1 \ge x_j \ge [x_j] \qquad (3.19)$$

**Step 3 Bounding**
Solve these two new linear programming problems and calculate the values of the integer decision variables. The linear programming problem with the infeasible solution is eliminated. The solution is optimal if there is no linear programming problem. Otherwise, continue the iteration from step 2.

## 3.4  PARETO OPTIMALITY

In the previous sections, we have discussed linear programming and integer programming optimization. This section is devoted to Pareto optimality for multi-objective optimization for its use in optimization using the genetic algorithms in Chapter 10. One way to find good solutions to multi-objective problems is with Pareto optimality, named after economist Vilfredo Pareto. Multi-objective optimization consists of two or more objective functions and constraints. Let us consider a simple multi-objective problem consisting of two objective functions and two constraints:

Minimize

$$f_1(\text{cost}) = 2000\,x_1 + 500\,x_2 \qquad\qquad (3.20)$$

$$f_2(\text{time}) = x_1 + 2\,x_2 \qquad\qquad (3.21)$$

Subject to the constraints

$$2\,x_1 + 2\,x_2 \geq 16 \qquad\qquad (3.22)$$

$$x_1 + 5\,x_2 \geq 28 \qquad\qquad (3.23)$$

Let us first plot the two constraints of Equation (3.22) and Equation (3.23) with axes of $x_1$ and $x_2$, as shown in Figure 3.2. The constraints cross the axes and corner points, and these create a feasible region of the constraints (shaded area). The feasible region of these constraints is called the design space of the problem (shaded area).

The next step is to convert the design space to a criterion space. In the criterion space, the axes are not $x_1$ and $x_2$, but $f_1(\text{cost})$ and $f_2(\text{time})$, which are the objective functions of this problem. Now, let us evaluate the corner points for both the objectives, $f_1(\text{cost})$ and $f_2$ (time), as shown in Figure 3.3 and create the criterion

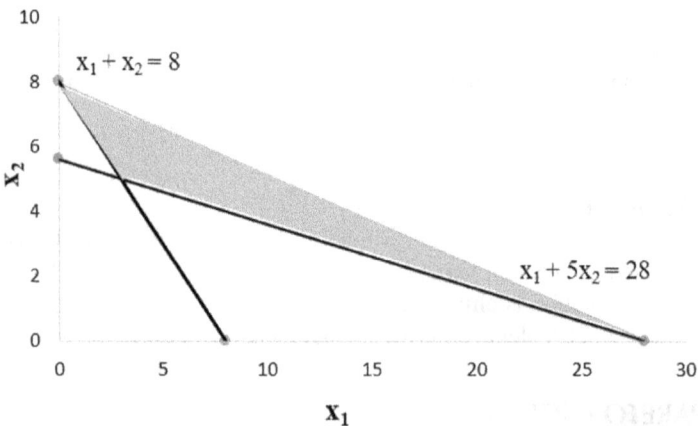

FIGURE 3.2  Feasible region for the constraints.

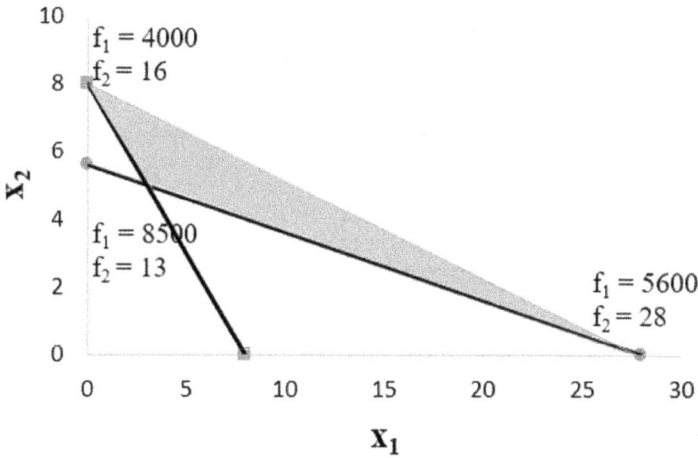

**FIGURE 3.3**   Design space.

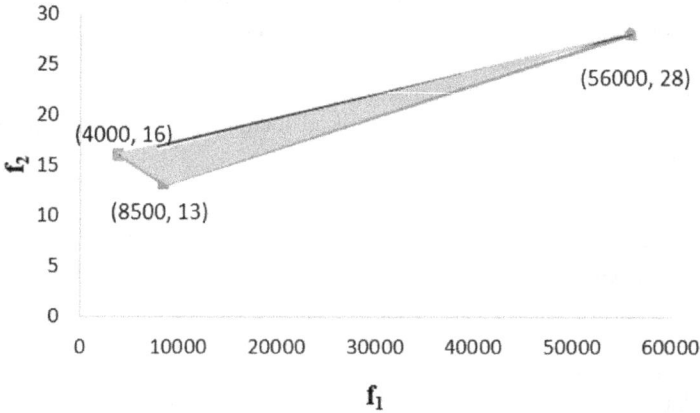

**FIGURE 3.4**   Criterion space.

space plotting $f_2$ (time) and $f_1$(cost) of the three corner points with axes of $f_2$(time) and $f_1$(cost), as shown in Figure 3.4.

We are trying to minimize time and cost. Note that in Figure 3.4, the point (8500, 13) in criterion space is the lowest value of $f_2$(time), and the point (4000, 16) is the lowest value of $f_1$(cost). The edge between them is called the Pareto front. Any point on this front is considered the Pareto optimal. By moving along this curve, you can minimize the cost at the expense of time or minimize the time at the expense of cost, but you cannot improve both at once. What this means is that there is no mathematical best point along the Pareto front. The slope of the Pareto curve in Figure 3.4 is -3/4500, which reduces to -1/1500.

**TABLE 3.8**
**Some Pareto Optimal Solutions**

| Cost | Time |
|------|------|
| 4000 | 16   |
| 5500 | 15   |
| 7000 | 14   |
| 8500 | 13   |

Using this slope, we can find some Pareto optimal solutions, and these are given in Table 3.8.

An important implication of the Pareto front is that any point in the feasible region that is not on the Pareto front is a bad solution. Either or both of the objectives can be improved at no penalty to the other, and an improvement that helps one objective without harming the other objective is called a Pareto improvement.

Many heuristic search algorithms have been developed to solve multi-objective problems, including genetic algorithms. Genetic algorithms are population-based search algorithms inspired by Darwin's evolutionary theory, i.e., survival of the fittest. Several Pareto genetic algorithms are also available to find Pareto optimal solutions minimizing computational efforts (Deb et al., 2002; Eskandari and Geiger, 2008). These algorithms can be used for the operational optimization of smart grids.

## 3.5 EXAMPLES OF LINEAR PROGRAMMING AND INTEGER PROGRAMMING PROBLEMS

**Example 3.1**
Electricity is to be supplied from a coal-based power plant as well as a diesel-based power plant. The power company's objective is to maximize the profit from the coal-based power plant as well as from the diesel-based power plant. The profits from sales at the coal-based power plant and the diesel-based power plant are Tk. 6.00 per kWh and TK. 4.00 per kWh, respectively. The production of coal-based electricity and diesel-based electricity are $0.3x_1$ GWh and $0.4 x_2$ GWh, respectively. Hence, the objective function is to maximize the profit from the sale of electricity and can be stated as:

Maximize

$$Z = 6x_1 + 4x_2 \tag{3.24}$$

The power company wants to search and evaluate three options: (1) a maximum capacity of 120 GWH of coal-based electricity, (2) a maximum capacity of 100 GWH of diesel-based electricity and (3) a combined sale option with a maximum

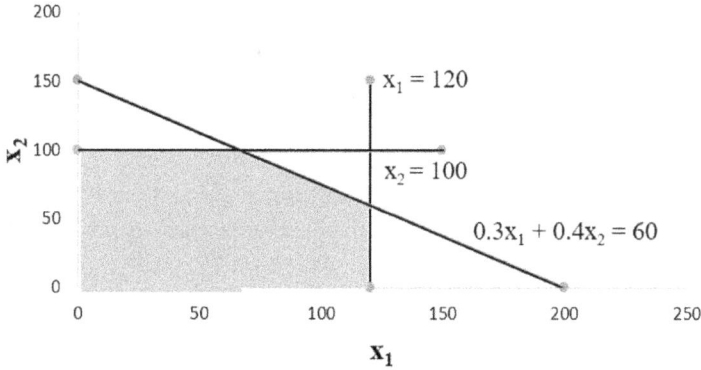

**FIGURE 3.5** Two-dimensional presentation of the linear programming problem.

capacity of 60 GWH, which consists of 30% coal-based electricity and 40% diesel-based electricity. Hence, the constraints can be stated as:

$$x_1 \le 120 \tag{3.25}$$

$$x_2 \le 100 \tag{3.26}$$

$$0.30x_1 + 0.40x_2 \le 60 \tag{3.27}$$

$$x_1 \ge 0, \quad x_2 \ge 0$$

The problem presented above has two decision variables, $x_1$ and $x_2$, and this two-dimensional problem can be solved graphically with $x_1$ and $x_2$ as the axes. The first step is to find the region of the decision variables $(x_1$ and $x_2)$ permitted by no nega-tivity and the constraints. This can be accomplished by drawing lines for $x_1 = 120$, $x_2 = 100$, and $0.3x_1 + 0.4x_2 = 60$ and $x_1 \ge 0$ and $x_2 \ge. 0$. The resulting region of the permitted values of the decision variables $(x_1$ and $x_2)$ is shown in Figure 3.5.

The next step would be to find the values of the decision variables that maxi-mize the objective function. The feasible values are the corner points of the region of permitted values of the decision variables. The line $Z = 6x_1 + 4x_2$ through the corner point (66.67, 100) gives the maximum value of the objective function $(Z = 967)$ and is the optimal solution.

**Example 3.2**
A diesel fuel supplier is planning to supply diesel and biodiesel in a local market. The power company's objective is to maximize the profit from the sale of both. The profits from the sale of biodiesel and diesel are Tk. 4.00 per liter and TK. 7.00 per liter, respectively. The production rates of biodiesel and diesel are $x_1$ liters and $x_2$

liters, respectively. Hence, the objective function is to maximize the profit from the sale of biodiesel and diesel and can be stated as:
    Maximize

$$Z = 4x_1 + 7x_2 \tag{3.28}$$

The fuel supplier wants to search and evaluate three options: (1) a maximum capacity of 150 liters of biodiesel, (2) a maximum capacity of 100 liters and (3) a combined sale option with a maximum capacity of 160 liters of biodiesel and diesel, which consist of 30% biodiesel and 40% diesel. Hence, the constraints can be stated as:

$$x_1 \leq 150 \tag{3.29}$$

$$x_2 \leq 100 \tag{3.30}$$

$$0.30x_1 + 0.40x_2 \leq 60 \tag{3.31}$$

$$x_1 \geq 0, \quad x_2 \geq 0$$

The problem presented above has two decision variables, $x_1$ and $x_2$, and this two-dimensional problem can be solved graphically with $x_1$ and $x_2$ as the axes. The first step is to find the region of the decision variables ($x_1$ and $x_2$) permitted by no negativity and the constraints. This can be accomplished by drawing lines for $x_1 = 150$, $x_2 = 100$, and $0.3x_1 + 0.4x_2 = 60$ and $x_1 \geq 0$ and $x_2 \geq$. 0. The resulting region of the permitted values of the decision variables ($x_1$ and $x_2$) is shown in Figure 3.6.

The next step would be to find the values of the decision variables that maximize the objective function. The feasible values are the corner points of the region of permitted values of the decision variables. The line $Z = 4x_1 + 7x_2$ through the

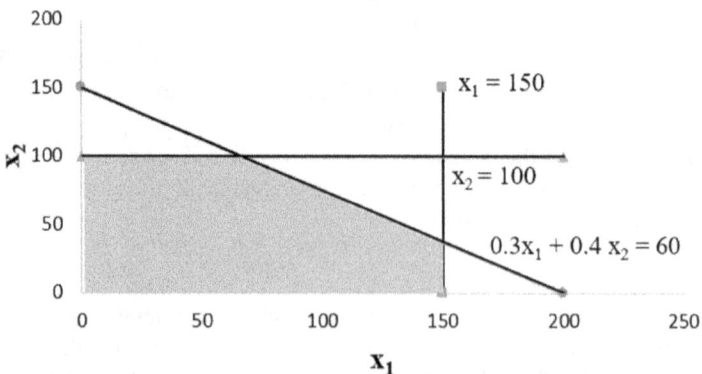

FIGURE 3.6  Two-dimensional presentation of the linear programming problem.

corner point (66.67, 100) gives the maximum value of the objective function ($Z = 967$) and is the optimal solution.

## Example 3.3

Two types of solar PVs are available for their use on a rooftop to provide electricity in houses: PVA and PVB. PVA is 34 kW with an area of 5 m², and PVB is 40 kW with an area of 7 m². The number of either type of PVs cannot be more than four. The total roof area available for placing these solar PVs is 35 m². We need to optimize the rooftop area for placing the solar PV and hence determine the number of each type of solar PV to be used. The objective of this placement of solar PV is to maximize energy conservation using solar electricity. The energy savings from the use of PVA and PVB are 87000 kWh per PV and 95000 kWh per PV, respectively. The number of solar PVAs and solar PVBs are $x_1$ and $x_2$, respectively. Hence, the objective function is to maximize the energy conservation from the use of solar PV and can be stated as:
    Maximize

$$Z = 87000x_1 + 95000x_2 \tag{3.32}$$

We need to search and evaluate three options: (1) the maximum number of PVAs is to be four, (2) the maximum number of solar PVBs is to be 4 and (3) the combined use option has a maximum allowed rooftop area of 35 m². Hence, the constraints can be stated as:

$$x_1 \leq 4 \tag{3.33}$$

$$x_2 \leq 4 \tag{3.34}$$

$$5x_1 + 7x_2 \leq 35 \tag{3.35}$$

$$x_1 \geq 0, \quad x_2 \geq 0$$

The problem presented above has two decision variables, $x_1$ and $x_2$, and this two-dimensional problem can be solved graphically with $x_1$ and $x_2$ as the axes. The first step is to find the region of the decision variables ($x_1$ and $x_2$) permitted by no negativity and the constraints. This can be accomplished by drawing lines for $x_1 = 4$, $x_2 = 4$, and $5x_1 + 7x_2 = 35$ and $x_1 \geq 0$ and $x_2 \geq 0$. The resulting region of the permitted values of the decision variables ($x_1$ and $x_2$) is shown in Figure 3.7.

The next step would be to find the values of the decision variables that maximize the objective function. The feasible values are the corner points of the region of permitted values of the decision variables. The line $Z = 67000 x_1 + 95000 x_2$ through the corner point (4, 2) gives the maximum value of the objective function ($Z = 712000$) and is the optimal solution.

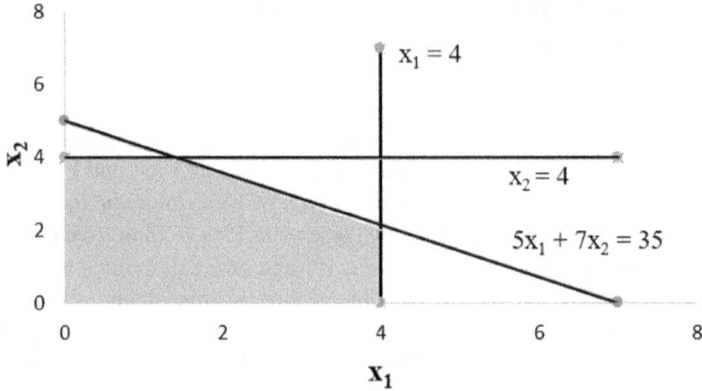

**FIGURE 3.7**   Two-dimensional presentation of the linear programming problem.

## Example 3.4

Energy conservation is one of the major concerns for the design of energy-efficient buildings (Das et al., 2017). The energy conservation of lighting loads using energy-efficient lights and solar PV to generate electricity is an option of recent interest in the design of energy-efficient buildings. To conserve energy and reduce the cost of electricity bills in a building, solar PVs are placed on the rooftop, and the CFL and tube lights are replaced by LED lights. To achieve this goal, the solar PVs that satisfy the energy requirements must be accommodated on the top of the building, and most importantly, the budget for this building electrification must be within the limits of the funds available. The problem is to maximize the energy conserved within the constraints of budget available, lighting loads required and the floor area available on the rooftop of a building.

Hence, the objective function is to maximize the energy conservation from the use of solar PV and can be stated as:

**Objective function**

Maximize the objective function:

$$Z = \sum_i C_{si} \times X_i \qquad (3.36)$$

where

$i = 1, 2, 3, 4$

$X_1$ = LED light, number of LED lights

$X_2$ = solar panel (low range), W

$X_3$ = solar panel (medium range), W

$X_4$ = solar panel (high range), W

$C_{s1}$ = energy saved per unit LED light, W/LED light

$C_{s2}$ = energy saved per unit energy supplied by solar panel (low range), W/W
$C_{s3}$ = energy saved per unit energy supplied by solar panel (medium range), W/W
$C_{s4}$ = energy saved per unit energy supplied by solar panel (high range), W/W

**Constraint 1:** Annual budget

$$\sum_i C_{bi} \times X_i \le B \tag{3.37}$$

where
  $i = 1, 2, 3, 4$
  $C_{b1}$ = cost of installation per unit LED light, TK/LED light
  $C_{b2}$ = cost of installation per unit W of solar panel (low range), TK/W
  $C_{b3}$ = cost of installation per unit W of solar panel (low range), TK/W
  $C_{b4}$ = cost of installation per unit W of solar panel (low range), TK/W
  B = annual budget, TK

**Constraint 2:** LED lights

$$X_1 \le L \tag{3.38}$$

where
  L = Number of LED lights

**Constraint 3:** Floor area for solar panel

$$\sum_j C_{fj} \times X_j \le A \tag{3.39}$$

where
  $j = 2, 3, 4$
  $C_{f2}$ = floor area per unit W of solar panel (low range), m²/W
  $C_{f3}$ = floor area per unit W of solar panel (medium range), m²/W
  $C_{f4}$ = floor area per unit W of solar panel (high range), m²/W
  B = total floor area of solar panel, m²

The data for this linear programming model are adopted from Das et al. (2017), and the linear programming model for solving using the software LiPS is shown in tabular format in Table 3.9.

The linear programming model is solved using LiPS. The optimal energy saved is Z = 3300 W with the budget Tk. 2,50,000. But X4 is not an integer and is greater than 1, as shown in Table 3.10. To avoid this non-integer value, the budget is readjusted to Tk. 2,20,400 after further calculation and the revised computer format is shown in Table 3.11.

**TABLE 3.9**
**The Optimization Problem Computer Format**

|              | X1   | X2    | X3     | X4     |      | RHS MAX |
|--------------|------|-------|--------|--------|------|---------|
| Objective    | 12   | 1000  | 1500   | 2000   | =    |         |
| Row 1        | 580  | 87000 | 130500 | 174000 | < =  | 25000   |
| Row 2        | 1    |       |        |        | < =  | 80      |
| Row 3        |      | 7.76  | 22.64  | 15.52  | < =  | 220     |
| Lower bound  | 0    | 0     | 0      | 0      |      |         |
| Upper bound  | INF  | INF   | INF    | INF    |      |         |
| Type         | CONT | CONT  | CONT   | CONT   |      |         |

**TABLE 3.10**
**Optimal Solution with Non-Integer Value of Solar PV**

| Variable | Value   | Obj. cost | Reduced cost |
|----------|---------|-----------|--------------|
| X1       | 80      | 12        | 0            |
| X2       | 0       | 1000      | 0            |
| X3       | 0       | 1500      | 0            |
| X4       | 509/435 | 2000      | 0            |

**TABLE 3.11**
**Optimization Problem Computer Format in the Revised Format**

|              | X1   | X2    | X3     | X4     |      | RHS MAX |
|--------------|------|-------|--------|--------|------|---------|
| Objective    | 12   | 1000  | 1500   | 2000   | =    |         |
| Row 1        | 580  | 87000 | 130500 | 174000 | < =  | 220400  |
| Row 2        | 1    |       |        |        | < =  | 80      |
| Row 3        |      | 7.76  | 22.64  | 15.52  | < =  | 220     |
| Lower bound  | 0    | 0     | 0      | 0      |      |         |
| Upper bound  | INF  | INF   | INF    | INF    |      |         |
| Type         | CONT | CONT  | CONT   | CONT   |      |         |

The linear programming model is again solved using LiPS. The optimal energy saved is the maximum energy saved, which is $Z = 2960$, and the final optimal solution is shown in Table 3.12

The optimal solution from Table 3.12 is $X1 = 80$, $X2 = 0$, $X3 = 0$, and $X4 = 1$, with an optimal value of energy saved of $Z = 2960$ W.

**TABLE 3.12**
**Final Optimal Solution**

| Variable | Value | Obj. cost | Reduced cost |
|----------|-------|-----------|--------------|
| X1 | 80 | 12 | 0 |
| X2 | 0 | 1000 | 0 |
| X3 | 0 | 1500 | 0 |
| X4 | 1 | 2000 | 0 |

**Example 3.5**

Textile manufacturing is an energy-intensive process (Kimutai et al., 2019), and the objective of this example is to develop a mathematical model that meets the finished product requirements at a minimum cost of energy used in the process, subject to different operational constraints. The result of this process will yield a unique set of values that represent the quantities of products to be processed to meet the monthly demands. Multi-stage linear programming is used in the development of the model.

Hence, the objective function is to minimize the energy conservation from the use of solar PV and can be stated as:

Minimize

$$Z = \sum_i C_{ei} \times X_i \qquad (3.40)$$

where

$i = 1, 2, 3, 4$

$X_1$ = number of units of product processed at stage 1 per month

$X_2$ = number of units of product processed at stage 2 per month

$X_3$ = number of units of product processed at stage 3 per month

$X_4$ = number of units of product processed at stage 4 per month

$X_5$ = number of units of product processed at stage 5 per month

$C_{e1}$ = energy cost per unit of product in spinning, TK/number of units

$C_{e2}$ – energy cost per unit of product in sizing/preparation, TK/number of units

$C_{e3}$ = energy cost per unit of product in rewinding, TK/number of units

$C_{e4}$ = energy cost per unit of product in weaving, TK/number of units

$C_{e5}$ = energy cost per unit of product in finishing, TK/number of units

**Constraint 1**: Demand for product produced in stage 1

$$X_1 - X_2 \geq D1 \qquad (3.41)$$

where

D1 = number of units of demand for product produced at stage 1 per month

**Constraint 2**: Demand for product produced in stage 4

$$C_4 \times X_4 - X_5 \geq D2 \qquad (3.42)$$

where

$C_4$ = share of the number of units of demand for product produced at stage 4 per month

D2 = number of units of demand for product produced at stage 4 per month

**Constraint 3**: Demand for product produced in stage 5

$$C_5 \times X_5 \geq D3 \qquad (3.43)$$

where

$C_5$ = share of the number of units of demand for product produced at stage 5 per month

D3 = number of units of demand for product produced at stage 5 per month

**Constraint 4**: Material balance 1

$$\sum_j C_{m1j} X_j = 0 \qquad (3.44)$$

where

$j = 2, 3, 4$

$C_{m12}$ = share of product in sizing/preparation, h/number of units

$C_{m13}$ = share of product in rewinding, h/number of units

$C_{m14}$ = share of product in weaving, h/number of units

**Constraint 5**: Material balance 2

$$\sum_j C_{m2j} X_j = 0 \qquad (3.45)$$

where

$j = 2, 3, 4$

$C_{m22}$ = share of product in sizing/preparation, h/number of units

$C_{m23}$ = share of product in rewinding, h/number of units

$C_{m24}$ = share of product in weaving, h/number of units

**Constraint 5**: Time

$$\sum_j C_{tj} X_j \leq T \qquad (3.46)$$

where

$j = 1, 2, 3, 4$

Ce1 = time needed per unit of product in spinning, h/number of units

Ce2 = time needed per unit of product in sizing/preparation, h/number of units
Ce3 = time needed per unit of product in rewinding, h/number of units
Ce4 = time needed per unit of product in weaving, h/number of units
Ce5 = time needed per unit of product in finishing, h/number of units
T = total time, h

The data for this linear programming model are adopted from Kimutai et al. (2019), and the linear programming model for solving using LiPS is shown in tabular format in Table 3.13.

The linear programming model is solved using LiPS. The optimal cost is the minimum cost, Z = Tk. 2,813,030, and the optimal solution is shown in Table 3.14.

**TABLE 3.13**
**The Optimization Problem Computer Format**

|  | X1 | X2 | X3 | X4 | X5 |  | RHS MAX |
|---|---|---|---|---|---|---|---|
| Objective | 50 | 7 | 40 | 15 | 48 | = |  |
| Row 1 | 1 | −1 | 0 | 0 |  | > = | 400 |
| Row 2 | 0 | 0.03 | −1 | 0.07 |  | = | 0 |
| Row 3 | 0 | 0.07 | 0 | −1 |  | = | 0 |
| Row 4 | 0 | 0 | 0 | 0.93 | −1 | > = | 600 |
| Row 5 | 0 | 0 | 0 | 0 | 0.96 | > = | 20000 |
| Row 6 | 0.007 | −0.007 | 0 | 0.013 | 0.0062 | < = | 720 |
| Row 7 | 1 | 0 | 0 | 0 | 0 | > = | 0 |
| Row 8 | 0 | 1 | 0 | 0 | 0 | > = | 0 |
| Row 9 | 0 | 0 | 1 | 0 | 0 | > = | 0 |
| Row 10 | 0 | 0 | 0 | 1 | 0 | > = | 0 |
| Row 11 | 0 | 0 | 0 | 0 | 1 | > = | 0 |
| Lower bound | 0 | 0 | 0 | 0 | 0 |  |  |
| Upper bound | INF | INF | INF | INF | INF |  |  |
| Type | CONT | CONT | CONT | CONT | CONT |  |  |

**TABLE 3.14**
**Optimal Solution**

| Variable | Value | Obj. cost | Reduced cost |
|---|---|---|---|
| X1 | 24159.38 | 50 | 0 |
| X2 | 23759.38 | 7 | 0 |
| X3 | 2326.043 | 40 | 0 |
| X4 | 23046.59 | 15 | 0 |
| X5 | 20833.33 | 48 | 0 |

The optimal solution from Table 3.14 is X1 = 24,159.4, X2 = 23,759.4, X3 = 2,326.04, X4 = 23,046.6, X5 = 20,833.3, with an optimal value of energy cost of Z = Tk. 2,813,030.

**Example 3.6**
Solve the following problems using the branch and bound method.
Maximize

$$Z = 9x_1 + 6x_2 + 5x_3 \tag{3.47}$$

subject to

$$2x_1 + 3x_2 + 7x_3 \le 17.5 \tag{3.48}$$

$$4x_1 \quad\quad + 9x_3 \le 15 \tag{3.49}$$

$$x_1 \text{ nonnegative integer}$$

**Solution**
The problem is solved in the following steps:

**Step 1**
The problem is solved using an ordinary linear program, and the results are:

$$x_1 = 15/4, x_2 = 10/3, \ x_3 = 0, Z = 215/4$$

**Step 2**
This problem does not satisfy the integer requirement. The original problem is divided into two mutually exclusive and exhaustive sub-problems, presented below:

(a)

Maximize

$$Z = 9x_1 + 6x_2 + 5x_3 \tag{3.501}$$

subject to

$$2x_1 + 3x_2 + 7x_3 \le 17.5 \tag{3.51}$$

$$4x_1 \quad\quad + 9x_3 \le 15 \tag{3.52}$$

$$x_1 \quad\quad\quad \ge 4 \text{ integer} \tag{3.53}$$

$$x_1, x_2 \ge 0$$

(b)

Maximize

$$Z = 9x_1 + 6x_2 + 5x_3 \qquad (3.54)$$

subject to

$$2x_1 + 3x_2 + 7x_3 \le 17.5 \qquad (3.55)$$

$$4x_1 \qquad + 9x_3 \le 15 \qquad (3.56)$$

$$0 \le x_1 \qquad \le 3 \text{ integer} \qquad (3.57)$$

$$x_1, x_2 \ge 0$$

It is clear that the problem presented in (a) is not a feasible solution and the solution of problem (b) is:

$$x_1 = 3, x_2 = 23/6, \ x_3 = 0, Z = 50$$

Since this problem does satisfy the integer requirement, it is the optimal solution.

**Example 3.7**
Solve the following problems using the branch and bound method.
    Maximize

$$Z = 9x_1 + 6x_2 + 5x_3 \qquad (3.58)$$

subject to

$$2x_1 + 3x_2 + 7x_3 \le 17.5 \qquad (3.59)$$

$$4x_1 \qquad + 9x_3 \le 15 \qquad (3.60)$$

$$x_1, x_2, x_3 \text{ nonnegative integers}$$

**Solution**
The problem is solved in the following steps:

**Step 1**
The problem is solved using an ordinary linear program, and the results are:

$$x_1 = 15/4, x_2 = 10/3, \ x_3 = 0, Z = 215/4$$

**Step 2**
Considering the variable $x_1$, we can create two sub-problems, (a) and (b). One solution is not feasible, and the other has the following solution:

$$x_1 = 3, x_2 = 23/6, \ x_3 = 0, Z = 50$$

**Step 3**
The problem in (b) is divided into two sub-problems, (c) and (d).

(c)

Maximize

$$Z = 9x_1 + 6x_2 + 5x_3 \tag{3.61}$$

subject to

$$2x_1 + 3x_2 + 7x_3 \le 17.5 \tag{3.62}$$

$$4x_1 \qquad + 9x_3 \le 15 \tag{3.63}$$

$$0 \le x_1 \le 3 \text{ integer} \tag{3.64}$$

$$x_2 \ge 4 \text{ integer} \tag{3.65}$$

$$x_2 \ge 0$$

(d)

Maximize

$$Z = 9x_1 + 6x_2 + 5x_3 \tag{3.66}$$

subject to

$$2x_1 + 3x_2 + 7x_3 \le 17.5 \tag{3.67}$$

$$0 \le x_1 \le 3 \text{ integer} \tag{3.68}$$

$$0 \le x_2 \le 3 \text{ integer} \tag{3.69}$$

$$x_2 \ge 0$$

Problem (c) is not a feasible solution. Problem (d) satisfies the requirement, and the optimal solution is:

$$x_1 = 3, x_2 = 3, \ x_3 = 0, Z = 45$$

## 3.6.  MARKAL MODEL

MARKAL (MARKet ALlocation) is a widely used bottom-up dynamic technique and, most importantly, a linear programming model developed by the Energy Technology Systems Analysis Program of the International Energy Agency (IEA). It can depict both the energy supply and energy demand sides of energy systems and can aid policy planners and managers in policy planning by providing optimization scenarios of different policy options.

### 3.6.1  MARKAL MODELING

MARKAL modeling is based on primary and secondary data and is an ideal scenario simulator and the perfect tool for policy analysis. The modeling framework requires the full spectrum of processes from the supply of primary fuels through to the conversion technologies used to meet end-user demand sectors. The energy carriers interconnect the conversion and consumption of energy, and the consumption of energy and demand for energy services may be disaggregated by sectors. The structure of MARKAL is shown in Figure 3.8, and the supply of electricity through energy conversion technologies to different demand sectors is shown in Figure 3.9.

The optimization routine in MARKAL selects from each of the sources, energy carriers and transformation technologies to produce the least-cost solution subject to a variety of constraints. As a result of this integrated approach, supply-side technologies are matched to energy service demands.

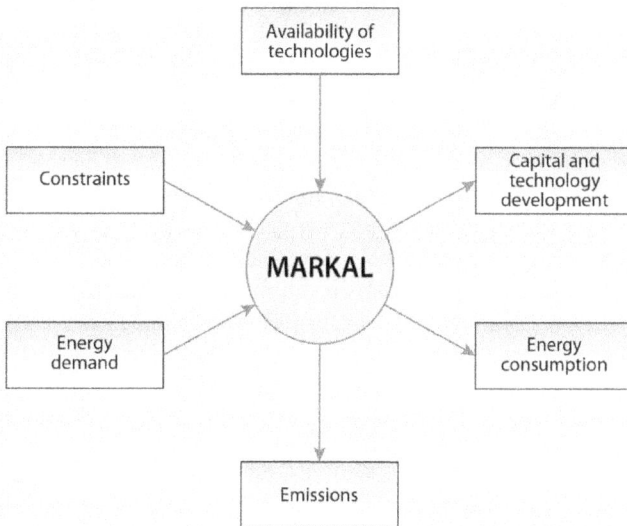

**FIGURE 3.8**   Structure of the MARKAL model.

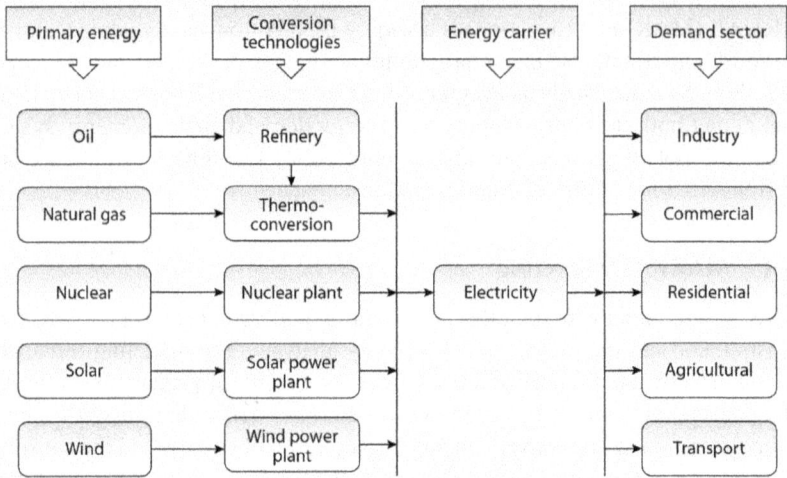

**FIGURE 3.9** Supply of electricity through energy conversion technologies to different demand sectors.

The steps in MARKAL modeling are as follows:

(1) Description of the integrated energy systems structure based on research reports and studies and primary and secondary data.
(2) Formulation of optimization of the integrated energy systems using MARKAL.
(3) Analysis of the scenarios of optimal energy systems and energy planning.

In MARKAL, all power plants connected to the grid are considered to match the centralized national grid, and it is assumed that consumption of electricity will never exceed generation levels, and the electricity demand will increase throughout the study horizon. The national average for transmission and distribution loss for electricity is taken into account. Currency can be specified in US dollars ($) or local currency, and primary and secondary fuel costs can be obtained from the Energy Information Agency. Capital, operating and variable cost, as well as technical efficiencies and availability factor for various technologies, are adopted as the Energy Technology Reference Indicator for the projection period. Transmission and distribution cost is not accounted for in the model, nor are seasonal and daily load fluctuations considered, and the power sector has no financial constraints due to active investments by the private sector.

## 3.6.2  SCENARIO ANALYSIS

The optimal scenarios that are developed and evaluated are: least-cost optimization of existing technologies with the addition of new committed technologies as planned by the government plus renewable; least-cost optimization of existing technologies plus renewable technologies with a strong policy impetus on renewables; and least-cost optimization of existing technologies plus solar PV and nuclear power.

Energy modeling and scenario development has multiple purposes: it provides a better scenario for current and future markets (supply, demand and prices); it facilitates a better design of energy supply systems in the short-, medium- and long-term; it ensures the optimal sustainable exploitation of scarce energy resources; and it provides scenarios for understanding the present and future interactions of energy and the rest of the economy, and the potential implications on environmental quality.

### 3.6.3   ENERGY PLANNING

Energy planning is required for sustainable development and effective environmental management due to the limitations of fossil fuel resources, the high capital costs of renewable energy development, and various concerns about energy supply and demand. Energy planning is not about predicting the future, but it is about the analysis and evaluation of a set of different possible futures, as well as the guidelines for optimal operational planning.

Some of the energy policies that can be optimized using MARKAL are:

(1) Policies for private and public investments and energy conservation.
(2) Policies for the optimal energy mix for integrated energy systems, including solar, wind and nuclear.
(3) National energy policies and policies for energy security for Vision 2041.

**Exercises**

**3.1** Given the linear programming problem:

Maximize

$$Z - 4x_1 + 3x_2$$

subject to

$$5x_1 + x_2 \leq 40$$

$$2x_1 + x_2 \leq 20$$

$$x_1 + 2x_2 \leq 30$$

$$x_1, x_2 \geq 0$$

$$x_1, x_2, x_3 \text{ nonnegative integer}$$

(a) Draw the feasible region in the resource space.
(b) Label the extreme points.
(c) Solve the problem for the maximum Z by inspection.

**3.2** Solve the following problems using the simplex method.

Maximize

$$Z = 2x_1 + x_2$$

subject to

$$x_1 + x_2 \leq 14$$

$$x_1 - 2x_2 \leq 4$$

$$x_1, x_2 \geq 0$$

**3.3** A power company supplies electricity to an isolated remote village. The options are (1) a coal-based power plant with a maximum capacity of 60 kW or (2) a diesel power plant with a maximum capacity of 80 kW. The microgrid proposed has a maximum capacity of 150 kW. The power company's objective is to maximize the profit from the operation of this microgrid. The profit from coal-operated electricity is Tk. 3.00 per kWh and the profit from diesel-operated electricity is Tk. 5.00 per kWh. Formulate this as a linear programming problem and solve using both the graphical and simplex methods.

**3.4** A power company supplies electricity to an isolated remote village. The options are (1) a coal-based power plant that can supply a maximum capacity of 60 kW or (2) a diesel power plant that can supply a maximum capacity of 80 kW. The microgrid is proposed to operate at 2% from coal and 3% from diesel of the microgrid of capacity 150 kW. The objective is to minimize the pollution from the operation of this microgrid. Formulate this as a linear programming problem and solve using both the graphical and simplex methods.

**3.5** An energy research plan is for reducing $CO_2$ emissions from the power grid system that consists of a variety of power-generating plants: coal-fired, natural gas, oil, and renewable energy (PV and wind). This is to be formulated as a linear programming problem and solved using the simplex method. The model is to be developed for a nation to meet a specified $CO_2$ emission target. Two $CO_2$ mitigation options may be considered in this linear programming modeling, i.e., fuel-balancing and fuel-switching. In order to reduce the $CO_2$ emissions by 30% by 2041, State Electric Supply Company plans to generate up to 30% of its electricity from renewable energy (RE), and the cost of electricity (COE) is expected to increase by Tk. 6.00 per kWh for the fuel-balancing option, while the fuel-switching option has a plan to generate 45% of electricity from RE, and the COE is expected to increase to Tk. 8.00 per kWh.

**3.6** The energy solutions of our past, present and future energy mixes are commonly formulated by participants in a complex decision-making process who decide important issues regarding limited fossil fuel resources, greater efficiency in producing electrical energy and reducing the emission levels of pollutants. These problems can be decided upon only by formulating and implementing the results of optimization by linear programming. The production of electricity in a grid is by a mix of fossil fuels such as oil, gas and coal, as well as renewable energy resources such as nuclear power, hydroelectric, solar and wind.

(a) Formulate a linear programming model to minimize the cost of electricity generation subject to the constraints of the availability of the fuels, the limits of renewable energy resources and the limit of $CO_2$ emissions from the power grid system.

(b) Present the linear programming model in tabular form for solving using the optimization software LIPS. Solve using the data of the electric grid in your country or using the data from the paper of Gruenwald and Oprea (2012).

(c) Present the optimal solution to this problem.

**3.7** An electrical grid consisting of an energy mix of fossil fuels such as oil, gas and coal, and renewable energy resources such as nuclear power, hydroelectric, solar and wind, needs to be upgraded by installing some new power plants of fossil fuels and renewable energy resources to meet the increasing demand for electricity.

(a) Formulate a linear programming model to minimize the cost of electricity generation, operation and maintenance, and the investment for new power plants subject to the constraints of the electricity demand, which ensures the electricity needed, and the operational constraint that states that the electricity produced by all the power plants should not exceed the full installed capacity. The lower bounds of operational constraints are emission constraints that limit $CO_2$ emissions from either existing power plants or new power plants, and renewable energy constraints that state that energy from renewable resources cannot exceed the availability of renewable energy resources.

(b) Present the linear programming model in tabular form for solving using LiPS. Solve using the data of the electric grid in your country or using the data from the paper of Farizal et al. (2016).

(c) Present the optimal solution to this problem.

**3.8** Solve the following problems using the branch and bound method.

Maximize

$$Z = 2x_1 + 3x_2 + 10x_3$$

subject to

$$-x_1 + 4x_2 + 7x_3 \leq 10$$

$$5x_1 - 2x_2 + 6x_3 \leq 10$$

$$0 \leq x_1, x_2, x_3 \leq 4$$

$x_1, x_2, x_3$ nonnegative integer

**3.9** Solve the following problems using the branch and bound method.

Minimize

$$Z = x_1 + 3x_2 + 5x_3$$

subject to

$$x_1 + 2x_2 \qquad \geq 7$$

$$3x_1 + x_2 + 2x_3 \geq 8$$

$$0 \leq x_1, x_2, x_3 \leq 4$$

$x_1, x_2, x_3$ nonnegative integer

# REFERENCES

Bala, B. K. (1998). *Energy and Environment: Modelling and Simulation.* Nova Science Publishers, USA

Beightler, C. S., Phllips, D. T., & Wilde, D. J. (1979). *Foundations of Optimization* (2nd ed.), Prentice Hall of India.

Das, S., Joshi, H. C., & Rao, D. K. (2017). Electrical energy conservation model using linear programming. *International Journal on Emerging Technologies.* 8(1): 250–259.

Deb, K., Pratap, A., Agarwal, S., & Meyarivan, T. A. (2002). Fast and elitist multiobjective genetic algorithm: NSGA-II. *IEEE Transactions on Evolutionary Computation.* 6: 182–197.

de Neufville, R. (1990). *Applied Systems Analysis: Engineering Planning and Technology Management.* McGraw-Hill.

Eskandari, H., & Geiger, C. D. (2020). Solving expensive multiobjective optimization problems: A fast Pareto genetic algorithm approach. *Journal of Heuristics.* 14(3): 203–241.

Farizal, F., Septia, W. E., Rachman, A., Nasruddin, & Mahila, T. M. I. (2016). Optimization of electrical energy generation schemes in the Java-Bali grid system with CO2 reduction consideration. *Makara Journal of Technology*. 20(2): 49–57.

Gruenwald, O., & Oprea, D. (2012, May 15–17). *Optimization model of energy mix taking into account the environmental impact*. 11th International Conference Energy – Ecology – Economy, Tatranske Matliare High Tatras, Slovak Republic.

Hillier, F. S., & Lieberman, G. J. (1974). *Introduction to Operations Research* (2nd ed.). Holden Day Inc.

Hooman, F. (2019). *Energy Systems Modeling: Principles and Applications*. Springer.

Kimutai, I., Maina, P., & Makokha, A. (2019). Energy optimization model using linear programming for process industry: A case study of textile manufacturing plant in Kenya. *International Journal of Energy Engineering*. 9(2): 45–52.

Munasinghe, M., & Meier, P. (2008). *Energy Policy Analysis and Modeling*. Cambridge University Press, Cambridge, UK.

Parikh, J. (1998). *Energy Models for 2000 and Beyond*. McGraw Hill.

Rao, S. S. (1984). *Optimization: Theory and Applications*. Wiley Eastern Limited.

Taha, H. A. J. (2017). *Operations Research: An Introduction* (10th ed.). Pearson.

# 4 Communication Techniques

## 4.1 INTRODUCTION

Power-system communications play an important role in the safe and efficient operation of electric power systems. The real-time automation and control of electric generation, transmission and distribution systems depend on reliable and secure communication networks. With the ever-expanding role of electric power systems, communications networks have started revolutionizing more computer and microprocessor-controlled devices. These networks and devices support the better utilization of extensive energy management systems (EMS) and incorporate information technology infrastructures. The use of SCADA (supervisory control and data acquisition) in distribution systems has become the norm for most new distribution equipment with electronic controls. A SCADA system enables operators in a control center to monitor the activity and status of field devices, such as reclosers, pulse closers, switches and other electronically controlled devices equipped with long-range communications capabilities. The future of developed, developing, and emerging countries in a global economy will depend even further on the availability and transport of electrical energy. It is believed that in the near future, the global consumption of electrical energy will grow at an unprecedented rate. Furthermore, security and sustainability have become major priorities for both industry and society. Consequently, the deployment of sustainable renewable energy sources is crucial for a healthy relationship between man and his environment. Conventional power systems with sophisticated information and communication technologies (ICT) are expected to evolve into a new grid paradigm called a smart grid. A smart grid can be thought of as a conventional power system with additional digital layers. In this chapter, we present the fundamentals of communication technologies. Later, in Chapter 10, we will present the current state-of-the-art communication techniques with reference to various research works reported in literature in the communication sector, as applicable to electrical power systems.

## 4.2 STANDARDS FOR COMMUNICATION AND INFORMATION EXCHANGE

The Institute of Electrical and Electronics Engineers (IEEE) has proposed a number of standards related to communications in power systems, including IEEE C37.1, IEEE 1379, IEEE 1547 and IEEE 1646 (Wang et al., 2011). IEEE C37.1 describes the functional requirements of IEEE on SCADA and automation systems, and IEEE 1379 recommends implementation guidelines and practices for communications and the interoperations of intelligent electronic devices (IEDs) and remote terminal units (RTUs) in an electric substation. IEEE 1547 defines and specifies an electric power system that interconnects distributed resources, and IEEE 1646 specifies the requirements in communication delivery times within and external to an electric substation.

Smart meters used in smart grids can be automated or advanced. An automated meter sends data occasionally, whereas an advanced meter sends data two ways frequently. There are several standards for information exchange. IEC 62056 and ANSI C12.22 are two sets of standards used for open communication for smart meters. IEC 62056 defines the transport and application layers for smart metering systems under a set of specifications called Companion Specification for Energy Metering (COSEM). ANSI C12.22 defines the sending and receiving of meter data both ways from external systems and can be used over any communication network.

## 4.3 COMMUNICATION TECHNOLOGIES

Communication systems are defined as the systems and processes are used to send information from a source to a destination efficiently and reliably, especially by means of electricity or radio waves. The transmission of information is known as communication, and the essential elements for communication from the source of an input signal to the receiver of the desired output signal are the transmitter, the channel and the receiver. The transmitter sends the input signal; the channel serves as the medium of transmission of the signal from the source of the signal to the receiver of the output signal; and the receiver extracts the desired message at the channel output. Figure 4.1 shows the block diagram of communication techniques. The main concern of this chapter is the communication medium, which consists of communication channels such as bandwidth and data rate, and information carriers such as wired and wireless information carriers and their interconnections.

### 4.3.1 TRANSMISSION MEDIUM

A transmission medium is a pathway that carries information signals from a transmitter to a receiver (plural: transmission media). This pathway may be a solid, a liquid or a gas. Typical pathways are copper or fiber-optic cables and the atmosphere

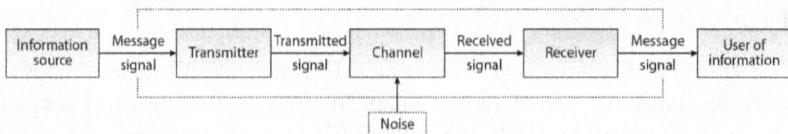

**FIGURE 4.1**   Block diagram of communication techniques.

or a vacuum. Information signals can be electrical, electromagnetic or optical. The transmission medium for sound received by ear is usually air. A vacuum or empty space is also a transmission medium for electromagnetic waves such as light and radio waves.

### 4.3.1.1 Information Transfer Using Sound Waves

Speech communication can be transferred between people over a limited distance (about 50 m). Other types of sound signals can transfer limited information (e.g., bells, sirens, guns, drums) up to several miles in rural settings but over a much shorter distance in busy cities.

### 4.3.1.2 Information Transfer Using Light Waves

Historically, line-of-sight visual signaling systems such as flags, semaphore, light, fire, smoke, and sunlight reflected by mirrors (heliograph) were used. These are effective for distances of up to about 20 km in clear weather and from high ground. Modern communication systems make use of optical fibers to carry optical signals.

### 4.3.1.3 Information Transfer Using Electrical Signals

The electric telegraph sent electrical currents along wires, as do local telephone networks today (but in a vastly more sophisticated form). Using amplifiers, electrical information signals can be transmitted over thousands of miles.

### 4.3.1.4 Information Transfer Using Electromagnetic Signals

It is possible to send signals through the air using electromagnetic radiation in the form of radio waves. These can connect people, mobile or stationary, on board a ship, on foot or in a vehicle. Electromagnetic information signals can be transmitted over tens of thousands of miles. At the receiver, a demodulator decodes (separates) the carrier and information signal so that the audio broadcast can be heard by the listener.

## 4.3.2 COMMUNICATION CHANNELS

Physical media between a source and a destination are termed communication channels. In the case of dedicated channels, as in a SCADA system, a single medium is generally used. Shared communication channels may consist of more than one medium, depending on how the signal travels. A communication channel may be a guided medium such as a copper cable or optical fiber or an unguided medium such as a radio link.

## 4.3.3 TERMS RELATED TO COMMUNICATION CHANNELS

### 4.3.3.1 Bandwidth

Bandwidth is the difference between the upper and lower cut-off frequencies of a communication channel. The bandwidth of the signal is essentially the difference

**TABLE 4.1**
**Frequency Band of Some Important Wireless Communications**

| Service | Frequency band |
| --- | --- |
| Standard AM broadcasting | 540–1600 kHz |
|     FM broadcasting | 88–800 MHz |
| Television (HF) | 54–72 MHz |
| TV | 76–88 MHz |
| VMF | 174–216 MHz |
| TV | 420–809 MHz |
| Cellular mobile radio | 896–901 MHz |
| | 840–935 MHz |
| Satellite communication | 5.925–0.425 MHz |
| | 3.2–4.2 GHz |

between the maximum frequency and the minimum frequency of the signal transmitted channel. In digital transmission, the term bit rate is often used to express the capacity (which some books refer to as bandwidth) of a channel. The bit rate is measured in bits per second (bps). The frequency bands of some important wireless communications are given in Table 4.1.

### 4.3.3.2  Available Bandwidth

Available bandwidth depends on the medium used, for example, copper, coaxial or fiber-optic cable, and the cable's specification. Typically, the bandwidth for copper cable is up to 1 MHz, for coaxial cable up to 100 MHz and 100 GHz for fiber-optic cable.

The available bandwidth for radio wave transmission is from 30 kHz to 300 GHz. Frequency and bandwidth allocation is strictly controlled by national governments. Satellite communication systems are allocated bandwidth at frequencies above 3 GHz.

### 4.3.3.3  Channel Bandwidth

The minimum acceptable bandwidth required for the transfer of meaningful information is essentially known as channel bandwidth. A transmission medium is divided into channels so that it can be used to send multiple streams of information simultaneously. The formula for the capacity of a transmission medium N is:

$$\text{Capacity of transmission medium} = \frac{\text{Available bandwidth}}{\text{Channel bandwith}} \quad (4.1)$$

### 4.3.3.4  Data Rate

Data rate is measured in units called bits/s or bps. For digital signals, the maximum data rate achieved by a transmission medium depends on the available bandwidth, the number of signal levels and the amount of noise present in the medium. Assuming the signal is binary encoded (consisting of two voltage levels) and the medium is noise-free, the data rate is expressed as:

$$\text{Maximum data rate} = 2 \times \text{available bandwidth} \qquad (4.2)$$

**Example 4.1**

(a) The maximum number of channels on a satellite communication is 1000, and each channel is allocated a bandwidth of 3 MHz. Determine the available bandwidth of the satellite communication.
(b) A cable should carry 35 channels, each with a bandwidth of 15 kHz. What is the minimum bandwidth the cable must have?
(c) What is the minimum bandwidth required for a fiber-optic cable to carry binary data at a rate of 50 Gbps?

**Solution**

(a) From Equation (4.1), we can write

$$\text{Available bandwidth} = \text{Channel bandwidth} \times \text{Maximum number of bandwidth}$$
$$= 0.003 \times 1000 = 3 \text{ GHz}$$

(**b**) From Equation (4.1), we can write

$$\text{Minimum bandwidth} = \text{Channel capacity} \times \text{Channel bandwidth}$$
$$= 35 \times 15 = 525 \text{ kHz}$$

(**c**) From Equation (4.2), we can write

$$\text{Maximum data rate} = 2 \times \text{available bandwidth}$$
$$50 = 2 \times \text{avaiable bandwidth}$$

Hence, the bandwidth is

$$\text{Bandwidth} = \frac{50}{2} = 25$$

## 4.3.4   TERMS RELATED TO COMMUNICATION SIGNALS

### 4.3.4.1  Gain

In electronics, gain is a measure of the ability of a subsystem (often an amplifier) to increase the voltage, current or power of a signal. It is usually defined as the ratio of the signal output to the signal input.

### 4.3.4.2 Power Gain

Originally, the decibel (abbreviated dB) was used to measure the intensity of sound. In communication systems, it is used to express the power gain of an amplifier or the power attenuation (loss) of a transmission medium.

The following equation converts a power ratio into decibels:

$$G_{dB} = 10\log_{10}\frac{P_{out}}{P_{in}} \tag{4.3}$$

Table 4.2 gives examples of the conversion from power loss/gain as a ratio to loss/gain in decibels:

### 4.3.4.3 Attenuation

When a signal propagates through a communication channel, its amplitude decreases with distance. In long-distance transmission, amplifiers (for analog signals) and repeaters (for digital signals) are installed at regular intervals to boost attenuated signals. For example, when transmitting digital signals in copper cables, repeaters are required every 10 km, whereas optical fiber can take a signal without significant attenuation over a distance as great as 100 km.

Attenuation occurs as signals propagate through electrical circuits, optical fibers, air (free space), etc. As the signal progresses, some of the energy is lost, and its amplitude decreases. The further it travels, the greater the loss in amplitude. Figure 4.2 shows the attenuation of the amplitude of a sinusoidal signal with distance. Attenuation can be thought of as a gain that is less than 1.

**TABLE 4.2**

**Examples of the Conversion from Power Loss/ Gain as a Ratio to Loss/Gain in Decibels**

| Ratio of loss/gain | Loss/gain in dB |
|---|---|
| 1000 | 30 |
| 100 | 20 |
| 10 | 10 |
| 1 | 0 |
| 0.1 | −10 |
| 0.01 | −20 |
| 0.001 | −30 |

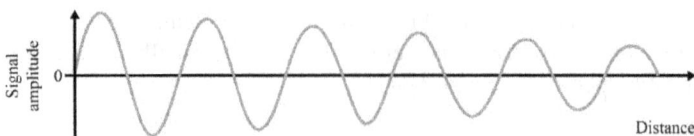

**FIGURE 4.2**   Attenuation of amplitude with distance.

## Example 4.2

(a) A 4 km communication link has input and output signals of 280 mW and 64 mW, respectively. Calculate the power loss/km in dB for the link.

(b) An amplifier has input and output signals of 20 mW and 5 W, respectively. Calculate the power gain in dB.

(c) Calculate the output power when a 20 mW signal is applied to a transmission path with a loss of -15 dB.

## Solution

(a) From Equation (4.3), we can write:

$$G_{dB} = 10\log_{10}\frac{P_{out}}{P_{in}} = 10\log_{10}\frac{64}{280} = 10\log_{10} 0.2286 = -6.41$$

Power loss per km is

$$\frac{loss}{km} = \frac{-6.41}{4} = -1.60$$

(b) From Equation (4.3), we can write:

$$G_{dB} = 10\log_{10}\frac{P_{out}}{P_{in}} = 10\log_{10}\frac{5000}{20} = 10\log_{10} 250 = 23.97$$

(c) From Equation (4.3), we can write:

$$-15 = 10\log_{10}\frac{P_{out}}{20}$$

$$10^{-1.5} = \frac{P_{out}}{20}$$

$$P_{out} = 10^{-1.5} \times 20 = 0.632\,mW$$

## Example 4.3

Loss per km of a 25 km link is 2.1 dB/km, and the gains of the input and output signals are 20 dB and 30 dB. Calculate the value of $P_{IN}$ for a value of $P_{out}$ of 5 mW.

## Solution

$$Loss\ on\ link = 25 \times -2.1 = -52.5dB$$

$$Overall\ gain/loss = 20 - 52.5 + 30 = -2.5$$

$$G_{dB} = -2.5 = 10\log_{10} \frac{5}{P_{in}}$$

$$10^{-0.25} = \frac{5}{P_{in}}$$

The input signal is:

$$P_{in} = \frac{5}{10^{-0.25..}} = 8.97 \text{ mW}$$

### 4.3.4.4 Noise

Electrical noise is an inherent problem in communication. When digital signals travel inside a channel, sometimes noise is sufficient to change the voltage level that corresponds to logic 0 to that of logic 1 or vice versa. Noise level is normally described by the signal-to-noise ratio (SNR) and measured in decibels (dB). SNR is defined by:

$$SNR_{dB} = 10\log_{10} \frac{\text{Signal power}}{\text{Noise power}} \qquad (4.4)$$

For example, if SNR = 20 dB, from Equation (4.4), it can be seen that the ratio of signal power to noise power is $10^{(20/10)} = 100$.

### 4.3.4.5 Signal Propagation Delay

The finite time delay that it takes for a signal to propagate from a source to a destination is known as propagation delay. In a communication channel, both the media and repeaters are used to amplify and reconstruct the incoming signals that cause delays. As some smart grid applications require real-time low latency communication capabilities, it is important to consider the propagation delay of a channel.

### 4.3.4.6 Distortion

Non-linearity is introduced in the signal by the components of the circuit. Although non-linear, the change is at least predictable, and its effects can be minimized.

### 4.3.4.7 Electrical Noise

Additional external signals are also added to the information signal during information transfer. Noise is responsible for the familiar static heard on the radio and the "snow" seen on television screens when displaying a weak signal. In general, noise limits the range over which electrical, radio or optical signals can be transmitted and received.

Noise in electrical signals is generated in one of three main ways: white noise caused by the random vibration of atoms, interference caused by electromagnetic

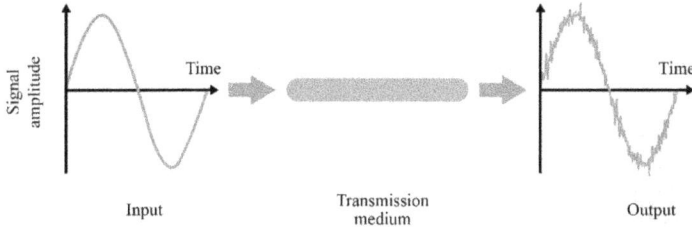

**FIGURE 4.3**   The effect of noise on a sinusoidal signal after passing through a transmission medium.

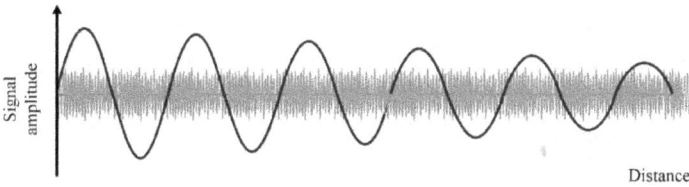

**FIGURE 4.4**   The effect of attenuation and noise on a signal as it travels along a transmission link.

radiation from lightning or electrical machinery, for example, and crosstalk picked up from nearby transmissions. The diagram in Figure 4.3 shows the effect that noise could have on a sinusoidal signal after passing through a transmission medium. (The effects of attenuation have been ignored for clarity.)

It is impossible to completely remove noise once it has been added to an analog signal. The effect of attenuation and noise on a signal as it travels along a transmission link is shown in Figure 4.4:

In the diagram, noise is not shown superimposed on the sinusoidal signal in order to more clearly show that the noise amplitude remains fairly constant while the signal amplitude decreases with distance. As a result, the ratio of signal amplitude to noise amplitude also diminishes with distance. The only way to alleviate this effect over a long-distance analog communication link is to amplify the signal at regular intervals on the link using repeaters or high-quality, low-noise amplifiers, which boost signal amplitude and filter out some of the noise.

### 4.3.4.8   Signal-to-Noise Ratio

The ratio of signal amplitude to noise amplitude is referred to as the signal-to-noise ratio (SNR), is expressed in dB and is calculated using a formula similar to that used for power gain in Equation (4.4):

$$ \text{SNR}_{dB} = 20 \log_{10} \frac{V_S}{V_N} \tag{4.5} $$

If the signal-to-noise ratio is below a certain level, then the information in the signal is degraded unacceptably. An SNR of 0 dB means that the signal power is equal to the noise power, and the signal is unrecoverable.

## Example 4.4

(a) The amplitudes of a signal and a noise in a transmission link are estimated at 5 V and 2 mV, respectively. Estimate the SNR.
(b) The noise output from a coaxial cable is 0.35 mW with no signal present. What signal power is required if the minimum acceptable SNR is 25 dB?
(c) A communications receiver requires an input signal amplitude of 3.2 V and an SNR of 36 dB. What is the maximum acceptable value of the noise amplitude?

## Solution

(a) From Equation (4.5), we can write:

$$SNR_{dB} = 20\log_{10}\frac{V_S}{V_N} = 20\log_{10}\frac{5000}{2} = 20\log_{10}250 = 47.96$$

(b) From Equation (4.4), we can write:

$$SNR_{dB} = 25 = 10\log_{10}\frac{P_S}{0.35}$$

Hence $P_S$ is

$$P_S = 10^{2.5}\times3.5 = 110.7 \text{ mW}$$

(c) From Equation (4.5), we can write:

$$SNR_{dB} = 36 = 20\log_{10}\frac{3.2}{V_N}$$

Hence, $V_N$ is:

$$V_N = \frac{3.2}{10^{1.8}} = 0.051 \text{ V}$$

## Example 4.5

(a) The noise power at the output of a cable link with no signal present is 43.9 μW. The combined noise and signal power is 69.8 mW. Calculate the SNR.

(b) A signal of power of 8.5 W is applied to the input of a cable. The attenuation in the cable is 4.2 dB /km. Calculate the maximum possible length of cable that will achieve the SNR calculated in part (a) without amplification.

**Solution**

(a) From Equation (4.4) we can write:

$$\text{Output signal power} = 69.8 - 0.0439 = 69.796 \text{ mW}$$

$$\text{SNR}_{dB} = 10\log_{10}\frac{69.796}{0.0439} = 32$$

(b) From Equation (4.4), we can write:

$$\text{Overall gain/loss} = 10\log_{10}\frac{69.796}{8500} = -20.86 \text{ dB}$$

For attenuation of 4.2 dB/km, we can write:

$$\text{Maximum length of cable} = \frac{20.86}{4.2} = 4.97 \text{ km}$$

### 4.3.4.9   Filters

Filters are used extensively in communication systems. Low-pass filters are used to modify the bandwidth of information signals to match the channel bandwidth, reconstruct audio signals that have been digitized for transmission and prevent interference between signals, for example, between the broadband and voice signals in a domestic telephone circuit. Band-pass filters are used to separate channels at the output of a frequency division multiplexing (FDM) system and prevent interference between radio stations in radio receivers.

### Example 4.6

Several audio signals are transmitted simultaneously over one pair of wires in an FDM cable transmission system. Each signal has a bandwidth of 4 kHz (including a guard band) and is allocated its own channel. Signal 1 is allocated the first channel and is unmodulated (i.e., baseband transmission). The other signals cannot be transmitted in the same way since all signals would then become inseparable. Instead, signals 2, 3, etc., are each modulated onto carrier waves, having different frequencies to produce modulated signals with frequencies in the ranges of 4–8 kHz, 8–12 kHz, etc. At the receiver, the different frequency bands are separated using a low-pass filter for signal 1 and band-pass filters for the other signals. Signals 2, 3, etc., are then demodulated to recover the original signals.

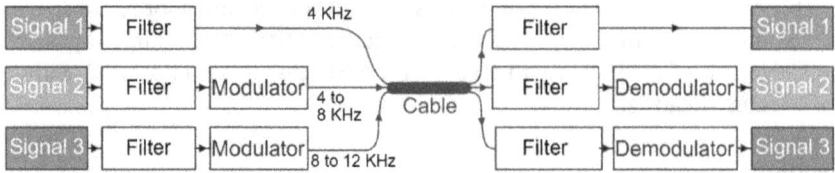

**FIGURE 4.5**   Block diagram of a system showing the first three channels.

(a) Draw the block diagram of the system showing the first three channels.
(b) Each pair of twisted cables has a bandwidth of 200 kHz. How many signal channels can be incorporated in this FDM system?

**Solution**

(a) The block diagram of the system with three channels is shown in Figure 4.5.
(b) The number of channels in the system for a bandwidth of 200 kHz with a channel bandwidth of 4 kHz is:

$$N = \frac{200}{4} = 50$$

### 4.3.4.10   Encoding and Decoding

Encoding is the process of transforming information from one format into another. Both the transmitter and receiver must be aware of the code; otherwise, the data cannot be retrieved. The encoder used for radio transmission is called a modulator. In it, a high-frequency carrier signal is combined with the audio signal, the information. One reason for using a high-frequency carrier is to propagate over large distances. The process of superimposing the audio signal over a high-frequency carrier wave is called modulation. The process of separating the audio signal from a modulated signal is called demodulation.

## 4.4   COMMUNICATION SYSTEMS

Communication systems play an important role in the operation of electric power systems. Real-time automation and the control of electric generation, transmission and distribution systems depend on communication systems. Communication systems may be either wired or wireless.

### 4.4.1   WIRED COMMUNICATION

Communication lines are used to interconnect points separated from each other, and the line communication may be in the form of an electrical signal or an optical signal. Early telephone circuits used two open-wire lines. With an open-wire circuit, care should be taken to avoid cross-coupling the electrical signals with adjacent circuits. Wired communications include power line carriers, twisted-pair copper cables, coaxial cables and fiber cables.

### 4.4.1.1 Power-Line Carrier

Power-line carrier (PLC) is a communication method that uses electrical wire to transmit data and electric power simultaneously. PLC uses a line-matching unit (LMU) to inject signals into a high-voltage transmission or distribution line. The injected signal is prevented from spreading to other parts of the power network by line traps. PLC transmits at a radio band frequency signal of between 10 kHz and 490 kHz over transmission lines. PLC with a power output of 150 W can be used up to a distance of 240 km.

Power-line networking (PLN) uses existing electrical wiring, whether in a building or in a utility grid, as network cables, meaning they also carry data signals. It is a means of extending an existing network into new places without adding new wires. The power line is transformed into a data line via the superposition of a low energy information signal to the power wave. Since the cycle of electricity is 50 or 60 Hz, data is transmitted at least 3 kHz to ensure the power wave does not interfere with the data signal. The technical challenge is that the power wiring is unshielded and untwisted, and the wiring acts as an antenna, so the wiring emits radio energy, causing interference to the existing users of the same frequency band. However, PLC does not offer a reliable solution for wide-area data transmission.

### Example 4.7
Some PLC equipment uses a carrier frequency of 100 kHz. If the value of the inductance in the line trap is 0.25 mH, what is the value of the capacitance required?

### Solution
The impedance of the line trap is:

$$Z = \frac{j\omega L \times 1/j\omega C}{j\omega L + 1/j\omega C} = \frac{L/C}{j(\omega L - 1/\omega C)}$$

At parallel resonance, that is when $\omega L = 1/\omega C$, $Z \to \infty$ and the line trap will block any frequency component at the frequency $f = 1/2\pi\sqrt{LC}$

When f = 100 kHz and L = 0.25 mH,

$$C = \frac{1}{(2\pi f)^2 L} = \frac{1}{(2\pi \times 100 \times 10^3) \times 0.25 \times 10^{-3}} = 10 \text{ nF}$$

### 4.4.1.2 Twisted Pair Copper Cable

Unshielded twisted pair (UTP) cables are used widely in telecommunication circuits. These usually consist of two twisted copper cables, each with an outer PVC or plastic insulator. Depending on the data rate (which is influenced by the cable material as well as the type of connectors), UTP cables are categorized into a number of categories (or CAT). For voice transmission, Category 1 UTP cables are used. However, they are not suitable for data transmission. For low-speed data transmission up to 4 Mbps, Category 2 UTP cables can be used. Category 3, 4 and 5 UTP cables provide data transmission rates of up to 10 Mbps, 16 Mbps and 100 Mbps, respectively. Category 5 is the most commonly used UTP cable type for data communication. For applications of much higher data rates, Category 5e, 6, 6e, 7 and 7a UTP cables are available. They can support data rates of up to 1.2 Gbps.

A digital subscriber line (DSL) allows the transmission of data over ordinary copper twisted pair telephone lines at high data rates to provide broadband services. A variant of DSL, asymmetric DSL (ADSL), is a commonly used technology for broadband services at homes.

Twisting the pairs of wires reduces crosstalk between adjacent circuits. High-grade twisted pair cables used on computer networks have a bandwidth in excess of 20 MHz. Twisted pair cables are only suitable for use over fairly short distances.

### 4.4.1.3   Coaxial Cable

In coaxial cables, a shielded copper wire is used as the communication medium. The outer coaxial conductor provides effective shielding from external interference and reduces losses due to radiation and skin effects. Bit rates up to 10 Mbps are possible over several meters. Coaxial cables are used in the home for feeding the signal from a satellite dish to a satellite TV receiver. They are also used in computer networks and long-distance telephone circuits to carry thousands of telephone signals over large distances. In the case of telephone circuits, several coaxial cables will be enclosed in a polythene sheath.

### Example 4.8

According to Shannon's capacity formula, the maximum channel capacity in bps is given by B $\log_2[1 + $ (signal power)/(noise power)], where B is the bandwidth of a channel in Hz. Compare the maximum channel capacity of twisted copper and coaxial cables. For copper cable, the bandwidth is 250 kHz, and the SNR is 20 dB. For coaxial cable, the bandwidth is 150 MHz, and the SNR is 22 dB.

### Solution

For twisted copper cable, from Equation (4.4), we can write:

$$\frac{\text{Signal power}}{\text{Noise power}} = 10^{(20/10)} = 100$$

Using Shannon's capacity formula, we can write:

$$\text{Maximum channel capacity} = 250 \times 10^3 \times \log_2 [1 + 100] \approx 1.7 \text{ Mbps}.$$

For coaxial cable, from Equation (4.4), we can write:

$$\frac{\text{Signal power}}{\text{Noise power}} = 10^{((22/10)} = 158$$

Using Shannon's capacity formula, we can write:

$$\text{Maximum channel capacity} = 150 \times 10^6 \times \log_2 [1 + 158] \approx 1.1 \text{ Gbps.}$$

### 4.4.1.4    Fiber-Optic Cable

Fiber-optic cable is used in local telephone networks to provide high-speed broadband at data rates in excess of 100Mbps. It is also used to carry high volumes of signals both in computer networks and in long-distance telephone communication systems.

Optical fiber transmission is used both inside substations and for long-distance transmission of data. They are often embedded in the stranded conductors of the shield (ground) wires of overhead lines. This cable is known as an optical ground wire (OPGW). OPGW cable contains a tubular structure with one or more optical fibers in it, surrounded by layers of steel and aluminum wire. Optical fibers may be wrapped around phase conductors or sometimes a standalone cable.

Optical fiber consists of three components: the core, the cladding and the buffer. The thin glass center of the fiber where the light travels is called the core. The outer optical material surrounding the core that reflects the light back into it is called the cladding. To protect the optical surface from moisture and damage, it is coated with a layer of buffer coating.

Compared to other communication media, fiber-optic cables have much higher bandwidth. They are less susceptible to signal degradation than copper wire, and their weight is less than a copper cable. Unlike electrical signals in copper wires, light signals from one fiber do not interfere with those of other fibers in the same cable. Further, optical fiber transmission is immune to external electromagnetic interference (EMI). This is important in power system applications since data transmission through the electrically hostile area of a substation is required.

Optical fiber is the most suitable data transmission medium for power system control, protection and monitoring function. Optical fiber products are used in communications for protection, monitoring, and control devices. Additional benefits of using optical fiber are that it is easy to terminate with a field connector, easy to test with visible light when checking a damage-resistant cable, and electrical isolation. The characteristics of optical fiber make it useful for low attenuation, high bandwidth, electromagnetic interface immunity and security (Kirkham et al., 1994).

The main disadvantages of optical fiber transmission include its cost, special termination requirements and vulnerability (it is more fragile than coaxial cable).

### 4.4.2    WIRELESS COMMUNICATION

Wireless communication does not require wires to interconnect points separate from each other. Communication may be in the form of different types of radio links. Wireless communications include radio communication, ultra-high frequency, microwave radio, cellular mobile and different types of satellite communications.

#### 4.4.2.1   Radio Communication

The substations of power networks are often widely distributed and located far from the control center.

For such long distances, the use of copper wire or fiber optics is costly. Radio links provide an alternative for communication between the control center and substations. Even though radio communication cannot provide the bandwidth offered by wired technology, the reliability, performance and running costs of radio networks have improved considerably over the past few years, making it an attractive option. Radio communication may be multipoint or point-to-point, operating typically at either ultra-high frequencies (between 300 MHz and 3 GHz) or microwave frequencies (between 3 and 30 GHz).

#### 4.4.2.2   Ultra-High Frequency

Ultra-high frequency (UHF) radio represents an attractive choice for applications where the required bandwidth is relatively low, and the communication endpoints are widespread over harsh terrain. It uses frequencies between 300 MHz and 3 GHz. Unlike microwave radio, UHF does not require a line of sight between the source and destination. The maximum distance between the source and destination depends on the size of the antennae and is likely to be about 10–30 km with a bandwidth of up to 192 kbps.

#### 4.4.2.3   Microwave Radio

Microwave radio operates at frequencies above 3 GHz and facilitates high-channel capacities and transmission data rates. Microwave radio is commonly used for long-distance communication systems. Parabolic antennae are mounted on masts and towers at the source to send a beam to another antenna situated at the destination tens of kilometers away. Microwave radio offers capacities ranging from a few Mbps to hundreds of Mbps. However, the capacity of transmission over a microwave radio is proportional to the frequency used; thus, the higher the frequency, the larger the transmission capacity but the shorter the transmission distance. Microwave radio requires a line of sight between the source and destination; hence, high masts are required. In the case of long-distance communications, the installation of tall radio masts will be the major cost of microwave radio.

#### 4.4.2.4   Cellular Mobile Communication

Cellular mobile technology offers communication between moving objects. To make this possible, a service area is divided into small regions called cells. Each cell contains an antenna that is controlled by a Mobile Telephone Switching Office (MTSO). In a cellular network, the MTSO ensures the continuation of communication when a mobile device moves from one cell to another. For example, assume that a mobile user in cell A communicates through the antenna located in the base station of cell A. If the user moves out of range of cell A and enters into the range of cell B, the MTSO automatically assigns a communication channel in cell B to the user without interrupting the communication session.

#### 4.4.2.5 Satellite Communication

Satellites have been used for many years for telecommunication systems and have also been adopted for SCADA systems. A satellite communication network can be considered a microwave network with a satellite acting as a repeater.

#### 4.4.2.6 Geostationary Orbit Satellite Communication

Currently, many satellites in operation are placed in geostationary orbit (GEO). A GEO satellite is typically 35,786 km above the equator, and its revolution around the Earth is synchronized with the Earth's rotation. The high altitude of a GEO satellite allows communications from it to cover approximately one-third of the Earth's surface (Hu and Li, 2001).

Even though GEO satellite-based communication offers technical advantages for long-distance communication, it still presents some drawbacks:

1. Transmitting and detecting a signal over the long distance between the satellite and the user is quite challenging.
2. The long distance traveled by the signal from the source to the destination results in an end-to-end delay or propagation delay of about 250 ms.

#### 4.4.2.7 Low Earth Orbiting Satellite Communication

Low earth orbiting (LEO) satellites are positioned 200–3000 km above the Earth and reduce propagation delay considerably. In addition to the low delay, the proximity of the satellite to the Earth makes the signal easily detectable, even in bad weather (Ekanake et al., 2012).

LEO satellite-based communication technology offers a set of intrinsic advantages such as rapid connection for packet data, asynchronous dial-up data availability, reliable network services, and relatively reduced overall infrastructure support requirements when compared to GEO. In addition, LEO satellite-based communication channels can support protocols such as TCP/IP since they support packet-oriented communication with relatively low latency.

#### 4.4.3 Multiplexing

Multiplexing is the process of permitting several independent users to share a common transmission medium (or link). The medium can be a cable (i.e., wire-pair, coaxial, optic fiber) or the atmosphere (or space), and the information pathway within this medium is called the channel. Multiplexing allows many channels to be accommodated within a single medium.

#### 4.4.3.1 Frequency-Division Multiplexing

Frequency-division multiplexing (FDM) is a form of signal multiplexing that assigns non-overlapping frequency ranges to each signal or "user" of the transmission medium. Each information signal is combined with a different high-frequency carrier signal. During transmission, each channel has access to a narrow range of frequencies all the time. A small frequency gap, known as a guard band, is left between channels to stop the signals from interfering with one another.

**Example 4.9**
Three radio stations are each allocated a bandwidth of 8 kHz. They broadcast radio waves into free space at carrier frequencies of 80, 90 and 100 kHz, respectively. The guard band between each channel is 2 kHz. Determine the frequency allocation (band) for each station.

Frequency band of station 1 is:

$$(80 \pm 4) = (76 - 84)\,\text{kHz}$$

Frequency band of station 2 is:

$$(90 \pm 4) = (86 - 94)\,\text{kHz}$$

Frequency band of station 3 is:

$$(100 \pm 4) = (96 - 104)\,\text{kHz}$$

### 4.4.3.2   Time-Division Multiplexing
Time-division multiplexing (TDM) is a type of digital (or, rarely, analog) multiplexing in which two or more signals or bitstreams take turns to access the channel. The transmission time is divided into time slots of fixed length, one for each signal. A byte of data from signal 1 is transmitted during time slot 1, signal 2 during time slot 2, etc.

A TDM frame consists of one timeslot from each signal. After the last of these, the cycle starts all over again with a new frame, starting with the second byte of data from signal 1. During transmission, each signal has access to the full range of frequencies available to the channel.

## 4.5   SUPERVISORY CONTROL AND DATA ACQUISITION (SCADA) AND SMART GRID

The present-day electricity grid has evolved as a result of rapid urbanization and infrastructure development. However, the growth of the electric power system is influenced by economic, political and geographic factors. Since energy resources are limited and demands for energy are increasing with time, this suggests a gradual transition to renewable resources such as solar and wind and the optimization of energy use. The automation of power systems is the solution to achieving this goal, and every sector of the power system, from generation to customer, is automated today to achieve the optimal use of energy and resources.

Modern SCADA systems essentially replace manual labor in operating electrical distribution tasks and manual processes in distribution systems with automated equipment. A SCADA system as an automation system is used to acquire data from instruments and sensors located at remote sites and to receive and transmit the data at the master station central site for either controlling or monitoring purposes.

High-quality electricity is a necessity in modern society for innumerable applications that require quality power, such as electronic manufacturing, microprocessors, and many sensitive devices. A smart grid can supply electricity from generation to consumer using digital technology to save energy, reduce costs and increase reliability. It connects everyone to abundant, affordable, clean, efficient and reliable electric power anytime, anywhere, addressing energy independence and global warming issues. The present revolution in communication systems, particularly stimulated by the internet, offers the possibility of much greater monitoring and control throughout the power system, and hence more effective, flexible and lower cost operation. The smart grid is an opportunity to use new ICTs to revolutionize the electrical power system.

### 4.5.1 SUPERVISORY CONTROL AND DATA ACQUISITION (SCADA)

SCADA systems are used for large-scale geographical distribution or generation systems and are applicable in large-scale renewable systems such as solar and wind. A SCADA system is the heart of large-scale renewable power generation as well as distribution management system (DMS) architecture. SCADA systems usually monitor and make slight changes to function optimally. They are considered closed-loop control systems and operate with comparatively little human interface.

A SCADA system is a centrally controlled master system that commands terminal RTUs, and these RTUs include relay devices, actuators and sensors, circuit power breakers, voltage regulators, etc. PLCs are used as control sensory devices. The SCADA system essentially consists of five components: an RTU, a communication system, a master station, a human–machine interface (HMI) and a database (Ancillotti et al., 2013). Figure 4.6 shows SCADA system architecture in modern power grids. Master terminal units (MTUs) are higher-level units, including supporting applications, HMIs, data storage and acquisition systems. RTUs or PLCs link the control system to field-sensing devices for collecting data from the field-sensing devices and passing commands from the control station to the field devices. Programmable automation controllers are used as the basic controlling unit.

A communication system transfers data between field data interface devices and control units and the computers in the SCADA system central host. The central host computer server or servers are sometimes called a SCADA center, master station or master terminal unit, where the operator monitors the system and makes control decisions to be conveyed to the field. The HMI is the interaction between the operator and the machine. Databases are for storing historical data, measuring trends and deriving forecasts. MTUs are higher-level units including supporting applications, HMIs, data storage and data acquisition. All automation systems essentially have these five components. A SCADA system allows a utility

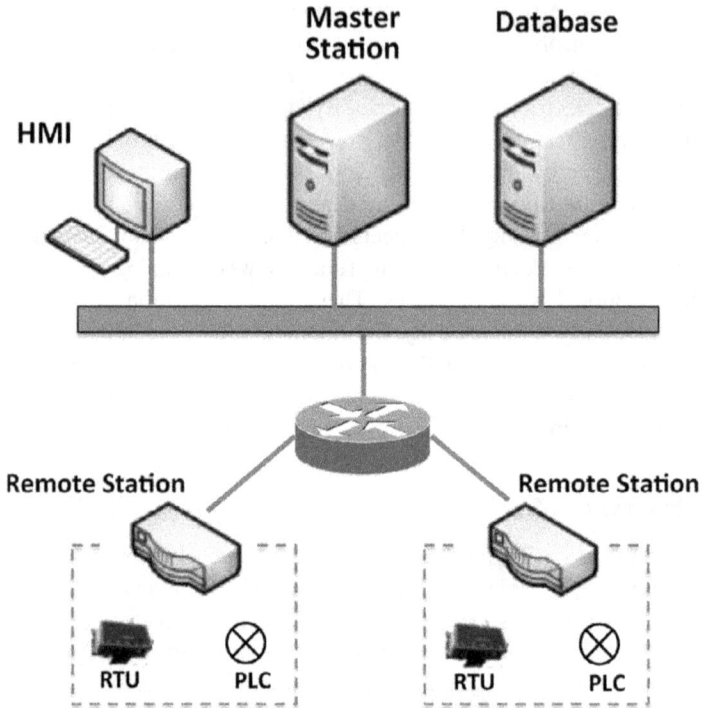

**FIGURE 4.6**    SCADA system architecture (Ancillotti et al., 2013).

operator to monitor and control field devices that are distributed among various remote sites.

The classification of the components based on the tasks assigned of the SCADA system architecture include the following:

(1) on-field devices, for example, RTUs, PLCs, IEDs and process automation controllers (PACs);

(2) monitoring and controlling equipment, for example, an HMI, a historian, a controller for the SCADA system and a real-time data processor; and

(3) communications, for example, Inter-Control Center Communications Protocol (ICCP), Odyssey Commutation Processor (OCP), Ethernet, wireless networks, serial network connections, and Modbus and DNP3 protocols.

The terminal controller unit is responsible for communicating, analyzing data, and displaying occurring events to the user, as well as the service provider. The devices are generally controlling and controlled devices that run on embedded operating systems to communicate data using various controlling protocols, such as Modbus and DNP3 protocols.

## 4.5.2 SMART GRID

Most of the present electric power grid infrastructures were built more than 50 years ago and are now a complex spiderweb of power lines and very old, with obsolescent technology and outdated communications. In addition, there is a growing demand for low carbon emissions, renewable energy resources, improved reliability, and security. It is expected that a smart grid should integrate renewable energy sources, especially wind and solar, with conventional power plants in a coordinated and intelligent way so that it not only improves system reliability and service continuity but also effectively reduces energy consumption and significantly reduces carbon emissions.

Smart grids are commonly used to refer to a managed electrical system and represent a new generation of standard power distribution grid. In a smart grid environment, heterogeneous communication technologies and infrastructures are involved. Communication infrastructure is critical for the successful operation of modern smart grids.

The present revolution in communication systems, particularly stimulated by the internet, offers the possibility of much greater monitoring and control throughout the power system, and hence more effective, flexible and lower cost operation. The smart grid is an opportunity to use new ICTs to revolutionize the electrical power system.

A smart grid is an electrical generation and distribution system that is fully networked, instrumented and automated. A smart grid uses digital technology to improve the reliability, security and efficiency (both economic and energy) of the electric system from large generation, through the delivery systems, to electricity consumers and a growing number of distributed generation and storage resources (U.S. Department of Energy, 2009). Figure 4.7 shows the smart grid

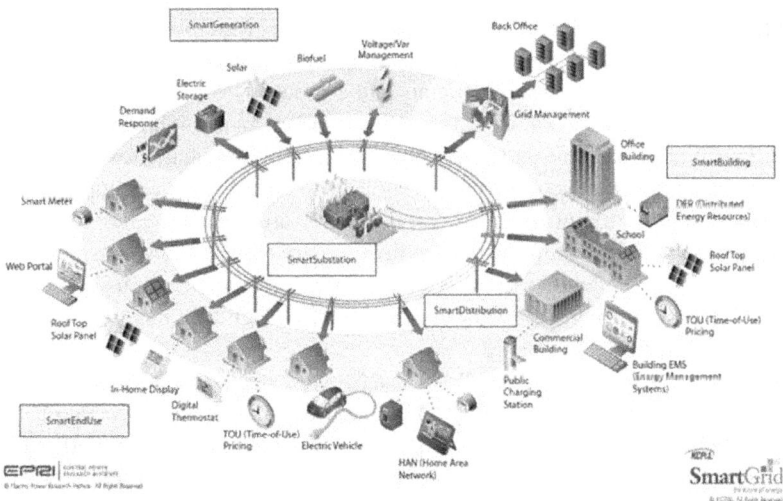

**FIGURE 4.7** Smart grid vision of an electric power research institute (EPRI, 2011).

vision of an electric power research institute, and the model set up of the smart grid includes smart generation, smart transmission, smart storage and smart sensors to isolate the fault (EPRI, 2011). It also shows that the integration of distributed resources into all levels of grid operations, including market trading, generation dispatch, distribution operations and consumer, creates an end-to-end smart grid from smart generation to smart consumption built around a major smart substation

The smart grid is the next generation of power distribution grid, and governments of many countries and companies are supporting research for smart grid applications. The smart grid uses two-way communications, digital technologies, advanced sensing and computing infrastructure and software abilities to provide improved monitoring, protection and optimization of all grid components, including generation, transmission, distribution and consumers (Baimel et al., 2016).

## 4.6   SMART METERING AND AUTOMATION

Smart grids are provided with broadband connections, and these are capable of using smart meters at household consumer premises and RTUs with digital intelligence. Broadband connections can transfer data at a faster rate, and this facility enables the remote monitoring of consumers' consumption at their households without visiting the premises, saving money and time. Broadband also introduces new functionality by collecting data both at the meters and the central system.

### 4.6.1   SMART METERS

Smart meters provide data on energy use to customers to control costs and consumption and send data to the utility for load factor control, peak load requirements, and the development of pricing strategies based on consumption information, etc. Automated data reading is an additional component of both smart meters and two-way communication between customers and utilities.

Smart meters provide knowledge to utility customers about how much they pay per kilowatt hour and how and when they use energy. This facilitates better pricing information and more accurate bills, in addition to ensuring faster outage detection and restoration by the utility. Additional features are demand response rates, tax credits, tariff options, and participation in voluntary rewards programs for reduced consumption. Other features include remote connect/disconnect of users, appliance control and monitoring, smart thermostat, enhanced grid monitoring, switching, and prepaid metering.

With governmental assistance, the large-scale deployment of smart meters has already started in many developing countries. Academic participation in the research and development of metering tools and techniques for network analysis has seen enhancement, and the use of smart meter outputs for voltage stability and security assessment and enhancement has been started.

## 4.6.2 SMART APPLIANCES

Smart appliances operate in response to signals sent by the electric utility. The appliances enable customers to participate in voluntary demand response programs that provide credits for limiting power use during peak demand periods or when the grid is under stress. An override function allows customers to control their appliances using the internet. Air conditioners, space heaters, water heaters, refrigerators, washers and dryers represent about 20% of total electric demand during most of the day and throughout the year. Grid-friendly appliances use a simple computer chip that can sense disturbances in the grid's power frequency and can turn an appliance off for a few minutes to allow the grid to stabilize during a crisis.

## 4.6.3 ADVANCED METERING INFRASTRUCTURE

SCADA and advanced metering infrastructure (AMI) systems were the first to introduce simple digital communication capabilities in power systems. However, smart grid applications need pervasive and real-time control of each grid component and not just of smart meters and substations. For this reason, a smart grid should incorporate a pervasive and scalable two-way communication infrastructure to enable more distributed command-and-control functionalities. One of the most important components of this communication infrastructure is the AMI, which can be used to interconnect the smart meters (i.e., electricity meters that incorporate networking and data management functionalities) installed at end customers' premises with other control systems and data aggregators. AMI systems can then contribute in several ways to the realization of the smart grid vision.

First, electric utilities can use AMI as data acquisition networks to monitor: (i) power quality, (ii) how much electricity is produced/stored by distributed energy resources units, and (iii) power consumption of household appliances. This large amount of metering data can be exploited to proactively identify failure conditions and anomalies and to take appropriate countermeasures or to implement sophisticated techniques to regulate electricity usage patterns (e.g., dynamic pricing or scheduling of residential loads). In a more general view, AMI networks can be foreseen to interconnect not only smart meters but also a variety of IEDs, which can be massively dispersed within smart grids. It is also important to note that AMI networks allow smart grids to collect a huge volume of heterogeneous data from a large number of sources. How to efficiently aggregate, store and analyze this data is the subject of intensive research.

The problem domains to be addressed within AMI implementations are relatively new to the utility industry; however, precedence exists for implementing large scale, network-centric solutions with high information assurance requirements. The defense, cable and telecom industries offer many examples of requirements, standards and best practices that are directly applicable to AMI implementations. The functions of AMI can be subdivided into three major categories (Momoh, 2012):

**Market applications:** Market applications serve to reduce/eliminate labor, transportation, and infrastructure costs associated with meter reading and maintenance, increase billing accuracy, and allow for time-based rates while reducing bad debts, and facilitate informed customer participation for energy management.

**Customer applications:** Customer applications serve to increase customer awareness about load reduction, reduce bad debt, improve cash flow, enhance customer convenience and satisfaction, and provide demand response and load management to improve system reliability and performance.

**Distribution operations:** Distribution operations curtail customer load for grid management, optimize networks based on data collected, allow for the location of outages and restoration of services, improve customer satisfaction, reduce energy losses, improve performance in the event of an outage with reduced outage duration and optimization of the distribution system and distributed generation management, and provide emergency demand response.

### 4.6.4　ELECTRICAL SUBSTATION AUTOMATION

In modern electric power systems, large central generators supply power into a high-voltage interconnected transmission network. The power is often transmitted over long distances and then stepped down through a series of distribution transformers to final circuits for the consumers. High availability and constant operation of an electrical substation have always been the focus of an electrical utility. Early substations consisted of mechanical relays and meters that barely supported recording and had no means of communication. With the introduction of microprocessor technology, digital protection and control devices have become more intelligent. New IEDs can collect and record information on the many different parameters of a system, process them in a fraction of a second, and make decisions on abnormal situations to send control commands to switches and breakers to clear the fault. This gives operators more flexibility on how and when to process the information to provide a fast recovery time from an interruption in the substation.

Traditionally, the distribution network has been passive with limited communication between elements. Some local automations are used, such as on-load tap changers and shunt capacitors for voltage control and circuit breakers, or auto-reclosers for fault management. These controllers operate with only local measurements, and wide-area coordinated control is not used. Over the past decade, automation of the distribution system has been increased in order to improve the quality of supply and support more distributed generation. The connection and management of distributed generation is increasing the shift from passive to active management of the distribution network. Network voltage changes and fault levels are increasing due to the connection of distributed generation. Without active management of the network, the costs of the connection of distributed generation will rise, and the connection of additional distributed generation may be limited. The connection of large intermittent energy sources and plug-in electric vehicles has led to an increase in the use of demand-side integration and distribution system automation.

Substation automation (SA) is not a new concept. People have started adopting changes in technology at substations. A substation automation system is a collection of hardware and software components that are used to monitor and control an electrical system, both locally and remotely. A substation automation system also automates some repetitive, tedious and error-prone activities to increase the overall efficiency and productivity of the system.

Smart substations form the key building block of a smart grid. A smart substation implies the creation of a highly reliable power system that rapidly responds to real-time events with appropriate actions to ensure uninterrupted power services to end-users. Substation automation has led to smart substations, and a SCADA system forms the entry towards substation automation.

### 4.6.4.1 Substation Automation Equipment

Traditionally, the secondary circuits of circuit breakers, isolators, current and voltage transformers and power transformers are connected to relays, and the relays are connected with multi-drop serial links to the station computer for monitoring and remote interrogation. However, the real-time operation of protection and voltage control systems is done through hardwired connections.

#### Connections of Substation Equipment

Two possible connections of the substation equipment are (Ekanake et al., 2012):

Connection 1: Secondary circuits of field equipment are hardwired to relay IEDs and bay controllers. The process bus is ring-connected.
Connection 2: Secondary circuits of field equipment are hardwired to the interfacing unit. The process bus is star-connected.

#### Levels of Substation Equipment

Although the connection may vary from design to design, generally, it may be at three levels (Ekanake et al., 2012):

(a) Station level: A redundant PC-based HMI enables local station control through a software package that contains an extensive range of SCADA system functions. The station level contains station-oriented functions that cannot be realized at bay level, e.g., an alarm list or event list related to the entire substation, and a gateway for communication with remote control centers. A dedicated master clock for synchronization of the entire system shall be provided.

(b) Bay level: A bay comprises a circuit breaker and associated isolators, earth switches and instrument transformers. At bay level, the IEDs provide all bay level functions, such as control, monitoring (e.g., status indications, measured values) and protection.

(c) Process level: This consists of all the switchyard devices that are hardwired using copper cables and use fiber-optic cables to connect the bay level IEDs used for control and protection.

In connection 1, analog signals are received from CTs and VTs, as well as status information, and are digitized at the bay controller and IEDs. In connection 2, analog and digital signals received from CTs and VTs are digitized by the interfacing unit. The process bus and station bus take these digital signals to multiple receiving units, such as IEDs, displays, and the station computer, that are connected to the Ethernet network.

To increase reliability, two parallel process buses are normally used. The station bus operates in a peer-to-peer mode. This bus is a LAN formed by connecting various Ethernet switches through a fiber-optic circuit. The data collected from the IEDs is processed for control and maintenance by SCADA system software that resides in the station computer.

The hardwiring of traditional substations required several kilometers of secondary wiring in ducts and on cable trays. This not only increased the cost but also made the design inflexible. In modern substations, inter-device communications are made via Ethernet and use the same communication protocol.

### 4.6.4.2   Types of Substation Automation

The electrical substation is of prime importance for electrical energy generation, transmission, and distribution systems, and there are four major types of substations: switchyard substations, customer substations, system substations and distribution substations (Leonardi et al., 2014).

(1) **Switchyard substation:** The switchyard substation at a generating station connects the generators to the utility grid and provides offsite power to the plant. Generator switchyards tend to be large installations that are typically engineered and constructed by the power plant designers and are subject to planning, finance and construction efforts different from those of routine substation projects.

(2) **Customer substation:** The customer substation functions as the main source of electric power supply for a particular business customer. The technical requirements and business case for this type of facility depend more on the customer's requirements than on utility needs.

(3) **System substation:** The system substation involves the transfer of bulk power across the network. Some of these stations provide only switching facilities (no power transformers), whereas others perform voltage conversion as well. These large stations typically serve as the endpoints for transmission lines originating from generator switchyards and provide the electrical power for circuits that feed transformer stations. They are integral to the long-term reliability and integrity of the electric system and enable large amounts of energy to be moved from the generators to the load

centers. System stations are strategic facilities and are usually very expensive to construct and maintain.

(4) **Distribution substation:** Distribution substations are the most common facilities in electric power systems and provide the distribution circuits that directly supply most customers. They are typically located close to the load centers, meaning that they are usually located in or near the neighborhoods that they supply and are the stations most likely to be encountered by the customers.

The roles of substations clearly indicate that they can be considered critical infrastructure, especially those in the transmission grid, interconnecting many systems. As such, they require proper physical and cyber protection to ensure uninterrupted and smooth operation.

### 4.6.4.3  Substation Automation System Components

The substation automation system uses a number of devices integrated into a functional array for the purpose of monitoring, controlling and configuring the substation. The components of the substation automation system are remote substation components and operations center (i.e., SCADA master station) components, and these are discussed here (Leonardi et al., 2014).

**Remote substation components**

The automation components of the substation are as follows.

(i)   Microprocessor-based IEDs provide inputs and outputs to the system while performing some primary control or processing activity. Common IEDs are protective relays, load survey or operator indicating meters, revenue meters, programmable logic controllers (PLCs), and power equipment controllers of various descriptions (McDonald, 2012).

(ii)  Devices dedicated to specific functions for substation automation systems are transducers, position sensors, and clusters of interposing relays (McDonald, 2012).

(iii) Dedicated devices often use an SA controller or interface equipment such as a conventional RTU as a means to connect into the SA system.

(iv)  A substation display or users station (local HMI), connected to or part of a substation host computer (local server) (McDonald, 2012).

(v)   Common communication connections to the outer world such as utility operations centers, maintenance offices or engineering centers. Most substation automation systems connect to a traditional SCADA system master station performing real-time requirements for operating the utility network from one or more operations centers. SA systems may also incorporate a variation of a SCADA RTU for this purpose, or the RTU function may appear in an SA controller or substation host computer. Other utility users/services are usually connected to the system through a firewalled DMZ, which is connected to the SCADA system.

**Operations center (SCADA master station) components**

In electric substation automation, the operations center (or master control center or SCADA master station) receives and processes data from several substations and takes appropriate measures for remote substation control (IEEE, 2008). The master station system may use an open and distributed architecture. There can also be multiple master stations, and accordingly, different topologies can be used to interconnect them for synchronizing grid operational data. Each master station (manned) is supported with a backup/emergency master station (unmanned) and is continuously synchronized with a primary master station database.

The main elements of a SCADA system master station (or SCADA master) are an HMI, application servers, a firewall, a communication front-end (to communicate with RTUs/data concentrators), and an external communication server/machine to machine (M2M) gateway (to communicate with other control centers). These elements are networked within the SCADA system master via a real-time dedicated LAN. The application servers are servers that support all energy management system (EMS) or distributed management system (DMS) applications. Redundancy is provided for the hardware and software elements of the SCADA master (e.g., redundant LAN) and substations (e.g., redundant critical computer), as well as for the M2M communication network.

### 4.6.4.4   Substation Automation Information Flow

Substation automation can be classified into five levels (McDonald, 2002): power system equipment at the bottom; IED implementation, IED integration, and SA applications in the middle; and utility enterprise at the top. There are three functional communication data paths from the substation to the utility enterprise:

   (i)  The operational data (e.g., volts, amps) path from the substation server to the SCADA system energy management, which is the most critical and utilizes one of the communication protocols supported by the SCADA system.
   (ii)  The remote access path from the substation server to the IEDs, which is a two-way network connection.
   (iii)  Finally, the rest of the data flow is nonoperational, and the nonoperational data paths to the utility's data warehouse are the IED nonoperational data, like event logs, from the SA to a warehouse.

Figure 4.8 shows the three functional communication data paths as well as the basic components of a substation automation system. The data flow is similar for distribution management or SCADA systems.

### 4.6.4.5   Substation Automation System Architecture

Figure 4.9 shows the smart integrated substation automation system architecture for remote monitoring, management, security and maintenance of unmanned

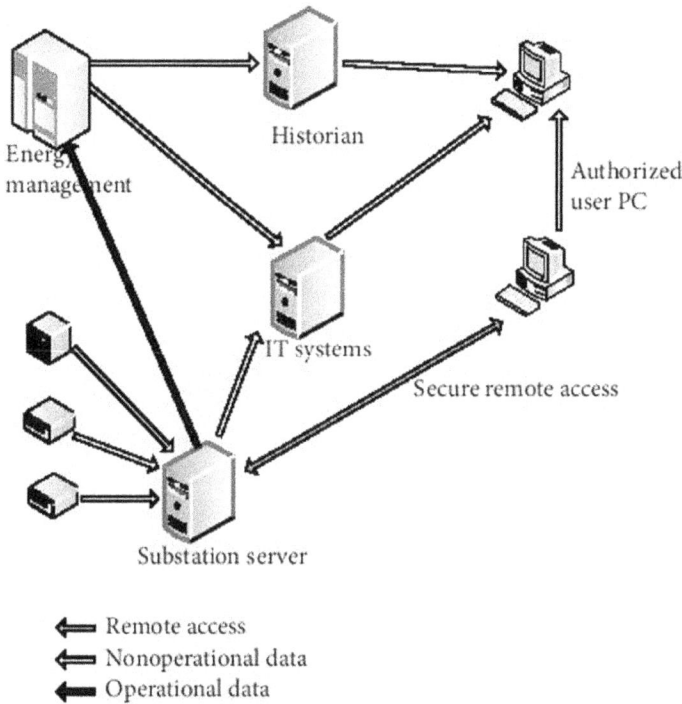

FIGURE 4.8   Substation data flow (McDonald, 2012).

energy substations and related sites. It takes full advantage of network-based architecture. The remote subsystems substation and operation center can be networked
via an M2M broadband communication network service to remotely monitor and
manage devices and machines. The communication links of the substation to the
authorized PC, cyber access management system, asset management system,
historian survey, distribution management system, energy management system,
physical access control system and video surveillance server are provided for the
remote monitoring, management, security, and maintenance of unmanned energy
substations and related sites

### 4.6.4.6   Smart Grid Control Center Applications

Smart electric transmission and distribution grid functionalities are centrally
controlled at the control center by several control centers or electric utility
applications, including SCADA systems, DMS, EMS, automated meter reading
(AMR), Network Integration System (NIS), and geographic information system
(GIS). The concepts of operations, system activities and performance, and operational functions of smart grid control center applications are discussed as follows
(Leonardi et al., 2014).

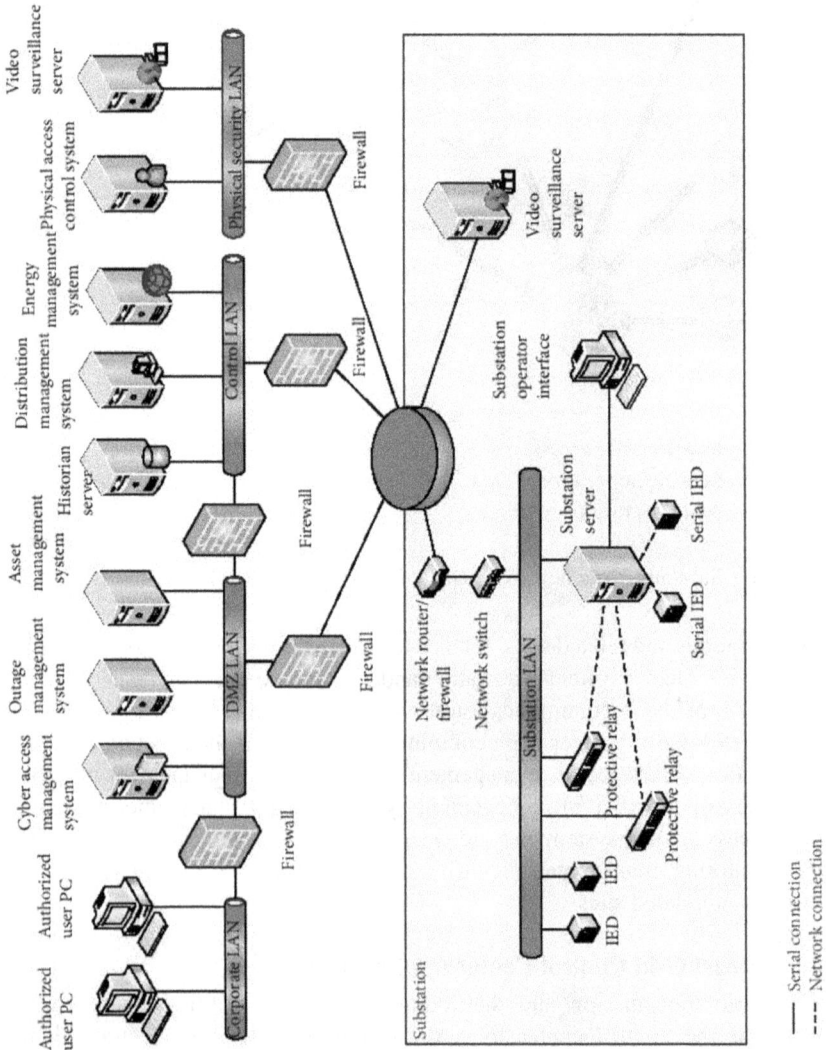

**FIGURE 4.9**  Smart integrated substation (McDonald, 2012).

### Concept of Operations

The SCADA manager, SCADA information security officer, SCADA system administrator, SCADA operator, SCADA engineer/developer, field maintenance worker and external user (via remote access) are involved in monitoring and control operations. Contractors, consultants, SCADA vendors (maintenance and emergency access), and managed security solution providers (MSSP) usually do the SCADA cyber security monitoring, which is outsourced to them by the utility. If the cyber security monitoring is not outsourced, it is done by the security incident manager.

The SCADA manager is responsible for ensuring that corporate policies are followed. The information security officer is responsible for ensuring the security policy is implemented, and audits are performed. The administrator is responsible for system activities such as maintenance, expandability and performance. The control center operator is responsible for operational functions such as electric substation monitoring and control. The maintenance workers do the fieldwork assigned to them by the control center operator, the scheduler or the dispatcher.

### System Activities and Performance

In addition to electric network management, the SCADA system serves as a source of important operating data for effective management of the utility's business. The performance of the SCADA system depends on its availability, maintenance, response time, security and expandability. The high-level availability of the SCADA system and its continuous operation are achieved by introducing reliability as well as redundancy in the hardware and software. In case of damage to the primary master station, for example, due to natural disasters, the backup/emergency master station takes over the system operation.

The system is in a normal state when the load and operating constraints are as assigned and the system performance is satisfactory. It switches to the emergency state when the system performance is not satisfactory. In the case of an emergency state, the response time becomes slow, and the system performance is degraded, but the basic functionalities such as alarm and status change operations are retained.

The system's access level is restricted for different groups of workers (i.e., access authorization). For example, operators have complete access to display and control functions for specific areas of responsibility (AoR), while maintenance staff may only have access to display functions.

The system maintenance task is hardware/software repair using diagnostic tools such as debugging, corrections, updates to the tasks with patch management, antivirus protection, tests, and preventive maintenance. The system can be expanded with new points, functions and equipment, depending upon functional and standardization requirements. The limitations and downtime are important factors to consider for expansion.

### Operational Functions

The operational functions of smart grid control center applications are SCADA, DMA and EMS. Real-time SCADA systems do data acquisition and processing, basic network monitoring, device and sequence control, network and device tagging, and alarms and events management. DMS applications (tools) do network topology monitoring, demand response and load management, load and generation forecasting, switching procedures, fault management, outage management, trouble call management, work management, crew management, customer information, and asset management. An EMS performs the remote and local control and supervision of transmission systems.

## 4.7   CYBER SECURITY

Cyber security means measures for the protection of computers, servers, mobile devices, electronic systems, networks and data from malicious attacks, while application security means measures for keeping software and devices free from threats.

### 4.7.1   CYBER SECURITY ISSUES

Several power utilities have installed SCADA/EMS systems around the globe that had limited security incorporated into the system solutions. However, the utilities have information and IT to tackle the security problem. The evolution of power system communication (PSC) systems and the limited concerns of cyber security in the past have created new cyber security issues, and the cyber security issues of SCADA/EMS systems and smart grids are described here (Ericsson, 2010).

When existing SCADA/EMS systems are being refurbished or replaced, information and IT security issues come into the picture. If a SCADA/EMS system is to be refurbished, the operational part of the system must be shielded from the administrative part so that the operational part is protected from the digital threats of an internet connection. When the SCADA/EMS system is replaced, the overall system structure should be reconsidered to incorporate IT security at all levels. To secure this state, if possible, the operational SCADA/EMS system and the administrative IT system should be decoupled, or a firewall configuration between the operational part and administrative part should be provided.

SCADA/EMS systems that are interconnected and integrated with external systems create new possibilities and threats. In fact, SCADA systems are now standardized to a great extent as off-the-shelf products and are increasingly connected to the internet for different purposes (e.g., remote access, remote maintenance). This implies that SCADA systems are being exposed to the same kind of vulnerabilities as ordinary office PC solutions based on Microsoft products (Ericsson, 2010; Gao et al., 2014). With all the devices being interconnected, the network of a SCADA system is becoming less isolated and, thus, more prone to attack (Creery and Byres, 2005).

The introduction of smart grid solutions means that cyber security and power system communication systems must be dealt with extensively. These parts together are essential for proper electricity transmission where information infrastructure is critical. The development of communication capabilities moving power control systems from islands of automation to totally integrated computer environments has opened up new possibilities and vulnerabilities (Ericsson, 2010).

### 4.7.2 INFORMATION SECURITY DOMAINS

A domain is a specific area where specific activities/business operations are performed and can be grouped together. It is logical to study a SCADA/EMS system in terms of domains. This concept in the application for power systems was introduced by Ericsson et al. (2007), and the security domains identified were:

(1) Public, supplier, maintainer domain.
(2) Power plant domain.
(3) Substation domain.
(4) Telecommunication domain.
(5) Real-time operation domain.
(6) Corporate IT domain.

#### 4.7.2.1 Purpose of Domain Concept

The purpose of the domain concept is to emphasize the involvement of all the stakeholders within a specific area. One domain may use hardware equipment or communications that are used by another domain, so the domains are typically interrelated. The domains described above may be different from one electric utility to another depending on the utility's operations and tasks. The proposed domains are chosen logically by the utility. It is, of course, up to each utility to choose and implement its domains. The ideas presented here are general and applicable to another set of security domains and their interdependencies.

#### 4.7.2.2 Authority and Security Policy

Security is treated within each domain, and there is always only one authority responsible for the security within the domain. Different interests and compliance with legislative and contractual requirements could make it necessary to define a security policy structure using different security domains inside a power utility. Within one security domain, there is only one security policy and only one authority responsible for the security policy inside the domain. The authority should guarantee a minimum security level for the systems in the domain. The security level of the individual systems must be classified and may actually vary.

When communicating across power utilities, organizations and other companies using communication networks, security domains should be recognized. For example, a power utility could define a security domain and related policies and procedures for its telecontrol activity to assure compliance with legislative or regulatory requirements. If similar definitions, procedures, policies, etc., are developed by other power utilities, it would be easier to discuss and define common rules for information exchange or the use of common resources in a communication network. However, there is no common definition of the terms "security" and "critical asset". There is also no common control system security policy or procedure, although groups such as ISA and NIST are working on generic policies and procedures.

A power utility should discuss and define its policy structure based on the topology and importance of resources in the telecontrol network itself. A power utility on a regional level, for example, must decide if all substations, all local control centers, and the regional control center should belong to the same security domain or be split into several domains. This is particularly true when the utility provides electric as well as gas or water products and services. This becomes more of an issue when utilities share equipment, such as RTUs. Furthermore, in WG D2.22, the information security domain model has been adopted and further used in the context of an information security framework (Ericsson et al., 2010).

### 4.7.2.3   Security Domain Model

An electric power utility (EPU) representing one security authority could define each domain according to the level of protection required by the organization. The domain model should be defined based on the results of a risk assessment process. Appropriate security controls must be assigned to the domains and inter/intra-connections. EPU systems and data networks supported by IT components, such as servers, client devices, data communication infrastructure, access and network management devices, operating systems and databases, must be mapped to the domain model as well. This model is suited for a defense-in-depth strategy against cyber risk.

Furthermore, an EPU needs to define its own selection of security controls for SCADA control systems. The controls must be appropriate for the EPU's regulatory regime and assessment of business risks. The security controls need to be defined within each domain, and the information flows between the domains based on the agreed risk assessments. For example, the corporate domain and business-critical domain controls will depend on an intra-business risk assessment, whereas the operational critical domain controls are likely to be interdependent risk assessments between other operators and possibly government agencies in addition to an intra-business risk assessment. Many types of IT components are required to support EPU control systems, and the lists of controls could be as follows (Torkilseng and Duchworth, 2008):

(1) System architecture security controls.
(2) IT support user security control.
(3) User access security control.

### 4.7.3   CYBER SECURITY: OBJECTIVES AND REQUIREMENTS FOR SMART GRID

Smart grids are designed to achieve multiple upgrades on the basis of innovative technologies and the collaboration of different organizations. These can provide easy access to devices and power data and may be the causes of existing and new attacks against the intelligent grid. Fully secured systems and apps do not exist, and smart grid technology cannot be bypassed.

### 4.7.3.1 Cyber Security Objectives

The National Institute of Standards and Technology (NIST) reported three high-level cyber security objectives (Wang and Lu, 2013; NIST, 2010; Lukszo et al., 2010):

(1) Availability: Ensuring services and information are timely and reliably accessed by users and attacks against utility companies and consumers are prevented.
(2) Integrity: Ensuring information authenticity and non-repudiation against modification or destruction attacks that can lead to inappropriate judgment for energy administration.
(3) Confidentiality: Ensuring information is defended against unauthorized access or disclosure.

The first two objectives are the most critical and important in a smart grid network from a reliability point of view, while the least crucial, confidentiality, is becoming more and more essential in the household area network (HAN) for communication with customers and in systems such as AMI networks (Wang and Lu, 2013).

### 4.7.3.2 Cyber Security Requirements

Cyber security requirements for a smart grid recommended by NIST involve physical security and cyber security that deals with critical parts such as communication or smart endpoints. These requirements are (Wang and Lu, 2013):

(1) Highly secured and efficient communication protocols.
(2) High resiliency and attack recognition.
(3) Access control, identification and authentication.

In terms of security, the importance is on HAN security because the customer's privacy is crucial, and LAN and WAN security has a very wide range. Usually, a HAN consists of four items: a gateway to connect the HAN with services in the LAN or WAN, network nodes or access points, a software management and operating system for the network, and smart endpoints (smart meters, appliances or info displays) (Deconinck, 2008).

### 4.7.3.3 Security Challenges

The customer's main concern refers to data protection exchanged between the utility company and the smart meter. Some of the security challenges in the transition to an intelligent grid are (Pearson, 2011):

(1) Ensuring customer security.
(2) Handling the problem of physical security.
(3) Dealing with a huge number of devices.
(4) Dealing with more stakeholders.
(5) Use of IP and commercial use of off-the-shelf hardware and software.

Under these circumstances, it is obvious that new techniques and technologies are needed to increase the cyber security of the smart grid. As more and more distributed information needs to be processed and protected, artificial intelligence techniques based on cellular automata, genetic algorithms, fuzzy logic, and artificial neural networks are finding applications for new solutions through local interconnections.

In line with the features, objectives, requirements, challenges and advantages of the future, smart grids are vulnerable to attacks by hackers for financial gains or others advantages.

### 4.7.4  ATTACKS AGAINST SMART GRID

Attacks against smart grid networks may cause potential risks and threats (Bîrleanu et al., 2019):

(1) New technologies can bring new issues.
(2) The grid is too complex, which can lead to weak points or unforced errors.
(3) Intelligent nodes can contribute to denial-of-service (DoS) attacks.
(4) The many network links can compromise the entire system.
(5) Interference with utility telecommunications can alter the generation, transmission and distribution systems.
(6) Bills can be manipulated for energy theft.
(7) Unauthorized access from a customer-end device or point.
(8) Mass load can be manipulated, which disturbs the bulk power grid.
(9) Equipment failure.
(10) Human error.
(11) Threats because of the weather, fire or animals.

In general, an attack has four steps:

(1) Investigation.
(2) Discovery.
(3) Vulnerability identification.
(4) Infiltration.

#### 4.7.4.1  Categories of Cyberattack
A large variety of attacks exist against smart grids and may be categorized depending on technologies used, layers, platforms and so on. Four aspects to

consider are (1) attacks based on smart grid security objectives, (2) attacks against utility companies, (3) attacks against customers, and (4) attacks regarding wide-area monitoring, protection and control systems (WAMPAC). WAMPAC is a technology based on phasor measurement units (PMUs) that ensures the real-time monitoring of current grid operations and real-time protection and control activities such as automatic generation control (AGC) and special protection schemes (SPS).

### 4.7.4.2    Cyberattacks Based on Three High-Level Cyber Security Objectives

Attacks based on three high-level cyber security objectives are (Bîrleanu et al., 2019):

(1) Attacks against availability: These are DoS attacks (jamming in substations, ARP10 spoofing, traffic flooding, buffer flooding) that are intended to corrupt, block or insert delays in the communications inside a smart grid.
(2) Attacks against integrity: These are designed to disrupt or modify data traffic inside a smart grid network, such as by false data injection, net metering, and sensor data manipulation.
(3) Attacks against confidentiality: These look to collect unauthorized data and information from network capabilities in the intelligence grid, such as by traffic analyzing and eavesdropping.

### 4.7.4.3    Cyberattacks against Utility Companies

When attacking utility companies, hackers typically use multiple attack schemes on several parts and resources of the utility organization in order to achieve their goals. Thus, attacks against utility companies are separated into (Bîrleanu et al., 2019):

(1) Physical attacks.
(2) Network attacks.
(3) System attacks.
(4) Wireless attacks.
(5) Application attacks.
(6) Social engineering attacks.

### 4.7.4.4    Cyberattacks against Customers

When attacking customers, hackers perform attacks against a HAN. These attacks include (Bîrleanu et al., 2019):

(1) Network attacks.
(2) Wireless attacks.
(3) Attacks against smart meters.
(4) Attacks against management devices.

### 4.7.4.5   Cyberattacks against WAMPAC Systems

Cyberattacks against WAMPAC systems in a smart grid are divided into three parts (Bîrleanu et al., 2019):

(1) Replay attacks: Through these attacks, hackers manipulate a phasor measurement unit (PMU) by stealing transit packets.
(2) Timing attacks: These attacks aimed at communication networks use packets that slow down or shut down the network
(3) Data integrity attacks: These attacks corrupt data in the forward or reverse path in the control flow.

### 4.7.5   Countermeasures for Attacks against Smart Grid

Countermeasures are the detection of malicious attacks and protection measures against malicious attacks. These malicious attacks can be avoided using a series of countermeasures and include (Bîrleanu et al., 2019):

(1) Using redundancy techniques for attacks targeting availability.
(2) Authorizing users to come up with modifications for attacks targeting integrity countermeasures based on access control.
(3) Providing the exact privileges necessary to users to implement actions for attacks targeting confidentiality (principle of least privilege).

However, these countermeasures should cover the entire smart grid and not just high-level cyber security objectives. Thus, the next countermeasures refer to defense mechanisms for protecting the power network of the smart grid, including cryptographic methods for securing communication in the network and security practices for securing the consumer and the utility companies.

For the power network in a smart grid, there are two countermeasures (Bîrleanu et al., 2019):

(1) Attack detection with four mechanisms: packet-based detection, signal-based detection, the proactive method (algorithms based on sending probing packets), and the hybrid method (a combination between the other three methods or other existing methods).
(2) Attack attenuation with three mechanisms: reconfiguration of the network architecture, rate-limiting on a possibly malicious set of packets, and packet-filtering after attack detection.

Cryptographic mechanisms for smart grid network communications are major countermeasures against attacks and are more effective than power network approaches. These mechanisms include (Bîrleanu et al., 2019):

(1) Encryption based on symmetric key cryptography (AES, DES).
(2) Encryption based on asymmetric key cryptography (RSA11).

(3) Authentication.
(4) Key management (PKI12 and symmetric key management).

### 4.7.5.1   Countermeasures to Secure a Customer and Their HAN on a Smart Grid

Countermeasures to secure a customer and their HAN on a smart grid can be encapsulated in a 12-point security practice (Bîrleanu et al., 2019):

(1) Threat modeling: potential threats must be found in order to implement countermeasures.
(2) Segmentation: minimize the impact and surface of attacks.
(3) Firewall rules: convenient rules must be utilized.
(4) Signing: digital signing for trusted applications in the grid.
(5) Honeypots: traps for hackers so that alerts come in time.
(6) Encryption: protecting sensitive data and information.
(7) Vulnerability analysis: centers for analyzing traffic and critical systems.
(8) Penetration testing: simulation of attacks to observe how efficient countermeasures are.
(9) Source code review: applications in the intelligent grid must present zero vulnerabilities.
(10) Configuration hardening: vulnerabilities scanner and benchmarking test, especially for smart endpoints.
(11) Strong authentication: at least two-factor authentication.
(12) Logging and monitoring: provides information for attack identification and for reassembling actions in case of natural failures.

### 4.7.5.2   Practices for Securing the Other End of Smart Grid Utility Companies

There is also a 12-point security practice for securing the other end of smart grid utility companies, which represents the ISO/IEC 27002 Code of Practice for Information Security Control and Management (Bîrleanu et al., 2019):

(1) Risk assessment: for the identification and prioritization of risks against the information grid.
(2) Security policy: to direct and evaluate the information security plans of utility companies.
(3) Organization of information security: an important factor in the successful and correct implementation of a functional information security plan.
(4) Asset management: to identify who is responsible for particular assets.
(5) Human resources security: information dissemination to the contractors or staff who must assume their responsibility regarding the security of the information that they just acquired.
(6) Physical and environment security: protecting critical grid systems and preventing unauthorized access.

(7) Communications and operations management: someone must be account-able and respond to the grid's information systems.

(8) Access control: principle of least privilege.

(9) Information systems acquisition, development and maintenance: utility companies must ensure that all information security requirements are included while a project is in the requirements step.

(10) Information security incident management: methods implemented for the timely identification and containment of information security incidents.

(11) Business continuity management: utility companies must hold up incidents that affect the continuity of their operations and actions

(12) Compliance with standards and legal requirements.

All these countermeasures to attacks against the smart grid are meant to prevent and protect the entire power network in case of intrusions and to alert and take action to combat these attacks.

## 4.8 ARCHITECTURE OF COMMUNICATION TECHNOLOGY FOR POWER SYSTEMS

SCADA systems are important for remotely monitoring, controlling and managing the quality/continuity and safety of facilities distributed at various remote sites. SCADA systems in smart grids are designed to monitor the state and control the behaviors of all the plants over all the power delivery systems. The information flow in a smart grid framework is quite complex and needs priority attention for power generation systems. The quantity of monitored data is vast compared to the data of command and control since a large number of devices are monitored. Clearly, the control, command and billing channels need to be highly secured, and the information needs to be processed in a timely manner for such action. Therefore, real-time, precise, robust and secure communication is key to monitoring and controlling purposes (Gao et al., 2014).

Communication infrastructure plays a vital role in the successful operation of modern smart grids and ensures optimal operation and coordination between the components of the smart grid from generation to consumer (Baimel et al., 2016). The smart grid uses two-way communications, digital technologies, advanced sensing and computing infrastructure for improved monitoring, protection and optimization. Two-way communications also allow energy consumers to receive real-time prices and bills, while the grid operator can receive real-time information about energy consumption. These reliable real-time information flows between the components are essential for the successful operation of smart grids.

### 4.8.1 COMMUNICATION INFRASTRUCTURE OF SCADA

The basic elements of a SCADA system are remote processes, field devices, RTUs, SCADA central host and dispatch consoles (usually called a master terminal unit or MTU), and communications paths.

A typical SCADA communication system generally consists of a master station and many other distributed RTUs. The RTUs are interconnected to the master station through a variety of communication channels. One of the greatest communication challenges is that the channel limits the speed of data acquisition and control. Furthermore, random noise on the channel is another challenge that has hindered SCADA communication.

Communication links in a SCADA-based control system are shown in Figure 4.10. In SCADA communication, system information can be automatically or manually exchanged between remote facilities and centralized control centers. RTUs in a remote plant exchange data with field devices or sensors, and the central host exchanges data with dispersed RTUs located at remote facilities via communications networks. The operator has a graphical interface that represents a plant or equipment displayed with their associated readings. As the data changes in the field, the operator graphical interface is updated accordingly.

There are three generations of SCADA system architectures. The first generation uses a WAN for communication between MTUs, which executes decision-making, and RTUs, which serve the end-users. The second generation uses local area networks (LANs) to communicate between MTUs and RTUs. The third generation uses a WAN and internet protocol (IP).

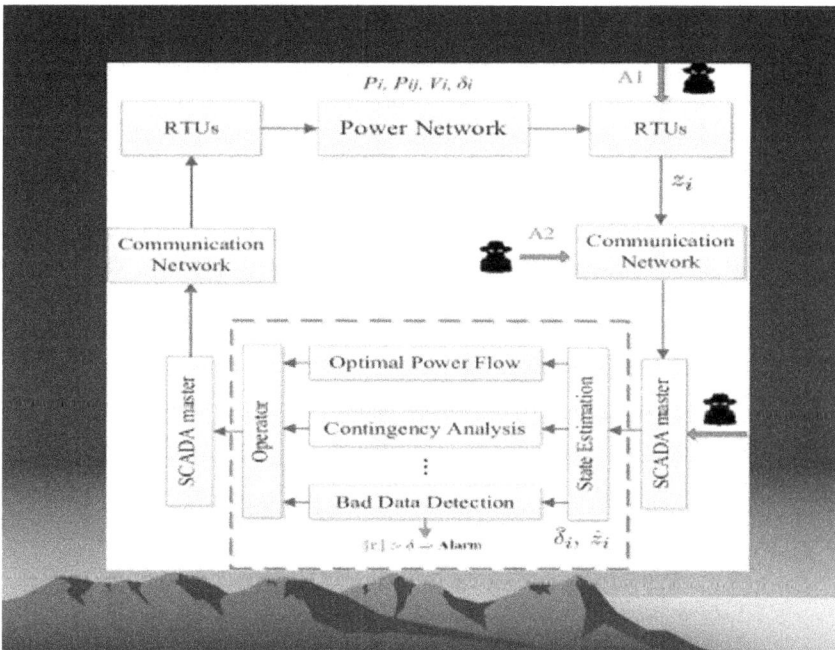

**FIGURE 4.10** Communication links in a SCADA-based control system.

## 4.8.2 COMMUNICATION INFRASTRUCTURE OF A SMART GRID

The smart grid communication infrastructure may be considered a hierarchical network with a three-tier architecture: (1) an access tier, (2) a distribution tier and (3) a core tier (Ancillotti et al., 2013). This architecture describes the communication facilities that should be provided at different tiers. Furthermore, it explains how the devices should be organized in network topologies and how small-scale networks could be interconnected to build a large communication system in order to map communication technologies. Figure 4.11 shows the reference architectural model of a smart grid communication system.

The tiers of the smart grid communication system are described below:

1. **Access tier**

   The communication networks in the access tier are responsible for real-time information flows between the consumers and the energy management systems and prove the more active role of end-consumers in the electricity market and smart grid management. Therefore, HANs are important candidates for the access tier and can be used to support the monitoring and control of electric devices at consumers' premises. Both low-range wired and wireless technologies can be used to build such networks. Wi-Fi is too expensive for most HAN devices and consumes too much power.

2. **Distribution tier**

   The communication networks used in the distribution tier are responsible for state estimation, real-time control of the distribution grid, supporting the interconnection of local area networks (i.e., HAN) with smart area communication backbones, and providing communications support implementing the data management services needed to efficiently handle the large amount of data collected in the distribution grid. The AMI network is the key component in the distribution tier since it interconnects smart meters with data aggregators and control systems in the distribution grid. Communication options include conventional programmable logic controller (PLC) technologies, point-to-point communication using cellular or medium-range wireless technologies, and multi-hop wireless technologies.

3. **Core tier**

   The core tier of a smart grid communication network is a wide area network (WAN). A high-capacity communication backbone is needed to deliver the huge amounts of data collected by AMI systems and field area networks (FANs) to remote control centers over long distances. The various communications are IP core networks and MPLS-based networks.

   The distinguishing features of communication network architectures are:
   (a) Low-power wireless communication for HANs.
   (b) Different types of gateways to control the communications of increasingly larger areas and provide interconnections of subsystems.
   (c) Cellular (3G) technologies for communication between multiple body are network (BAN) gateways and a neighborhood area network (NAN).

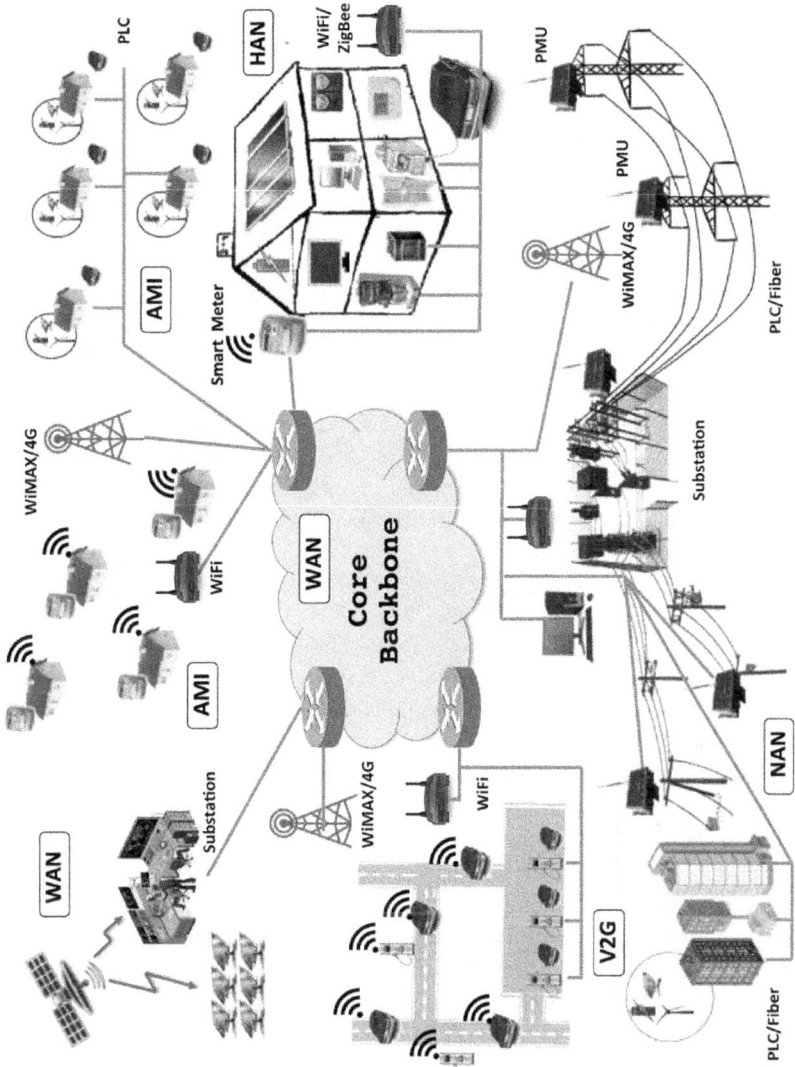

**FIGURE 4.11** An example of an end-to-end communication infrastructure for smart grids (Ancillotti et al., 2013).

The communication infrastructure of a smart grid can be categorized into three types of networks: a home area network (HAN), a neighborhood area network (NAN) and a wide area network (WAN).

A HAN is used and operated within a small area of tens of meters, usually a house or small office. The data transmission rate of a HAN is low as compared with either a NAN or a WAN and is hundreds of bits per second. In a typical application, a HAN is a broadband internet connection shared between multiple users using a wired or wireless modem. In smart grid applications, all smart home devices that consume energy and smart meters are connected to a HAN and can be implemented using Zigbee or Ethernet technologies.

A NAN is used and operated within an area of hundreds of meters, usually a few urban buildings. Several HANs can connect to form one NAN. The data transmission rate of a NAN is up to 2 Kbps and can be implemented using PLC, Wi-Fi and cellular technologies.

A WAN is used and operated within a vast area of tens of kilometers and usually consists of several NANs and local data centers (LDCs). The communication of all smart grid components includes an operator control center, conventional and renewable energy generation, and transmission and distribution. A WAN has a very high data transmission rate up to a few Gbps and can be implemented using Ethernet networks, WiMAX, 3G/LTE and microwave transmissions.

## 4.9  MULTI-AGENT SYSTEMS

A multi-agent system (MAS) contains multiple intelligent, interconnected collaborating agents for solving a problem beyond a single-agent ability. Agents are intelligent entities that act flexibly and autonomously to make wise decisions based on their intelligence and experience.

Agents can interact with other agents via some kind of agent communication language. Agents also can perceive and react to their environment. Lastly, multi-agent systems are proactive in that they are able to exhibit goal-oriented behavior by taking initiatives (Gupta et al., 2013; Nagata et al., 2012; Lu et al., 2009). Information is exchanged among agents to take appropriate actions, and the process is repeated constantly to monitor the system so that management of the system can be achieved instantly. The monitoring and measurement schemes of electric power systems can be enhanced using a multi-agent system.

A MAS is a computational platform where several agents or actors act together to achieve the desired goal, and the performance of the system depends on the activities of all the agents. In the context of power systems, multi-agent technologies can be applied in a variety of applications, such as to perform power system disturbance diagnosis, power system restoration, power system secondary voltage control and power system visualization.

A smart grid combines advanced intelligence systems, control techniques and sensing methods into an existing utility power network. For controlling smart grids, various control systems with different architectures have already been developed. The MAS-based control of power system operations has been shown to overcome the limitations of time required for analysis, relaying and protection, transmission switching, communication protocols, and management of plant control. These systems provide an alternative for fast and accurate power network control. Various intelligent agents have been used for the control of smart grids where a MAS is most effective to control various aspects of smart grids.

Several researchers (Gupta et al., 2013; Nagata et al., 2012; Pipattanasomporn et al., 2009) reported the capability of a multi-agent system as a technology for managing microgrid operation. Merabet et al., (2014) presented the platforms used for the implementation of a MAS for the control and operation of smart grids. Roybal and Jeffers (2013) reported that combining the methods of agent-based and system dynamics modeling minimizes complexity, improves realism, and increases speed for power system dynamic frequency simulations.

**Agent attributes**

A MAS is essentially a software system consisting of multiple autonomous components, the intelligent agents that interact with each other and react to environmental changes in order to accomplish a task provided (e.g., to control a physical resource or solve a complex problem in a distributed manner) (Wooldridge, 1999). An agent of a MAS may be defined as an entity with attributes considered useful in a particular domain. In this framework, an agent is an information processor that performs autonomous actions based on information. Common agent attributes include:

- Autonomy: goal-directedness, proactive and self-starting behavior.
- Collaborative behavior: the ability to work with other agents to achieve a common goal.
- Knowledge-level communication ability: the ability to communicate with other agents with language resembling human speech rather than typical symbol-level program-to-program protocols.
- Reactivity: the ability to selectively sense and act.
- Temporal continuity: persistence of identity and state over long periods.

Since agents are autonomous entities and each agent has unique objectives and responsibilities when working in collaboration towards achieving an overall goal, the specifications of the agents in a MAS are essential for assigning responsibilities and their roles in a MAS, and these are needed to simplify the modeling using a collaborative diagram that defines the interaction among the agents and their environment. The specifications of the agents in a MAS are presented in subsection 4.9.1 (Pipattanasomporn et al., 2009).

In the simplest case, agents are reactive objects that can only respond to signals from the environment. In more advanced systems, agents can also be proactive in the sense that they are programmed to execute different actions in order to achieve a global goal. In addition, agents can cooperate and coordinate with each other in order to find the best sequence of actions that can be addressed to achieve the desired goal. To meet this requirement, agents can employ various artificial intelligence techniques, including machine learning, fuzzy logic, neural networks or genetic algorithms for local decision-making. An agent can show a high degree of flexibility and autonomy because it can dynamically change its behavior to achieve a desired goal. It is also subject to an important limitation. Specifically, given the scale of the system to control, the agent can only observe and measure small portions of the grid (i.e., the environment is only partially observable). This implies that an agent cannot use global knowledge to make its decisions. Finally, it is useful to note that the environment that is observable by a MAS can be either physical (e.g., transmission lines, generators of renewable energy, electrical appliances, etc.) and observed through sensors; or virtual (e.g., databases, computing facilities, other agents) and observable through programming interfaces (McArthur et al., 2007).

### Applications of a multi-agent system

Smart grids are complex systems consisting of a large number of heterogeneous and interdependent components operating in dynamic environments. Therefore, decentralized management is considered the only feasible approach to control the operations of electric grids over multiple timescales (Amin, 2001; Hauser et al., 2005; Farangi, 2010). However, with decentralized control, it is difficult to achieve global optimization objectives, such as the efficiency and stability of the entire grid. On the other hand, recently, the emerging field of autonomic distributed computing has produced innovative middleware technologies to build intelligent distributed systems that are capable of managing, repairing, optimizing and protecting themselves without human intervention (Brazier et al., 2009).

Various approaches have been proposed to implement autonomic management capabilities in distributed environments. MASs are the most popular among these approaches in the smart grid research community (Rehtanz, 2003; Amin and Wollenberg, 2005; McArthur et al., 2007). This has also been demonstrated by a large number of MASs designed for a variety of smart grid applications including power system restoration (Nagata et al., 1995; Nagata and Sasaki, 2002), fault diagnosis (Hossack et al., 2003 and 2003: McArthur et al., 2004; Deshmukh et al., 2005), management of distributed energy resources (Lu and Chen, 2009; Baran and El- Markabi, 2007; Jiang, 2006), demand-side management (Ramchurn et al., 2011), management of energy storage systems (Vytelingum et al., 2010; Wei et al., 2010), optimization of EV operations (Vandael et al., 2010 and 2011), substation automation (Buse et al., 2003), distribution control (Pipattanasomporn et al., 2009), network monitoring (Nagata et al., 2005; Deshmukh et al., 2005) and visualization (Cristaldi et al., 2003), electricity market simulation (Vytelingum et al., 2010; Reddy and Veloso, 2011), profiling of power generation and energy usage patterns (Mallick er al., 2011), and management of microgrids and virtual power plants (VPPs) (Dimeas and Hatziargyriou, 2005; Jiang, 2006; Colson and Nehrir, 2009; Meng et al., 2010; Xu and Liu, 2011).

**Advantages and challenges of agent-based technologies**

The smart grid integrates advanced sensing technologies, control methods and integrated communications into the present electricity grid at both transmission and distribution levels. The smart grid is expected to :

1. be self-healing;
2. be consumer-friendly;
3. be attack-resistant;
4. provide power quality for 21st century needs;
5. accommodate all generation and storage options;
6. enable markets; and
7. optimize assets and operate efficiently.

In essence, in a multi-agent system:

- Each agent has incomplete capabilities to solve a problem.
- There is no global system control.
- Data is decentralized.
- Computation is asynchronous.

From the above discussion, we can observe that the main advantage of agent-based technologies in smart grids is to provide a decentralized management solution based on autonomous local decisions that can ensure a high level of flexibility and robustness. Thus, we can outline three main advantages of multi-agent systems as follows (Roche et al., 2010):

1. They have a view of local environment and have limited knowledge. This property is a requirement in the case of a very large system and is a decisive reason why MASs are chosen for grid energy management if the view of the agents can be limited to their neighbors in the grid.
2. They are flexible and fault-tolerant. An interesting aspect of the flexibility of a MAS is the possibility to add new functionalities without having to completely redesign the system. Fault tolerance can be observed when part of the MAS fails due to a broken component or a cut communication channel. A MAS can continue to work even without some parts of its structure and can adapt to it.
3. They are well-suited to distributed problems. The two previous characteristics enable this third one. MASs are well-suited for solving difficult problems where consumption can be distributed among several agents.

Besides these, another advantage is that MAS architecture is not dependent on a particular technology. Different programming languages can be used and can interact together if proper messaging tools are used.

Although there are already many smart grid applications where MAS technologies have been investigated, there are also unsolved technical issues that must be addressed in order to use this technology in real-world deployments. There are some technical and management challenges. We outline three of the most important technical challenges in subsections 4.9.2 to 4.9.5 (Ancillotti et al., 2013). Three of the most important management challenges are as follows (Gazanfroudi et al., 2017):

1.  A MAS can be applied to decentralized energy management methods that decrease the computational burden of the system. There are some challenges to implementing these systems.
2.  Another challenge is the decentralized energy management of systems related to the optimum decisions of these systems.
3.  MASs create platforms which enable agents to act autonomously and this increases the reliability of the systems. There is also another challenge in designing the platforms to implement these systems.

While local decision-makers can decrease the computational burden of the system, their decisions are not global. Hence, the trade-off between local decisions and global optimization is one of the research areas in these systems. Furthermore, cyber security is another important challenge in smart grids.

### 4.9.1 Specifications of the Agents of Multi-Agent Systems

The specifications of a control agent, a distributed energy resource (DER) agent, a user agent, and a database agent in an intelligent distributed autonomous power system (IDAPS) MAS are defined as:

1.  **Control agent:** The responsibilities of a control agent are monitoring system voltage and frequency to detect contingency situations or grid failures and sending signals to the main circuit breaker to isolate the IDAPS microgrid from the utility when an upstream outage is detected; receiving electricity price ($/kWh) signals from the main grid obtained from AMI; and publishing them to the IDAPS entities.
2.  **DER agent**: The responsibilities of a DER agent are storing associated DER information, monitoring and controlling DER power levels and connect/disconnect status. (DER information includes DER identification number, type (e.g., solar cells, microturbines, fuel cells), power rating (kW), local fuel availability, cost function or price at which users agree to sell, and DER availability, i.e., planned maintenance schedule.)
3.  **User agent**: The user agent acts as a customer gateway that makes the features of an IDAPS microgrid accessible to users. Responsibilities include providing users with real-time information on entities residing in the IDAPS system; monitoring electricity consumption by each critical and non-critical load; and allowing users to control the status of loads based on the user's predefined priority.

4. **Database agent:** The data agent serves as a data access point for other agents as well as users. Responsibilities include storing system information and recording messages and data shared among agents.

## 4.9.2 FUNCTIONAL ARCHITECTURES OF MULTI-AGENT SYSTEMS

A number of different functional architectures have been proposed in the literature for building multi-agent systems, but it is not yet decided which one of those approaches is the most suitable for smart grid applications (Rehtanz, 2003). It is important to note that flexible, extensible and open architectures allow the agents to be easily added or removed, but closed architectures are preferred where agent interactions are fixed at design time. The architectural model most commonly adopted in power systems is multi-layered architecture. For instance, a three-layer architecture is proposed for managing distributed energy resources (Lu and Chen, 2009). In this model, the bottom layer consists of agents managing physical resources (e.g., energy generators and power storage systems). The middle layer includes agents that provide high-level management services (e.g., fault diagnosis, protection and restoration, optimization of power parameters, etc.) to the agents connected to the physical resources. Finally, the top layer includes the agents handling the user interfaces. To improve the scalability of MAS solutions, many studies have proposed to group agents, especially those operating on physical resources, into clusters (e.g., a microgrid cluster or a VPP cluster). Indeed, clustering and coalition formation are common techniques in multi-agent systems for reducing architecture complexity. For instance, coalitions are used to integrate in a more efficient way electric vehicles into the electricity grid. In this case, an aggregator agent, called a coalition server, is used to hide the details about the individual vehicles and to present a group of vehicles as a single resource to grid operators.

Alternative two-layer architecture models are considered for power system restoration and for the management of the distribution grid (Nagata and Sasaki, 2002; Nagata et al., 2005; Garcia et al., 2010). More precisely, two types of agents are considered: equipment agents and facilitator agents (Nagata and Sasaki, 2002; Nagata et al., 2005). The former control physical resources, such as transmission lines, transformers and phase controllers, while the latter are used to promote the cooperation of the equipment agents that are associated with them. For instance, a facilitator agent is responsible for controlling an entire electric substation (Nagata et al., 2005). A more elaborate architecture has been described that defines multiple types of facilitators that perform different tasks such as device control, data acquisition and transfer, data analysis and data querying (Buse et al., 2003). The use of facilitator agents is also proposed (Hossack et al., 2003). However, in this case, the facilitator agent is responsible for maintaining a list of services (or resources) that other agents in the system can offer (or control). With the support of a name server agent, which maintains the names and locations of each agent (e.g., IP addresses), a facilitator allows other agents to dynamically enter or leave the system and register or deregister their locations and capabilities.

### 4.9.3 INTEROPERABILITY

When developing a multi-agent system, it must be a standardized agent model so that the agents can easily cooperate irrespective of their different capabilities and functions or platforms used. For these reasons, a significant body of work is focused on the formalization of:

(a) Agent specification languages, which are standards for defining message types and agent interaction models. Today, Foundations for Intelligent Physical Agents' Agent Communication Language (FIPA-ACL) is used by MAS developers as the de facto standard for message exchange and interaction protocols.
(b) Ontologies, which define a common vocabulary of terms and concepts that agents can exchange and interpret.

Currently, the trend is to implement application-specific ontologies. However, interoperability between MASs using different ontologies is difficult to implement, even if they are operated on the same platform or are based on similar concepts (Chatterson et al., 2005). A solution to this problem is to use a two-layer model for ontology specification. Specifically, the ontology in the top level is called the upper ontology (Chatterson et al., 2005), which is responsible for defining the basic concepts that are common to most smart grid applications (e.g., substation, switch, voltage, etc.), and is used as a template for defining more specific ontologies for different applications. The existing object-oriented standards for information exchange in power engineering applications such as IEC 61970-301 (2011) can be easily used as reference models when developing an upper ontology for smart grid applications.

### 4.9.4 MAS MODELING AND MAS SIMULATION PLATFORMS

Large-scale testing is essential for demonstrating the robustness and reliability requirements of MAS technologies for power grid applications. A valuable option to minimize the costs and risks associated with testing new technologies in real power grids is to use realistic simulation tools that simultaneously model electric power scenarios and the behaviors of computer communication protocols.

Agent-oriented approaches have emerged as a powerful technology that can support the design and development of complex systems, and the potentiality of these agent-based approaches to solving complex problems enables the implementation of systems that are flexible and autonomous. The flexibility in solving complex problems comes with challenges. To overcome these challenges, researchers have searched for suitable methods to support the development of agent-based complex systems and developed two methods: semi-formal methods and formal methods. Semi-formal methods use mainly Unified Modeling Language (UML) extensions, and formal methods are based on mathematical notations (Merabet et al., 2014).

The concept of the platform is related to the implementation of the MAS and is the environment that manages the lifecycle of agents and wherein to certain services. Several multi-agent platforms exist wherein the agents have access to certain services, and some examples of popular and regularly updated platforms include ZEUS, AgentBuilder and JADE. JADE, developed in Java, runs on all operating systems, including all required components that control the MAS, and has a very specific architecture for building so-called standard agents (Merabet et al., 2014).

### 4.9.5 IMPLEMENTATION PLATFORMS OF MULTI-AGENT SYSTEMS

Software is now available for the modeling of multi-agent systems (McArthur et al., 2007). The Java Agent Development Framework (JADE) is a common platform for the MAS modeling of smart grid applications (Bellifemine et al., 2007). Most available MAS implementations are prototypes and are used for lab experiments to demonstrate the potentiality of the MAS approach.

Recently, some multi-agent systems have been tested for operational electric grids. A multi-agent system called Protection Engineering Diagnostic Agents (PEDA) has been used by a transmission system operator in the UK to interpret SCADA-related data (Hossack et al., 2003) and to provide online diagnostic information and alarms (Davidson et al., 2006). A multi-agent system can be implemented using JADE to integrate a group of electric vehicles into the electric grid and to support both demand response and vehicle-to-grid (V2G) services.

Alternative technologies are becoming available for building large-scale and interoperable power system applications, including service-oriented architectures (SOA) (Pagani and Aiello, 2012), grid and cloud computing (Fang et al., 2012; Irving et al., 2004; Rusitschka et al., 2010) and web services (Zhu, 2003). However, agent-based technologies appear to be the most suitable solution for distributed control applications.

### Exercises

**4.1** (a) A coaxial cable has a bandwidth of 1.2 MHz. How many channels, each having a bandwidth of 8 kHz, can be accommodated on the cable?

(b) A communications medium has 150 channels. The available bandwidth of the medium is 2.4 MHz. Calculate the maximum bandwidth of each channel.

(c) What is the maximum data rate for carrying binary data on a fiber-optic cable with a bandwidth of 80 GHz?

**4.2** (a) An amplifier has input and output signal powers of 36 μW and 145 mW, respectively. Calculate the power gain of the amplifier in dB.

(b) A 4 km communication link has input and output signal powers of 1 W and 96 mW, respectively. Calculate the power loss/km in dB for the link.

(c) The output power from a transmission line is 250 mW. If the overall loss is -32 dB, calculate the input power.

**4.3** (a) The amplitude of signal and noise in a transmission link is estimated at 3 V and 4 mV, respectively. Estimate the signal-to-noise ratio (SNR).

(b) The noise output from a coaxial cable is 85 μW with no signal present. What signal power is required if the minimum SNR is 33 dB?

(c) A communications receiver requires an input voltage amplitude of 5 V and an SNR of 32 dB. What is the maximum acceptable value of the noise amplitude?

## REFERENCES

Amin, M. (2001). Toward self-healing energy infrastructure systems. *IEEE Computer Applications in Power*. 14(1): 20–28.

Amin, S., & Wollenberg, B. (2005). Toward smart grid power: Power delivery for the 21st century. *IEEE Power and Energy Magazine*. 3(5): 34–41.

Ancillotti, E., Bruno, R., & Conti, M. (2013). The role of communication systems in smart grids: Architectures, technical solutions and research challenges. *Computer Communications*. 36: 1665–1697

Baimel, D., Tapuchi, S., & Baimel, N. (2016). Smart grid communication technologies. *Journal of Power and Energy Engineering*. 4: 1–8.

Baran, M., & El-Markabi, I.M. (2007). A multiagent-based dispatching scheme for distributed generators for voltage support on distribution feeders. *IEEE Transactions on Power Systems*. 22(1): 52–59.

Bellifemine, F. L., Caire, G., & Greenwood, D. (2007). *Developing Multi-Agent Systems with JADE*. Wiley and Sons.

Bîrleanu, F. G., Anghelescu, P., Bizon, N., & Pricop, E. (2019). Cyber security objectives and requirements for smart grid. In Kabalci, E., & Kabalci Y., *Smart Grids and Their Communication Systems*. Springer.

Brazier, F. M. T., Kephart, J. O., Farunak, H. V. D., & Huhus, M. N. (2009). Agents and service-oriented computing for automatic computing: A research agenda. *IEEE Internet Computing*. 13(3): 82–87.

Buse, D. P., Sun, P., Wu, Q. H., & Fitch, J. (2003). Agent-based substation automation. *IEEE Power and Energy Magazine*. 1(2): 50–55. Prentice Hall.

Catterson, V., Davidson, E., & McArthur, S. (2005). Issues in integrating existing multi-agent systems for power engineering applications. In Proceedings of the 13th International Conference on Intelligent Systems Application to Power Systems. 05: 1–6.

Colson, C. M., & Nehrir, M. H. (2009). *A review of challenges to real-time power management of micro grids*. In Proceedings of the IEEE PES 09.

Creery, A., & Byres, E. J. (2005). Industrial cybersecurity for power system and SCADA networks. Paper No. PCIC-2005-34.

Cristaldi, I., Monti, A., Ottoboni, R., & Ponci, F. (2003). *Multi-agent based power systems monitoring platform: A prototype*. In Proceedings of the IEEE Power Tech Conference on System Sciences. Volume 2.

Davidson, E. M., McArthur, S. D. J., McDonald, J. R., Cumming, T., & Watt, I. (2006). Applying multi-agent system technology in practice: Automated management and analysis of SCADA and digital fault recorder data. IEEE Transactions on Power Systems. 21(2): 559-567

Deconinck, G. (2008). *An evaluation of two-way communication means for advanced metering in Flanders (Belgium)*. IEEE Instrumentation and Measurement Technology Conference. 900–905.

Deshmukh, A., Ponci, F., Monti, A., Riva, M., & Cristaldi, L. (2005). *Multi-agent system for diagnostics, monitoring and control of electric systems*. In Proceedings of the IEEE Conference on Intelligent Systems Application to Power Systems, 05. 1–6.

Dimeas, A. L., & Hatziargyriou, N. D. (2005). Operation of a multiagent system for microgrid control. *IEEE Transactions on Power Systems*. 20(3): 1447–1455.

Ekanayake, J. B., Liyanage, K., Wu, J., Yokoyama, A., & Jenkins, N. (2012). *Smart Grid: Technology and Applications*. John Wiley and Sons.

EPRI. (2011). EPRI smart grid demonstration initiative two - update. Figure 2 on page 7. www.epri.com/

Ericsson, G. N. (2010). Cyber security and power system communication – Essential parts of a smart grid infrastructure. *IEEE Transactions on Power Delivery*. 25(3): 1501–1507.

Ericsson, G., Trokilseng, A., Dondossola, G., Jansen, T., Smith, J., Holstein, D., Vidrascu, A., & Weiss, J. (2007). Security for information system and intranets in electric power systems. *Technical Brochure* (TB) 317 CIGRE, 2007.

Ericsson, G., Trokilseng, A., Dondossola, G., Pietre-Cambacédes, L., Duckworth, S., Bartels, A., Tritschler, M., Kropp, T., & Weiss, J. (2010). Treatment of information security for electric power utilities (EPUs). *Technical Brochure* (TB) 317 CIGRE, 2010.

Fang, X., Misra, S., Xue, G., & Yang, D. (2012). Managing smart grid information in the cloud: Opportunities, model, and applications, *IEEE Network*. 26 (4): 32–38.

Farangi, H. (2010). The path of the smart grid. *IEEE Power and Energy Magazine*. 8(1): 18–28.

Gao, J., Liu, J., Rajan, B., Nori, R., Fu, B., Xiao, Y., Liang, W., & Chen, C. L. P. (2014). SCADA communication and security issues. *Security and Communication Networks*. 7: 175–194.

Garcia, A. P., Oliver, J., Gosch, D. (2010). *An intelligent agent-based distributed architecture for smart-grid integrated network management*. In: 1st IEEE Workshop on Smart Grid Networking Infrastructure. Denver, Colorado. 1013-1018.

Gazafroudi, A. S., De Paz, J. F., Prieto-Castrillo, F., Villarrubia, G., Talari, S., Shafie-khah, M., & Catalão, J. P. S. (2017). *A review of multi-agent based energy management systems*. In: De Paz J., Julián V., Villarrubia G., Marreiros G., Novais P. (eds) Ambient Intelligence – Software and Applications – 8th International Symposium on Ambient Intelligence (ISAmI 2017). Advances in Intelligent Systems and Computing, vol. 615. Springer.

Gupta, R., Jha, D. K., Yadave, V. K., & Kumar, S. (2013). *A multi-agent framework for operation of a smart grid. Energy and Power Engineering*. 5: 1330-1336.

Hauser, C., Bakken, D., & Bose, A. (2005). A failure to communicate: Next generation communication requirements, technologies and architecture for the electric power grid. *IEEE Power and Energy Magazine*. 3(2): 47–55.

Hossack, J. A., Menal, J., McArthur, S. D. J., & McDonald, J. R. (2003). A multiagent architecture for protection engineering diagnostic assistance, *IEEE Transactions on Power Systems* 18 (2): 639–647.

Hu, Y., & Li, V. O. K. (2001). Satellite-based internet: A tutorial. *IEEE Communications Magazine*, 39(3): 154–162.

IEC 61970-301. (2011). *Energy management system application program interface (EMS-API) – part 301: Common information model (CIM) base*.

IEEE. (2008). *IEEE standard for SCADA and automation systems C37.1- 2007.*

Irving, M., Taylor, G., & Hobson, P. (2004). Plug in to grid computing, *IEEE Power and Energy Magazine* 2(2): 40–44.

Jiang, Z. (2006). *Agent-based control framework for distributed energy resources microgrids.* In Proceedings of the IEEE/ACM IAT 06. 646–652.

Kirkham, H., Johnston, A. R., & Allen, G. D. (1994). Design considerations for a fiber optic communications network for power systems. *IEEE Transactions on Power Delivery.* 9(1): 510–518.

Leonardi, A., Mathioudakis, K., Wiesmaier, A., & Zeiger, F. (2014). *Advances in Electrical Engineering.* Hindawi.

Lu, M., & Chen, C. (2009). *The design of multi-agent based distributed energy system.* In Proceedings of the IEEE SMC09. 2001–2006.

Lu, J., Da, X., & Qian, A. (2009). *Research on smart grid in China.* Proceedings of Transmission & Distribution Conference and Exposition: Asia and Pacific, 2009, pp. 1–4.

Lukszo, Z., Deconinck, G., & Weijnen, M. P. C. (2010). *Securing Electricity Supply in the Cyber Age: Exploring the Risks of Information and Communication Technology in Tomorrow's Electricity Infrastructure.* Springer.

Mallik, R., Sarda, N., Kargupta, H., & Bandyopadhyay, S. (2011). *Distributed data mining for sustainable smart grids.* In Proceedings of the ACM SustKDD 11. 1–6.

McArthur, S., Davidson, E., Catterson, V. M., Dimeas, A., Hatziargyriou, N. D., Ponci, F., & Funabashi, T. (2007). Multi-agent systems for power engineering applications – Part II: Technologies, standards and tools for building multi-agent systems, *IEEE Transactions on Power Systems* 22 (4) (2007) 1753–1759.

McArthur, S. D. J., Davidson, E. M., Hossack, J. A., & McDonald, J. R. (2004). *Automating power system fault diagnosis through multi-agent system technology.* In Proceedings of the 37th Hawaii International Conference on System Sciences. 22 (4) (2007) 1753–1759.

McDonald, J. D. (2002). Substation automation basics - The next generation. *Electric Energy Transmission and Distribution Magazine.*

McDonald, J. D. (2012). *Electric Power Substations Engineering.* CRC Press, Boca Raton, Fla, USA.

Meng, F., Akella, R., Crow, M., & McMillin, B. (2010). *Distributed grid intelligence for future microgrid with renewable energy sources and storage.* In Proceedings of the NAPS10.

Merabet, G. H., Essaaidi, M., Talei, H., Abid, M., Khalil, N., Madkour. M., & Benhaddou, D. (2014). *Applications of multi-agent systems in smart grids: A survey.* International Conference on Multimedia Computing and Systems (ICMCS), 088–1094.

Momoh, J. (2012). *Smart Grid: Fundamentals of Design and Analysis.* IEEE Press and John Wiley and Sons.

Nagata, T., Sasaki, H., & Yokoyama, R. (1995). Power system restoration by joint usage of expert system and mathematical programming approach. *IEEE Transactions on Power Systems.* 1–4.

Nagata, T., & Sasaki, H. (2002). A multi-agent approach to power system restoration. *IEEE Transactions on Power Systems.* 17(2): 457–462.

Nagata, T., Fujita, H., & Sasaki, H. (2005). *Decentralized approach to normal operations for power system network.* In Proceedings of the 13th International Conference on, Intelligent Systems Application to Power Systems 05. 1–6.

Nagata, T., Ueda, Y., & Utatani, M. (2012). *A multi-agent approach to smart grid operations.* 2012 IEEE International Conference on Power System Technology (POWERCON), 2012 pp. 1–5.

NIST. (2010). The smart grid interoperability panel – cyber security working group: Guidelines for smart grid cyber security. *NISTIR* 7628: 1–597.

Pagani, G. A., & Aiello, M. (2012). Service orientation and the smart grid state and trends. *Service Oriented Computing and Applications* 6(3): 267–282.

Pearson, I. L. G. (2011). Smart grid cyber security for Europe. *Energy Policy* 39: 5211–5218.

Pipattanasomporn, M., Feroze, H., & Rahman, S. (2009). *Multi-agent systems in a distributed smart grid: Design and implementation.* In Proceedings of IEEE/PES Power Systems Conference and Exposition. 1–6.

Ramchurn, S. D., Vytelingum, P., Rogers, A., & Jennings, N. (2011). *Agent-based control for decentralized demand side management in the smart grid.* In: Proceedings of AAMAS 11. 5–12.

Reddy, P., & Veloso, M. (2011). *Learned behaviors of multiple autonomous agents in smart grid markets.* In Proceedings of AAAI 11. 1–8.

Rehtanz, C. (2003). *Autonomous Systems and Intelligent Agents in Power System Control and Operation.* Springer.

Roche, R., Blunier, B., Miraoui, A., Hilaire, V., Koukam, A. (2010). *Multi-agent systems for grid energy management: A short review.* In IECON 2010 – 36th Annual Conference on IEEE Industrial Electronics Society, November 2010.

Roybal, L. G., & Jeffers, R. F. (2013). *Using system dynamics to define, study, and implement smart control strategies on the electric power grid.* The 31st International Conference of the System Dynamics Society, July 2013.

Rusitschka, S., Eger, K., & Gerdes, C. (2010). *Smart grid data cloud: A model for utilizing cloud computing in the smart grid domain.* In Proc. of IEEE Smart Grid Comm. 10: 483–488.

Torkilseng, Å., & Duchworth, S. (2008). Security frameworks for electric power utilities – Some practical guidelines when developing frameworks including SCADA/control system security domains. *Electra.* CIGRE, Dec.

U.S. Department of Energy. (2009). *Smart Grid System Report*, July 2009, www.oe.energy. gov/sites/prod/files/oeprod/DocumentsandMedia/SGSRMain_090707_lowres.pdf (accessed on 4 August 2011)

Vandael, S., Boucké, N., Holvoet, T., & Deconinck, G. (2010). *Decentralized demand side management of plug-in hybrid vehicles in smart grid.* In Proceedings of ACM AAMAS 10. 67–74.

Vandael, S., Creamer, K., Boucke, N., Holvoet, T., & Deconinck, G. (2011). *Decentralized coordination of plug-in hybrid vehicles for imbalance reduction in smart grid.* In Proceedings of ACM AAMAS 2011. 803–810.

Vandael, S., Holvoet, T., & Deconinck, G. (2011). *A multi-agent system for managing plug-in hybrid vehicles as primary reserve capacity in a smart grid.* In Proceedings of the Second International Workshop on Agent Technologies for Energy Systems (ATES 2011). 11. 13–20.

Vytelingum, P., Ramchurn, S., Voice, T. D., Rogers, A., & Jennings, N. R. (2010). *Trading agents for the smart electricity grid.* In Proceedings of AAMAS 10. 897–904.

Wang W., & Lu, Z. (2013). Cyber security in the smart grid: Survey and challenges. *Computer Networks.* 57: 1344–1371.

Wang W., Xu, Y., & Khanna, M. (2011). A survey on the communication architectures in smart grid. *Computer Networks*. 55: 3604–3629.

Wei, C., Hu, H. Chen, Q., & Yang, G. (2010). *Learning agents for storage devices management in the smart grid*. In Proceedings of IEEE Conference on CiSE 10. 1–4.

Wooldridge, M. (1999). Intelligent agents. In G. Weiss (Ed.). *Multiagent Systems: A Modern Approach to Distributed Artificial Intelligence*, MIT Press.

Xu, Y., & Liu, W. (2011). Novel multiagent based load restoration algorithm for microgrids. *IEEE Transactions on Smart Grid*. 2(1): 152–161.

Zhu, J. (2003). Web services provide the power to integrate. *IEEE Power & Energy Magazine* 1(6): 40–49.

# 5    Modeling of Energy Demand, Supply and Price

## 5.1   INTRODUCTION

Energy demand is increasing globally in order to meet the needs of increasing populations and economic growth. Fossil fuels are still the main sources of energy and contribute to global warming, and furthermore, they are being depleted rapidly. This has prompted the transition to environmentally friendly renewable energy resources to ensure energy security and to reduce the contribution to global warming. System dynamics modeling methods for understanding the dynamics of energy supply and demand, energy price and contributions to global warming from energy production are presented in this chapter.

## 5.2   ENERGY DEMAND

We use energy in different forms, such as biomass or gas for cooking, electricity for lighting and air-conditioning, petrol for driving cars, and diesel for running buses for transport and tractors for cultivation. Therefore, energy is demanded in different forms. Global demand is increasing rapidly and is related to GDP.

The energy demand sector estimates and projects the demands for energy for different uses and total energy demand, which mainly depends on fast-growing populations and economic growth. It also depends on income and location, such as rural and urban (Bala, 1998).

The electric power system is one of the most important energy sectors, and the demand analysis and load forecast of this sector are well-established. Accurate metered consumption data are available. However, errors in demand forecasts are more serious since large investments and long construction delays are associated with the expansion capacity of the electric power system. A forecast that is too low may lead to a short-term solution or load-shedding, which affects economic growth and industrial development. On the other hand, overestimations will lead to the tying-up of scarce investment of capital and other resources.

The terms demand and load forecast are used interchangeably to indicate the magnitude as well as the structure of the requirements of both electrical power and energy (kW and kWh, respectively). The commonly used unit of electrical energy

DOI: 10.1201/9781003218401-5

is kilowatt hour (kWh). The rate of energy per unit time is power and is expressed in kilowatts (kW).

The load factor is the ratio of average to maximum (peak) power over an interval of time and can be on a daily, annual, etc., basis. It is important because the size or capacity of power system components are determined to a great extent by their capability to handle peak power flows. Since the majority of customers require their maximum kW only during a short peak period during the day, the load factor is important to determine the peak load and is also a measure of the intensity of capacity use. A load forecast may be made either in terms of the peak power in kW or the total energy consumed in kWh. In general, the kW peaks of different customers are different. The diversity factor for a group of customers is a measure of the divergence of the spreading over time of the individual peak loads and permits the computation of the combined peak load of the group in terms of the disaggregate peak values.

## 5.2.1 MODELING OF ENERGY DEMAND

The assessment of long-term energy demand is a prerequisite for effective energy planning to support development and research. Two commonly used approaches to predict energy demand are: bottom-up approach of using LEAP (Long-range Energy Alternatives Planning) (Lazarus et al, 1995; Bala, 2006; Bala et al., 2014) and causal loop and stock–flow approach of system dynamics (Bala et al., 2017; Nail, 1992; Vogstad, 2004); while artificial neural networks (ANNs) are an emerging approach.

### 5.2.1.1 LEAP Model

LEAP is a tool that models energy and environmental scenarios and is used to project demand situations. Energy demand data are assembled in a hierarchy based on four levels denoted by names, sector, subsector, end uses and device. At level 1, the economy is divided into sectors. At level 2, each sector is divided into subsectors. At level 3, each subsector is divided into end uses. At level 4, each end-use is further divided into devices used. The average energy requirements or energy intensities of each device can also be estimated at this level. Energy intensity is expressed in energy units per activity. The four-level demand structure is shown in Figure 5.1.

The demand program computes energy demands over time by multiplying activity measures by their energy requirements. Both the activities and unit energy requirements may vary over time. In a LEAP demand sector, each of the major energy-consuming sectors – residential, industrial and transportation – are represented separately, and energy demands are compiled using a bottom-up approach. The energy demand in each sector is computed using the following relation:

$$E_{st} = \sum_{j} A_{st} \times I_{st} \qquad (5.1)$$

Where $E_{st}$ = energy demand, $A_{st}$ = activity and $I_{st}$ = energy use intensity.

```
/-Agriculture ──────── +-Rice ───────── +- Cultivation ──────── +- Tractor
          |                    |                  |                  \- Traditional plow
          |                    |           \- Irrigation ───────── +- Diesel pump
          |                    |                                     \- Electric pump
          |              +-Wheat ──────── +- Cultivation ──────── +- Tractor
          |                    |                  |                  \- Traditional plow
          |                    |           \- Irrigation ───────── +- Diesel pump
          |                    |                                     \- Electric pump
          |              +-Jute ──────────── Cultivation ──────── +- Tractor
          |                    |                                     \- Traditional plow
          |              \-Sugercane ──── +- Cultivation ──────── +- Tractor
          |                                      |                  \- Traditional plow
          |                               \- Irrigation ───────── +- Diesel pump
          |                                                         \- Electric pump
 \-Rural population ─ All ─────────────── +- Cooking ─────────── +- Stove (wood)
                                                 |                 +- Stove (crop)
                                                 |                 \- Stove (animal)
                                          +- Electric lighting ──── Bulb
                                          \- Kerosene lighting ── Kupi
```

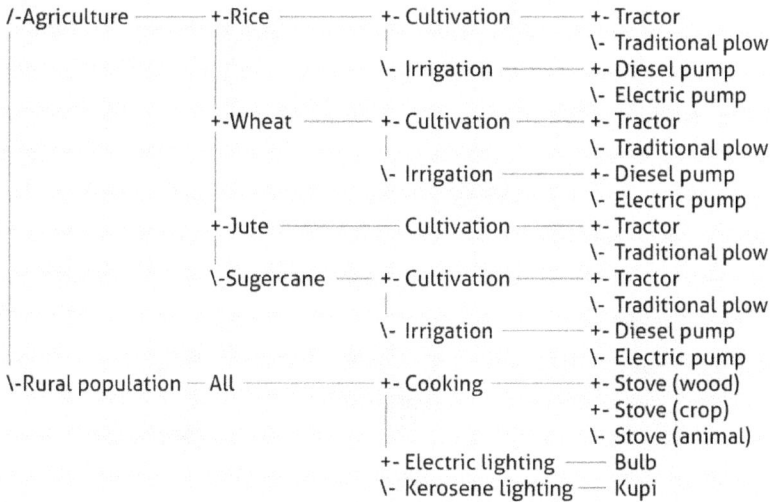

**FIGURE 5.1**  Four-level demand tree structure.

For example, the residential electricity demand is projected from the share of the number of households in rural and urban areas, the end-use devices used and their energy intensities for the future increase in households and end-use devices. However, energy intensity may also change.

### 5.2.1.2  System Dynamics Model

System dynamics models are formulated based on causal loop diagram and stock–flow diagram approaches. In the causal loop diagram and stock–flow diagram approach of system dynamics, the problem is described using cause–effect relationships and causal loop diagrams generating the dynamics of energy demand, and the model structure is formulated in the form of stock–flow diagrams that are essentially formulations of the model into integral finite difference equations (Bala et al., 2017). The causal loop diagram and stock–flow diagram approaches of system dynamics methodology are discussed in Chapter 2.

Consider a simple system dynamics model of electricity demand. In this model, per capita consumption of electrical energy depends on GDP and the total energy consumption is computed from per capita energy consumption and population. Per capita consumption of electrical energy depends on GDP, which in turn influences per capita consumption of electricity. Population also increases with time, resulting in an increase in electricity demand. The dynamic hypothesis is that electricity demand increases with time. Electrical energy demand is modeled first by drawing the causal loop diagram and then developing the stock–flow diagram, and finally, the energy demand is simulated.

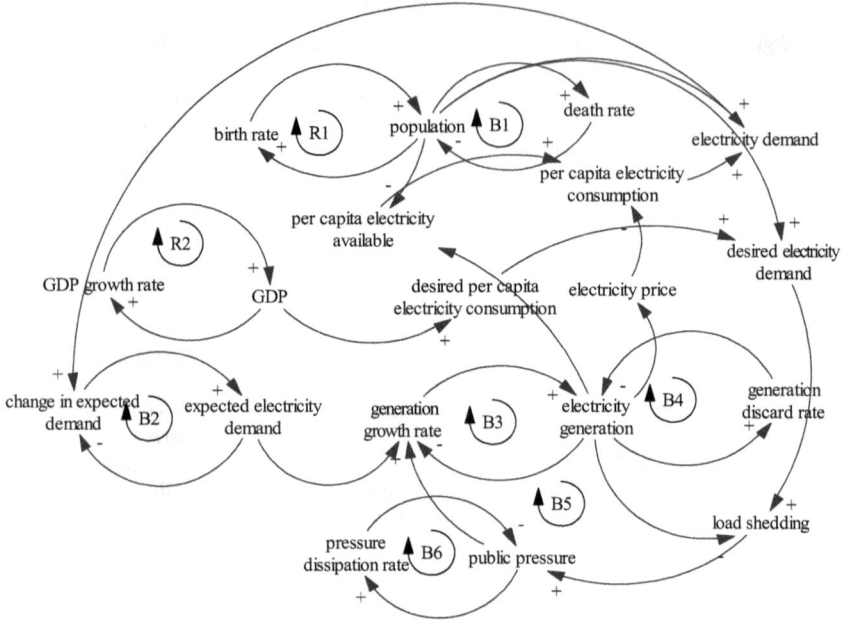

**FIGURE 5.2**    Causal loop diagram of electrical energy demand model.

The causal loop diagram of an electricity demand model is shown in Figure 5.2. There are six negative feedback loops and two positive feedback loops, and these eight feedback loops generate the dynamics of the energy demand. Population and birth rate form the positive feedback loop R1, and GDP and GDP growth rate form the re-enforcing feedback loop R2. Population and death rate forms a balancing feedback loop B1, and expected energy demand and change in expected energy demand form a balancing feedback loop B2. Electricity generation forms two negative feedback loops B3 and B4 with electricity generation rate and generation discard rate, respectively. Public pressure, electricity generation rate, electricity generation and load-shedding form the balancing feedback loop B5, while public pressure and dissipation rate form the negative feedback loop B6.

Electricity demand is computed from per capita electricity consumption and population. A stock–flow diagram of the electricity demand model is shown in Figure 5.3.

The fundamental equations that correspond to the major variables shown in Figure 5.3 are as follows:

Expected demand increases by change in expected demand and is expressed as:

expected_demand(t) = expected_demand(t - dt) +
(change_in_expected_demand) * dt
INIT expected_demand = 8.0E+10

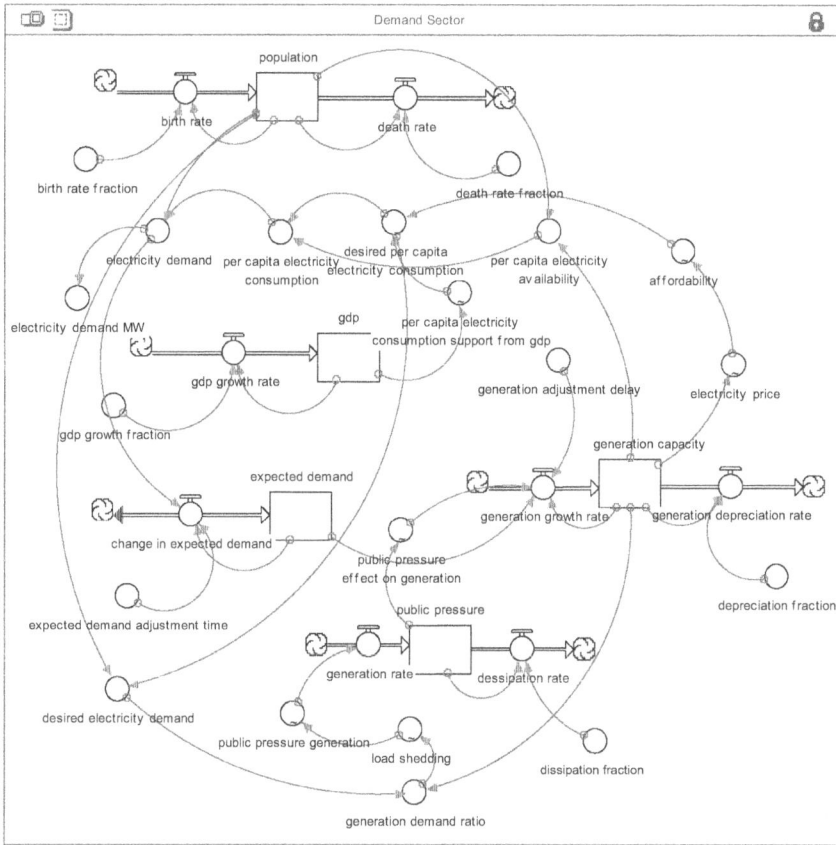

**FIGURE 5.3**   Stock–flow diagram of an electrical energy demand model.

Change in expected demand depends on difference between electricity demand and expected demand, and demand adjustment time. It is expressed as:

**INFLOWS**
change_in_expected_demand = (expected_demand - electricity_demand)/ expected_demand_adjustment_time
expected_demand_adjustment_time = 5

Electricity demand is computed from population and per capita electricity consumption and is expressed as:

electricity_demand = population * per_capita_electricity_consumption
electricity_demand_MW = electricity_demand/(0.6 * 8760 * 1000)

Population is increased by birth rate and decreased by death rate and is expressed as:

population(t) = population(t - dt) + (birth_rate - death_rate) * dt
INIT population = 163E+6

Birth rate is computed from population and its birth fraction and is expressed as:

**INFLOWS**
birth_rate = population * birth_rate_fraction
birth_rate_fraction = 0.044

Death rate is computed from population and its death fraction and is expressed as:

**OUTFLOWS**
death_rate = population * death_rate_fraction
death_rate_fraction = 0.015

Per capita electricity consumption is the minimum of either per capita electricity consumption or per capita electricity availability and is expressed as:

per_capita_electricity_consumption = MIN(desired_per_capita_electricity_consumption, per_capita_electricity_availability)

Per capita electricity availability is computed from electricity generation capacity and population and is expressed as:

per_capita_electricity_availability = generation_capacity/population

Desired per capita electricity consumption is computed from per capita electricity consumption support from GDP and affordability and is expressed as:

desired_per_capita_electricity_consumption = per_capita_electricity_consumption_support_from_gdp * affordability

Per capita electricity consumption support from GDP depends on GDP and is expressed graphically as:

per_capita_electricity_consumption_support_from_gdp = GRAPH(gdp)
(0.00, 205), (50.0, 218), (100, 240), (150, 268), (200, 295), (250, 325), (300, 348), (350, 363), (400, 388), (450, 403), (500, 428)

GDP is increased by the growth in GDP and is expressed as:

gdp(t) = gdp(t - dt) + (gdp_growth_rate) * dt
INIT gdp = 221.4

Growth in GDP is computed from GDP and its growth fraction and is expressed as:

**INFLOWS**
gdp_growth_rate = gdp * gdp_growth_fraction
gdp_growth_fraction = .071

Affordability of electricity consumption is a function of electricity price and is expressed graphically as:

affordability = GRAPH(electricity_price)
(0.00, 1.01), (5.00, 1.08), (10.0, 1.12), (15.0, 1.15), (20.0, 1.18), (25.0, 1.21), (30.0, 1.22), (35.0, 1.24), (40.0, 1.26), (45.0, 1.28), (50.0, 1.29)

Electricity price depends on electricity generation capacity and is expressed graphically as:

electricity_price = GRAPH(generation_capacity)
(0.00, 9.50), (1e+011, 14.0), (2e+011, 19.5), (3e+011, 23.5), (4e+011, 28.5), (5e+011, 35.0), (6e+011, 39.0), (7e+011, 41.5), (8e+011, 45.5), (9e+011, 48.5), (1e+012, 50.0)

Electricity generation is increased by electricity generation growth rate and decreased by generation depreciation rate and is expressed as:

generation_capacity(t) = generation_capacity(t - dt) + (generation_growth_rate - generation_depreciation_rate) * dt
INIT generation_capacity = 4.8E+10

Electricity generation growth rate depends on difference between expected electricity demand and electricity generation capacity, electricity generation adjustment delay and the effect of public pressure for electricity generation. It is expressed as:

**INFLOWS**
generation_growth_rate = ((expected_demand - generation_capacity)/generation_adjustment_delay) * public_pressure_effect_on_generation
generation_adjustment_delay = 10

Public pressure effect on generation depends on public pressure and is expressed graphically as:

public_pressure_effect_on_generation = GRAPH(public_pressure)
(0.00, 1.00), (0.1, 1.04), (0.2, 1.10), (0.3, 1.14), (0.4, 1.17), (0.5, 1.18), (0.6, 1.20), (0.7, 1.21), (0.8, 1.23), (0.9, 1.24), (1, 1.25)

Electricity generation depreciation rate is computed from electricity generation capacity and depreciation fraction and is expressed as:

**OUTFLOWS**
generation_depreciation_rate = generation_capacity * depreciation_fraction
depreciation_fraction = 0.00085

Public pressure increases from insufficient and irregularity in electricity generation and it is dissipated with time. It is expressed as:

public_pressure(t) = public_pressure(t - dt) + (generation_rate-dissipation_rate) * dt
INIT public_pressure = 0.0

Public pressure generation rate is simply the public pressure generation and is expressed as:

**INFLOWS**
generation_rate = public_pressure_generation

Public pressure generation depends on load-shedding and is expressed graphically as:

public_pressure_generation = GRAPH(load_shedding)
(0.00, 0.005), (0.1, 0.135), (0.2, 0.32), (0.3, 0.445), (0.4, 0.585), (0.5, 0.72), (0.6, 0.835), (0.7, 0.89), (0.8, 0.94), (0.9, 0.975), (1, 1.00)

Public pressure dissipation rate is computed from public pressure and dissipation fraction and is expressed as:

**OUTFLOWS**
dissipation_rate = public_pressure * dissipation_fraction
dissipation_fraction = 0.0001

Load-shedding depends on electricity demand ratio and is expressed graphically as:

load_shedding = GRAPH(generation_demand_ratio)
(0.00, 0.995), (0.1, 0.73), (0.2, 0.565), (0.3, 0.36), (0.4, 0.24), (0.5, 0.185), (0.6, 0.095), (0.7, 0.075), (0.8, 0.035), (0.9, 0.01), (1, 0.005)

Electricity generation demand ratio is the ratio of electricity generation capacity to desired electricity demand and is expressed as:

generation_demand_ratio = generation_capacity/desired_electricity_demand

Desired electricity demand is computed from population and per capita electricity consumption and is expressed as:

desired_electricity_demand = population * desired_per_capita_electricity_consumption

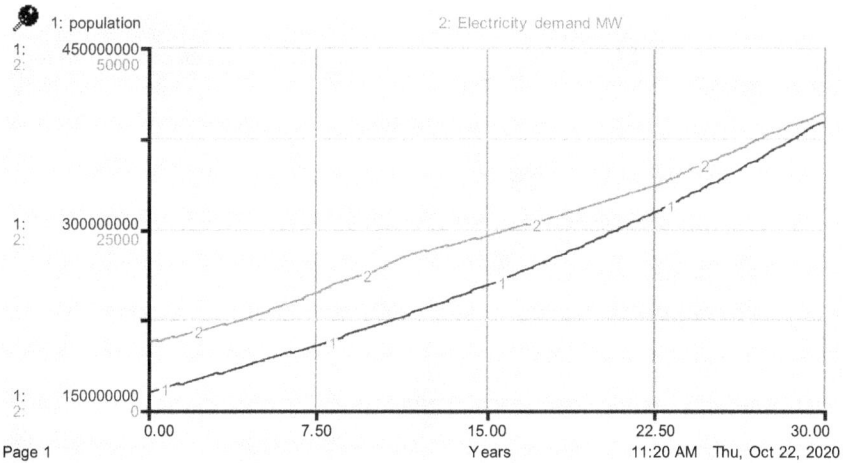

**FIGURE 5.4**   Simulated population and electrical energy demand.

Figure 5.4 shows simulated population and electrical energy demand. Population and energy demand increases with time. Energy demand increases with time because of increase in population and GDP with time.

### 5.2.1.3   Artificial Neural Network Model

Electricity consumption prediction has been considered an effective measure that helps power grid designers and planners build robust, adaptive, efficient and economical smart grids (Zhao and Magoulès, 2012). An artificial neural network (ANN) is an alternative method to project energy demand. The recent decade has brought an alternative and qualitatively new tool called a neurotechnique for solving complex problems that enables the user to conduct very fast and simple simulations. Several studies have been reported on the neural network modeling of electric energy demand (Song et al., 2020; Adhiswara et al., 2019; Panklib et al., 2015). Making electricity consumption forecasting based on a neural network has been a popular research topic in recent years, and the back-propagation neural network (BPNN) algorithm has been recognized as a mature and effective method (Song et al., 2020).

Neurocomputing techniques are shaped after biological neural functions and structures. Therefore, they are popularly known as artificial neural networks. As for their biological counterparts, the functions of ANNs are developed not by programming them but by exposing them to carefully selected data on which they can learn how to perform the required processing task. In such a modeling approach, there is no need to formulate an analytical description of the process. Instead, a black-box process model is constructed by interacting the network with representative samples of measurable quantities characterizing the process.

### Structure of an ANN Model

An independent multilayer ANN model of electric energy demand can be developed to project electric energy demand. The network of the model may be several layers and have a large number of simple processing elements called neurons. The input layer of the model consists of neurons that correspond to the input variables, and the output layer has a neuron that represents the final output in the model. The use of the number of hidden layers in an ANN model depends on the degree of complexity of the problem and on the application of the network. There are no fixed rules for determining the number of hidden layers and nodes. In general, one hidden layer has been found to be adequate, and in some cases, a slight advantage may be achieved by using two hidden layers (Hecht-Nielsen, 1989). A large number of neurons can represent the system more precisely, but complications arise in attaining proper training (Zhang et al., 2002). All inputs are standardized between 0.00 and 1.00. The learning rate initially is set at a value, for example, 0.10, and momentum is chosen in order to prevent overtraining. The performances of ANN models are usually evaluated by comparing the root mean square error (RMSE) and correlation coefficient ($r^2$) of predicted and observed values. Figure 5.5 shows an independent multilayer ANN model of a typical electric energy demand model. The input layer of the model consists of three neurons that correspond to three input variables, and the output layer has one neuron that corresponds to one output. The model has three layers and the number of neurons in the hidden layer is four. The input variables are GDP, population, and peak electric demand load.

### Training of an ANN Model

ANNs can modify their behavior in response to their environment. This factor, more than any other, is responsible for the interest they have received. Unlike mathematical models, the structure of an ANN model itself cannot represent a system's

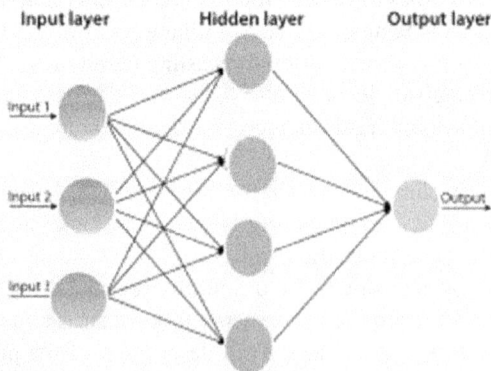

**FIGURE 5.5** Structure of a typical ANN model.

behavior unless it is properly trained. The objective of training the network is to adjust the weights of the interconnecting neurons of the network so that the application of a set of inputs produces the desired set of outputs. Initially, random values are used as weights. For reasons of brevity, one input-output set can be referred to as a vector. Training assumes that each input vector is paired with a target vector representing the desired output; together, these are called a training pair. A network is usually trained over a number of training pairs.

A wide variety of training algorithms have been developed, each with its own strengths and weaknesses. The ANN electric demand models are trained by a backpropagation algorithm so that the application of a set of inputs would produce the desired set of outputs. The steps of the training procedure are summarized as follows: (i) an input vector is applied, (ii) the output of the network is calculated and compared to the corresponding target vector, (iii) the difference (error) is fed back through the network, and (iv) weights are changed according to an algorithm called a delta rule (Wasserman, 1989) that tends to minimize the error. The vectors of the training set are applied sequentially. This procedure is repeated over the entire training set as many times as necessary until the error is within some acceptable criteria or until the outputs do not significantly change anymore. After the end of the training, simulations are done with the trained model to check the accuracy of training achieved.

Panklib et al. (2015) reported that the prediction of the ANN model is excellent and can be used to predict the potential of electric energy demand at different locations, and maintained that the ANN model yields more accurate load forecasts than the regression-based model using the same exogenous variables. Akhwanzada and Tahar (2012) reported that system dynamics simulation model output is very close to the multilayer perceptron (MLP) ANN model. Comparisons between the observed and simulated electric energy demands were also found to be very good. Thus, if the model is adequately trained, it can appropriately represent electric energy demand and can predict electric energy demand very well.

## 5.3   ENERGY SUPPLY

Fossil fuels and hydroelectric power are the main sources of commercial energy, and commercial energy is expected to penetrate even rural areas in order to raise the quality of life. Energy demand is met by energy supply. The first law of thermodynamics states that energy can be neither created nor destroyed. Therefore, we can meet this demand from different energy sources: fossil fuels (coal, gas and crude oils), nuclear fuel, and renewable energy sources (e.g., biomass, solar, wind).

Various energy supplies are considered in estimation, projection, modeling and policy analysis, and these are conventional energy sources, nuclear energy and renewable energy sources. Not all of these resources are fundamental and important for all countries; rather, the energy sources required vary by country. From the Bangladesh perspective, for electricity generation, natural gas is the main energy resource, although the construction of nuclear power plants and a gradual transition to renewable energy sources such as solar and wind are important.

Demand for electrical energy for industrial growth and to meet household requirements is increasing rapidly. The construction of nuclear power is a candidate for supplementing these energy requirements with clean and less costly energy, and the gradual transition to renewable energy to ensure energy security and to reduce contributions to global warming is advocated.

### 5.3.1  MODELING OF ENERGY SUPPLY

An assessment of the long-term projection of energy supply from different resources is needed for optimal uses of different energy resources to meet the energy demand that is required for effective energy planning to support development and research. Consider a simple electricity supply model. Upgrading the available energy resources or transitioning to renewable energy involves a delay in the initiation and construction of the energy supply project. The dynamic hypothesis is that the electricity supply would increase with time to match energy demand. Electrical energy supply is modeled first by drawing a causal loop diagram and then by developing a stock–flow diagram. Finally, the energy supply is simulated.

The causal loop diagram of a simple electrical energy supply model is shown in Figure 5.6. There are two power plants for supplying electricity, a natural gas

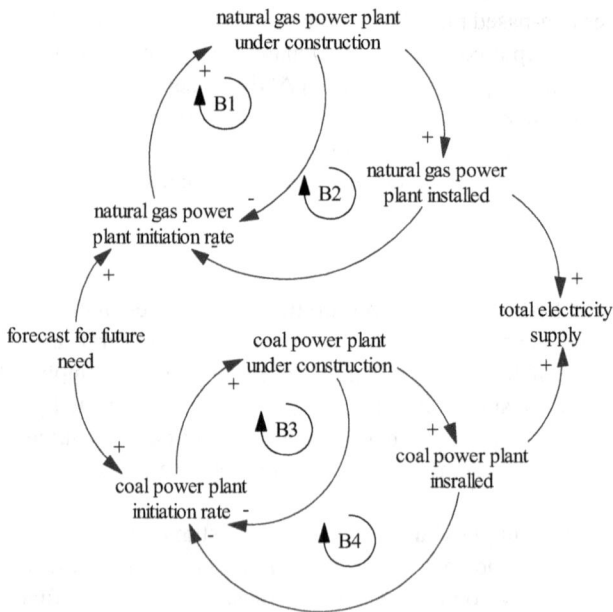

**FIGURE 5.6**  Causal loop diagram of electrical energy supply model.

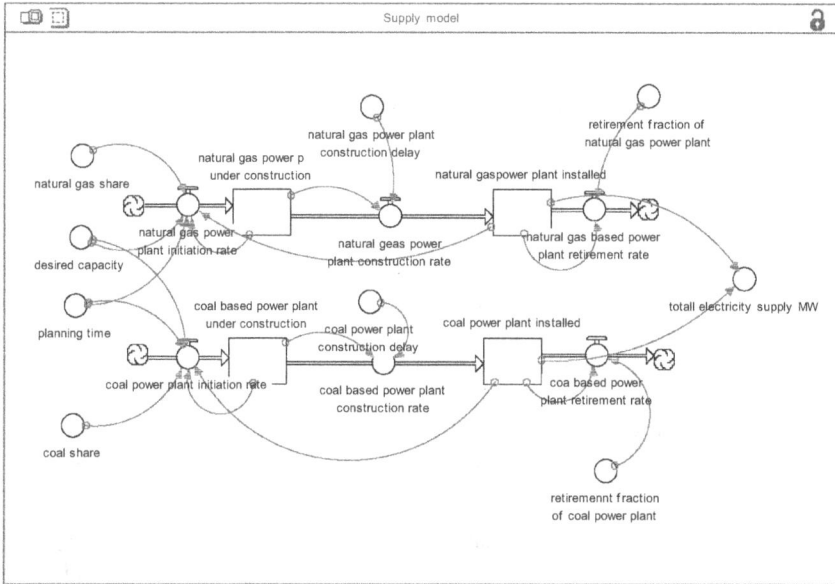

**FIGURE 5.7** Stock–flow diagram of electrical energy supply model.

power plant and a coal-based power plant. Each of these power plants forms two negative feedback loops. Power plant under construction and power plant initiation rate forms one negative feedback loop, and power plant installed, power plant initiation rate and power plant under construction forms another negative feedback loop. Natural gas power plant forms negative feedback loops B1 and B2, while coal-based power plant forms negative feedback loops B3 and B4.

Electricity supply is from two types of power plant: natural gas power plant and coal-based power plant. The stock–flow diagram of an energy supply model is shown in Figure 5.7. The fundamental equations that correspond to the major variables shown in Figure 5.7 are as follows:

**Coal-based power plant**

Coal-based power plant installed is increased by coal-based power plant construction rate and decreased by coal-based power plant retirement and is expressed as:

coal_power_plant_installed(t) = coal_power_plant_installed(t - dt) + (coal_based_ power_plant_construction_rate - coa_based_power_plant_retirement_rate) * dt
INIT coal_power_plant_installed = 2000

Coal-based power plant construction rate depends on coal-based power plant under construction and coal-based power plant construction delay and is expressed as:

**INFLOWS**

coal_based_power_plant_construction_rate  =  coal_based_power_plant_under_
construction/coal_power_plant_construction_delay
coal_power_plant_construction_delay = 3.5

Coal-based power plant retirement rate is computed from coal-based power plant installed and its retirement fraction and is expressed as:

**OUTFLOWS**

coal_based_power_plant_retirement_rate = coal_power_plant_installed * retire-
ment_fraction_of_coal_power_plant
retirement_fraction_of_coal_power_plant = 0.001

Coal-based power plant under construction is increased by coal-based power plant initiation rate and decreased by coal-based power plant construction and is expressed as:

coal_based_power_plant_under_construction(t) = coal_based_power_plant_
under_construction(t - dt) + (coal_power_plant_initiation_rate - coal_based_
power_plant_construction_rate) * dt
INIT coal_based_power_plant_under_construction = 2000

Coal-based power plant initiation rate is computed from desired coal-based power plant capacity, coal-based power plant installed, coal-based power plant under construction and planning time and is expressed as:

**INFLOWS**

coal_power_plant_initiation_rate = (desired_capacity * coal_share -
coal_based_power_plant_under_construction-coal_power_plant_installed)/
planning_time
desired_capacity = 50000
coal_share = 0.2
planning_time = 30

**Natural gas-based power plant**

Natural gas-based power plant installed is increased by natural gas-based power plant construction rate and decreased by natural gas-based power plant retirement and is expressed as:
natural_gaspower_plant_installed(t) = natural_gaspower_plant_installed(t - dt) +
(natural_gas_power_plant_construction_rate - natural_gas_based_power_plant_
retirement_rate) * dt
INIT natural_gaspower_plant_installed = 5000

Natural gas-based power plant construction rate depends on natural gas-based power plant under construction and natural gas coal-based power plant construction delay and is expressed as:

**INFLOWS**
natural_gas_power_plant_construction_rate = natural_gas_power_plant_under_
construction/natural_gas_power_plant_construction_delay
natural_gas_power_plant_construction_delay = 3.5

Natural gas-based power plant retirement rate is computed from natural gas-based
power plant installed and retirement fraction and is expressed as:

**OUTFLOWS**
natural_gas_based_power_plant_retirement_rate = natural_gaspower_plant_
installed * retirement_fraction_of_natural_gas_power_plant
retirement_fraction_of_natural_gas_power_plant = 0.001

Natural gas-based power plant under construction is increased by natural gas-based
power plant initiation rate and decreased by natural gas-based power plant construction and is expressed as:

natural_gas_power_plant_under_construction(t) = natural_gas_power_plant_
under_construction(t - dt) + (natural_gas_power_plant_initiation_rate - natural_
gas_power_plant_construction_rate) * dt
INIT natural_gas_power_plant_under_construction = 5000

Natural gas-based power plant initiation rate is computed from desired natural gas-
based power plant capacity, natural gas-based power plant installed, natural gas-
based power plant under construction and planning time and is expressed as:

**INFLOWS**
natural_gas_power_plant_initiation_rate = (desired_capacity * natural_gas_
share - natural_gas_power_plant_under_construction - natural_gaspower_plant_
installed)/planning_time
natural_gas_share = 0.8

**Total electricity supply**
Total electricity supply is the sum of electricity supply from coal-based power
plant and electricity supply from natural gas-based power plant and is expressed as:

total_electricity_supply_MW = coal_power_plant_installed + natural_gaspower_
plant_installed

Figure 5.8 shows the simulated total energy supply, energy supply from natural
gas power plant and coal-based power plant. Total energy supply from natural gas
power plant and coal power-based power plant is planned to meet the demand,
which increases with time. Energy demand increases with time because of increases
in population and economic development.

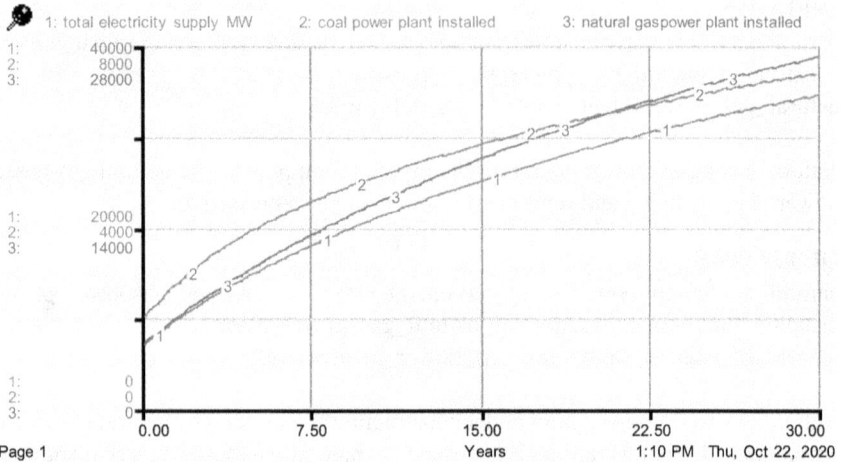

**FIGURE 5.8**   Simulated total electricity supply, coal power plant installed and natural gas power plant installed.

## 5.4   ENERGY BALANCE

Energy balance is the balance of energy demand and energy supply and may be the balance of specific energy or total energy systems. Energy balance essentially represents historic energy statistics in tabular format and represents alternative energy scenarios of consumption and production. It provides and projects energy balance in terms of energy deficit or energy surplus, and the projection of energy balance is needed in order to ensure energy security. It also influences energy price.

The basic scheme of energy balance is illustrated in Table 5.1. The columns of the table represent the different forms of energy such as electricity, gas and oil, and the rows represent the different stages of energy production, transformation and consumption (Meier, 1986). Each column must satisfy the following identity:

$$\text{Domestic production} + \text{imports} - \text{exports} + \text{stock change} - \text{conversion inputs} \atop + \text{conversion outputs} = \text{domestic consumption} \tag{5.2}$$

Note that exports, losses and inputs to energy conversion always have negative signs.

Energy accounting is needed to follow a somewhat analytical pattern relative to the energy supply situation. Whenever supply was abundant or cheap, little need was felt to devote financial and human resources to set up appropriate and detailed statistics. The energy crisis demands precise energy accounting, and scarce resources must be assessed to determine capacities, demand patterns and possible areas of management of energy resources. It is essential to have a knowledge of energy supply and consumption not only on a single product basis but also in the context of the national overall energy picture. Hence, energy balances are designed

**TABLE 5.1**
**Energy Balance**

|                        | Hydroelectricity | Electricity | Gas  | Oil  |
| ---------------------- | ---------------- | ----------- | ---- | ---- |
| Domestic production    | 90               |             |      |      |
| Imports                |                  |             | 220  | 280  |
| Stock changes          |                  |             | −20  | 20   |
| Gross supply           | 90               |             | 200  | 300  |
| Electricity generation |                  |             |      | −100 |
| Inputs                 | −90              |             |      |      |
| Generation             |                  | 60          |      |      |
| T and D                |                  | −12         |      |      |
| Total consumption      |                  | 48          | 200  | 200  |
| Consumption by sector  |                  |             |      |      |
| Households             |                  | 28          | 200  |      |
| Industry               |                  | 20          |      | 200  |

and applied. This energy balance has proved to be particularly valuable for analysis and is very useful for a variety of functions, such as: (i) showing the national energy situation in a comprehensive format by presenting various forms of energy in a common unit (e.g., gigajoules), (ii) checking the internal consistency of data, (iii) tracing energy flows from production through conversion to consumption, and (iv) showing losses in the various conversion and end-use sectors.

Until the mid-to-late seventies, energy balance existed practically only for developed countries. The concepts of energy balance were oriented predominantly to their needs and concentrated mainly on the supply and demand characteristics of commercial forms of energy (e.g., production, international trade, stock changes and apparent consumption of hard coal, ignite, crude and refined oils including their derivatives, natural gas and electricity). With increasing energy costs, planners started to ask for statistical information on how much and where energy is being used and at which efficiency in order to explore the feasibility of tapping alternative sources or to identify areas where energy management could be improved. Thus, efforts were made to achieve a more detailed breakdown of the data on the conversion industry and the various end-uses of energy.

In developing countries, national overall energy balances started to become an issue after the mid-seventies, and a comprehensive and more sophisticated statistical tool was needed to support evolving energy planning concepts. However, the first problem was that in many countries, even data on commercial forms of energy didn't exist, and even if it did exist, it was inadequate. For non-commercial or traditional forms of energy, the lack of data was even worse, a situation that still prevails today. It was, therefore, not possible to generate even the most aggregate type balance covering just commercial forms of energy.

A number of countries have already moved towards establishing a statistical system for their energy planning activities. Efforts are being made to mobilize institutional, human and financial resources for installing and maintaining such systems in national statistical offices or in other institutions such as energy ministries. Once this point is reached, discussions about the precise nature of energy accounting concepts become meaningful.

First, a clear notion of what kind of data is needed for which type of analysis must be developed in order to allocate scarce resources in the most efficient way. For countries with a relatively high share of commercial fuels, it might then be possible to adopt and modify slightly an accounting system and energy balance format derived from the prevailing formats in developing countries. However, for the majority of countries, a statistical system would have to be tailored to their national characteristics, and to some extent, their planning objectives. This should not have to preclude compatibility with concepts used at an international level. In fact, international concepts might have to be made more flexible in order to accommodate such features.

### 5.4.1 Approaches to Energy Balancing

The design of energy balances must be preceded by a thorough evaluation of the analytical and policy objectives pursued. In addition, the analyst will have to be careful while taking into consideration the scarce resources that a country can allocate to energy statistics. The basic approaches to designing balances are discussed, and this consideration is particularly relevant for developing countries to assess all aspects of the energy supply and consumption picture, including the future energy strategy. The following two concepts for compiling and presenting overall balances covering both primary and secondary forms of energy can be used for different approaches to planning.

#### 5.4.1.1 Top-Down Approach

This concept basically departs from the supply side (i.e., production, imports, exports and stock changes) and then shows the allocation of the available energy to the consuming sectors. From its focus and logical structure, it is primarily, but not necessarily exclusively, oriented towards historical analyses. This is still the prevailing concept used by international institutions, as well as by most countries, basically because it is perceived as less demanding in terms of data.

#### 5.4.1.2 Bottom-Up Approach

In its more developed variant, this approach starts out by looking at the various levels of useful energy for functions such as cooking, heating (process heat at high, medium and low temperatures), lighting and transportation. It considers the means (i.e., the technical apparel) by which energy is used (e.g., cook stoves, electrical appliances, type of automobile, etc.) and the respective efficiency in converting the potential energy into useful forms. The principle is that it follows up the ladder through transmission and distribution losses, conversion efficiencies (e.g., in generating electricity) to the point where the total energy requirement of a country or region can be portrayed. This is not just a statistical instrument for ex-post analyses, but an accounting concept with a forward-looking character, A bottom-up balance can be a useful tool for demonstrating to the policy-making bodies the gap between existing local supplies and overall (maybe minimum) demand. As a consequence, it can show how much imported energy would be required in the short term and which indigenous resources could possibly be developed and utilized in the long term (e.g., through a reforestation program, enhanced cooking stove efficiencies, improvements in the transmission lines to avoid excessive losses).

Energy balances for various regions within a country are not considered widely as the data requirements are very significant (e.g., specification of imports and exports on a regional basis, detailed end-use information). However, particularly if one looks at biomass for energy, the aspect of geographical distribution can be extremely important, and sometimes decisive in the assessment of the supply and demand situation. Whereas the national overall energy balance may show a satisfactory picture, the concentration of the population and the locality and accessibility of the energy of some regions may not coincide. Transportation over long distances may not be feasible because of the low bulk densities of fuels involved. A de facto scarcity could thus only be detected by regionally disaggregating the available data. Such an approach, together with demographic and socio-economic information (e.g., population densities, land tenure rights, transportation facilities), could contribute substantially to a meaningful analysis and projection of the biomass supply and demand.

Although both top-down and bottom-up concepts can be used for compiling regional balance, the latter is better suited to an energy planning approach that wants to consider local characteristics.

### 5.4.2   Energy Balance Format

When international organizations started to consider the compilation of energy balances for their member countries in a standardized format using common energy units such as joules, it was thought that it would be a big step forward if simple and highly aggregated balances could be produced. The current state-of-the-art energy balance format is that of the UN Statistics Division, set out in Table 5.2. It is currently being generated for more than 46 developing countries and areas around the world.

Although the quality of currently available data remains the principal concern, the need for refining and extending internationally used formats is generally recognized. The energy balance format for Bangladesh makes further provisions to disaggregate energy data for biomass in the UN format. The Bangladesh energy balance format is set out in Table 5.3. This energy balance incorporates

**TABLE 5.2**
**UN Statistical Office Energy Balance Sheet**

| Commodities transactions | Hard coal, lignite and peat | Briquettes and cokes | Crude petroleum and NGL | Light petroleum products | Heavy petroleum products | Other petroleum products | LPG and other petroleum gases | Natural gases |
|---|---|---|---|---|---|---|---|---|
| 1. Production of primary energy | | | | | | | | |
| 2. Imports | | | | | | | | |
| 3. Exports | | | | | | | | |
| 4. Marine/ aviation bunkers | | | | | | | | |
| 5. Stock change | | | | | | | | |
| 6. Total energy requirement | | | | | | | | |
| 7. Energy converted | | | | | | | | |
| 8. Briquetting plants | | | | | | | | |
| 9. Coke ovens and coke plants | | | | | | | | |
| 10. Gasworks | | | | | | | | |
| 11. Blast furnaces | | | | | | | | |
| 12. Petroleum refineries | | | | | | | | |
| 13. NGL processing plants | | | | | | | | |
| 14. Electric power plants | | | | | | | | |
| 15. Heating plants | | | | | | | | |
| 16. Other conversion industries | | | | | | | | |
| 17. Net transfers | | | | | | | | |
| 18. Consumption by the energy sector | | | | | | | | |
| 19. Losses in transport and distribution | | | | | | | | |
| 20. Cons. for non-energy uses | | | | | | | | |
| 21. Statistical differences | | | | | | | | |
| 22. Final consumption | | | | | | | | |
| 23. By industry and construction | | | | | | | | |
| 24. Iron and steel industry | | | | | | | | |
| 25. Chemical industry | | | | | | | | |

| Derived gases | Nuclear, hydro and geothermal electricity | | | Electricity | Total commercial energy | | Traditional fuel | Total energy | |
|---|---|---|---|---|---|---|---|---|---|
| | Conventional fuel equivalent | Physical energy input | | | Conventional fuel equivalent | Physical energy input | | Conventional fuel equivalent | Physical energy input |

(*Continued*)

**TABLE 5.2 (Continued)**
**UN Statistical Office Energy Balance Sheet**

| Commodities transactions | Hard coal, lignite and peat | Briquettes and cokes | Crude petroleum and NGL | Light petroleum products | Heavy petroleum products | Other petroleum products | LPG and other petroleum gases | Natural gases |
|---|---|---|---|---|---|---|---|---|
| 26. Other industry and construction | | | | | | | | |
| 27. By transport | | | | | | | | |
| 28. Road | | | | | | | | |
| 29. Rail | | | | | | | | |
| 30. Air | | | | | | | | |
| 31. Inland and coastal waterways | | | | | | | | |
| 32. By households and other cons. | | | | | | | | |
| 33. Households | | | | | | | | |
| 34. Agriculture | | | | | | | | |
| 35. Other consumers | | | | | | | | |

a more detailed representation of biomass fuels, and the right half of the energy balance sheet covers the different categories of biomass fuels used in Bangladesh. These biomass fuels are essentially agricultural residues, tree residues, fuelwood and dung.

Future energy demand and supply and energy balance need to be modeled for energy planning. Figure 5.9 shows the reference energy system representation of the energy balance derived from the energy supply model and energy demand model presented earlier in subsections 5.2.1.2 And 5.3.1.

## 5.5 ENERGY PRICE

In the past, the energy prices in many countries were determined on the basis of socio-economic considerations and revenues from the sales of electricity to meet operating and investment costs. More recently, there has been an increasing emphasis on and interest in economic principles for more efficient consumption and production of energy (Munasinghe, 1990). Muunasinghe and Meier (2008) have reported an economic framework for the efficient pricing of electricity.

The earliest works on the system dynamics modeling of price are the modeling of the price of agricultural commodities by Meadows (1970) based on the concept of expectation adjustment introduced by Nerlove (1958), and the modeling of the

| Derived gases | Nuclear, hydro and geothermal electricity | | | Total commercial energy | | | | Total energy | |
| --- | --- | --- | --- | --- | --- | --- | --- | --- | --- |
| | Conventional fuel equivalent | Physical energy input | Electricity | Conventional fuel equivalent | Physical energy input | Traditional fuel | Conventional fuel equivalent | Physical energy input |

price of energy by Sterman (1981) based on price adjustment. Jäger et al. (2009) also reported a system dynamics model of the German electricity market. Energy price models are concerned with non-linear interactions between energy prices, energy supply and energy demand. The price of energy is adjusted dynamically to suit the expectations and demand–supply balance (Mutingia et al., 2017).

### 5.5.1 PRICE BASED ON EXPECTATIONS

Price setting is one of the most difficult formulation challenges in economic modeling (Sterman, 2000). In many economic models, price P is formulated as an equilibrium price $P_e$ and adjusted by a function of a current demand–supply balance.

$$P = P_e \times f(\text{demand supply}) \qquad (5.3)$$

Evidence suggests that expectations about prices are strongly conditioned by past prices and can be modeled by some form of expectations such as exponential smoothing (Sterman, 2000):

$$P_e = \text{SMOOTH}(\text{price, Expection adjustment time}) \qquad (5.4)$$

**TABLE 5.3**
**Total Energy Balance of Bangladesh (in Peta Joules)**

| Description | Commercial energy | | | | | | Biomass fuels | | | | | Total energy |
|---|---|---|---|---|---|---|---|---|---|---|---|---|
| | Natural gas | Crude oil | Petroleum product | Coal | Electricity | Total commercial fuels | Agric residues | Tree residues | Fuelwood | Dung | Total biomass | |
| **I. SUPPLY** | | | | | | | | | | | | |
| Primary | 163.4 | x | 2.7 | x | 3.3 | 169.4 | 316.6 | 22.5 | 88.2 | 71.7 | 499.0 | 668.4 |
| Production | x | 53.4 | 48.0 | 12.3 | x | 113.7 | x | x | x | x | x | 113.7 |
| Imports | x | x | -6.3 | x | x | -6.3 | x | x | x | x | x | -6.38 |
| Exports | x | -5.9 | -6.8 | 0.1 | x | -12.6 | x | x | x | x | x | -12.6 |
| Stock exchange | | | | | | | | | | | | |
| Total primary supply | 163.4 | 47.5 | 37.6 | 12.4 | 3.3 | 264.2 | 316.6 | 22.5 | 88.2 | 71.7 | 499.0 | 763.2 |
| Primary percent | 21.4 | 6.2 | 4.9 | 1.6 | 0.4 | 34.5 | 41.5 | 2.9 | 11.6 | 9.4 | 65.4 | 99.9 |
| **II. Transformation** | | | | | | | | | | | | |
| Refinery | -1.0 | -47.5 | 44.1 | x | x | -4.4 | x | x | x | x | x | -4.4 |
| Thermal power | -69.3 | x | -8.8 | x | 24.4 | -53.7 | x | x | x | x | x | -53.7 |
| Losses and own use | -9.9 | x | -4.0 | x | -8.3 | -22.2 | x | x | x | x | x | -22.2 |
| Total final supply | 83.2 | x | 68.9 | 12.4 | 19.4 | 183.9 | 316.6 | 22.5 | 88.2 | 71.7 | 499.0 | 682.9 |
| **III. Consumption** | | | | | | | | | | | | |
| 1. Domestic | 9.3 | x | 23.6 | x | 4.9 | 37.8 | 243.0 | 22.5 | 67.3 | 71.7 | 404.5 | 442.3 |
| 2. Industrial | 14.0 | x | 7.0 | 9.5 | 10.0 | 40.5 | 73.6 | x | 19.1 | x | 92.7 | 133.2 |
| 3. Commercial | 3.1 | x | x | 0.4 | 3.6 | 7.1 | x | x | 1.8 | x | 1.8 | 8.9 |
| 4. Transport | x | x | 25.0 | 2.5 | x | 27.5 | x | x | x | x | x | 27.5 |
| 5. Agriculture | x | x | 11.0 | x | 0.9 | 11.9 | x | x | x | x | x | 11.9 |
| 6. Others | x | x | 0.3 | x | x | 0.3 | x | x | x | x | x | 0.3 |
| 7. Non-energy use | 56.8 | x | 2.0 | x | x | 58.8 | x | x | x | x | x | 58.8 |
| Total final Consumption | 83.2 | x | 68.9 | 12.4 | 19.4 | 183.9 | 316.6 | 22.5 | 88.2 | 71.7 | 499.0 | 682.9 |
| Final energy percent | 12.2 | x | 10.1 | 1.8 | 2.8 | 26.9 | 46.4 | 3.3 | 12.9 | 10.5 | 73.1 | 100.0 |

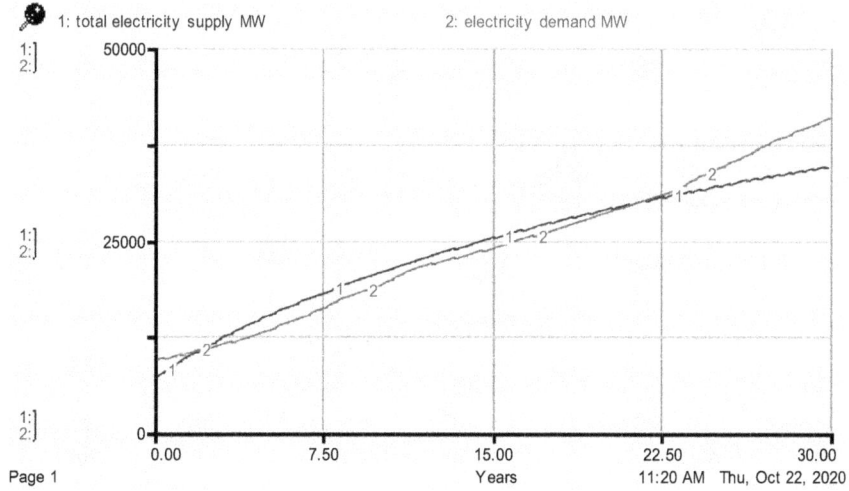

**FIGURE 5.9**   Simulated total energy supply and total energy demand.

## 5.5.2   PRICE BASED ON PRICE MARKUP AND COST

Consider the simple model that links price, investment and consumption (Munasinghe and Meier, 2008):

$$Q_d = \alpha \lambda^\beta \tag{5.5}$$

$$I = \lambda (Q_d - Q_0)^\delta \tag{5.6}$$

$$\lambda = \sigma \frac{I}{Q_d} + \pi \tag{5.7}$$

where $\lambda$ = price
$\quad Q_d$ = consumption
$\quad I$ = investment
$\quad$ and $\alpha$, $\beta$, $\lambda$, $\delta$, $\sigma$, $\pi$ and $Q_0$ are constants.

Since investment requirement is a function of the growth in demand, supply requirement may be assumed to be proportional to the investment requirement. Equation (5.7) can be expressed as:

$$\lambda = \sigma \frac{k_1 Q_s}{Q_d} + \pi \tag{5.8}$$

The first term of the right-hand side of Equation (5.8) is the price markup, and the second term is the cost. The price markup $\lambda_{markup}$ can be rewritten as:

$$\lambda_{markup} = k_2 \frac{1}{\left(\dfrac{Q_d}{Q_s}\right)} \tag{5.9}$$

For reference price markup $\lambda_{markup\ reference}$, the ratio of price markup to price markup reference can be expressed as:

$$\frac{\lambda_{markup}}{\lambda_{markup\ reference}} = k \frac{1}{\left(\dfrac{Q_d}{Q_s}\right)} \tag{5.10}$$

Equation (5.10) shows that the ratio of the price markup $\lambda_{markup}$ to price markup reference $\lambda_{markup\ reference}$ is related to the ratio of demand $Q_d$ to supply $Q_s$ non-linearly.

### 5.5.3 MODELING OF ENERGY PRICE

Two approaches to price determination have been discussed in earlier subsections. Here, the system dynamics models of price based on these two approaches are formulated in the form of causal loop diagrams, and a model based on price markup expectations and cost has been simulated for electricity price.

#### 5.5.3.1 Indicated Price Model

Price determination consists of two feedback loops: a price adjustment loop and a price discovery loop (Sterman, 2000). The price adjusts to the indicated price forming the negative feedback price adjustment loop, but the indicated price based on the current price forms a positive feedback discovery loop. The indicated price is computed from the price and the effect of demand–supply balance on price. The effect of demand–supply balance on price is computed from the following relationship (Sterman, 2000):

$$\text{Demand supply effect} = \left(\frac{\text{Demand}}{\text{Supply}}\right)^s \tag{5.11}$$

where $s > 0$

The causal loop diagram of a price determination model is shown in Figure 5.10. There are two feedback loops. Price and change in price form the balancing adjustment feedback loop B1, and price, indicated price and change price form the re-enforcing discovery adjustment feedback loop R1 (Sterman, 2000). These two loops simulate the dynamics of price. Several studies have been reported on

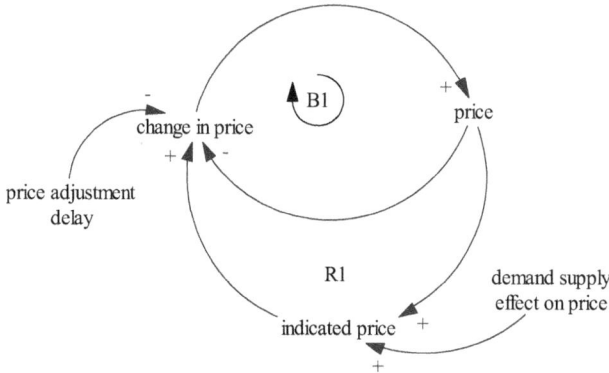

**FIGURE 5.10**   Causal loop diagram of price determination model.

the determination of the price of palm oil and electricity based on this approach (Abdulla et al., 2014; Sahara et al., 2015; Ford, 1997).

### 5.5.3.2   Indicated Price Markup Model

As discussed in subsection 5.5.2, the price of energy can be determined by the cost plus the price markup, which is affected by the demand–supply balance, and this influences the energy consumption. In this approach, the price determination consists of two feedback loops: a price markup adjustment loop and a price markup discovery loop (Bala et al., 2017). The price markup adjusts to the indicated price markup forming the negative feedback price markup adjustment loop, but the indicated price markup based on the current price markup forms a positive feedback discovery loop. The indicated price markup is computed from the price markup and the effect of demand–supply balance on price markup. The effect of demand–supply balance on price markup is computed from the following relationship (Sterman, 2000):

$$\text{Demand supply effect} = \left(\frac{\text{Demand}}{\text{Supply}}\right)^{s} \tag{5.12}$$

where s>0

Finally, the price should not be less than the cost, which is given as:

$$\text{Price} = \text{MAX}(\text{cost},(1 + \text{price makup}) \times \text{cost}) \tag{5.13}$$

Figure 5.11 shows the price determination model from price markup, including demand and supply ratio and cost. Price markup forms a negative feedback loop B1 with the change in price markup and indicated price markup, and the change in price

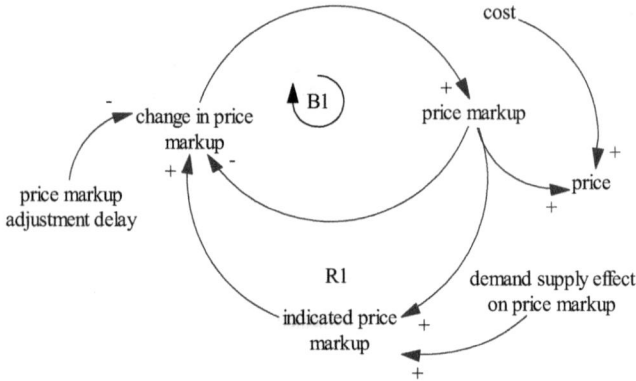

**FIGURE 5.11**   Causal loop diagram of price determination from price markup and cost.

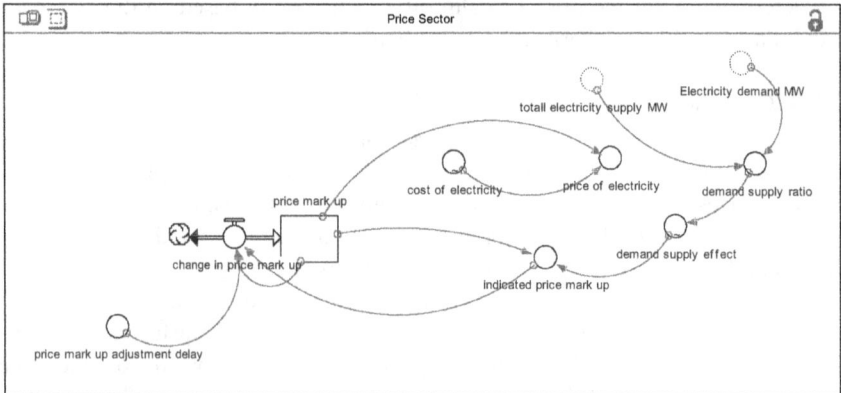

**FIGURE 5.12**   Stock–flow diagram of price determination model.

markup forms a positive feedback loop R1 (Bala et al., 2017). This price markup model is identical to the price model of Sterman (2000). Several studies have been reported on the determination of the price of perishable fruits and vegetables and rice based on this approach (Bala et al., 2017; Teimoury et al., 2013).

A stock–flow diagram of the price determination from price markup and cost is shown in Figure 5.12.

The fundamental equations that correspond to the major variables shown in Figure 5.12 are as follows:

Price markup is increased by change in price markup and is expressed as:

price_markup(t) = price_markup(t - dt) + (change_in_price_markup) * dt
INIT price_markup = 0.1

Change in price markup is computed from indicated price markup, price markup and price markup adjustment delay and is expressed as:

## INFLOWS
change_in_price_markup = (indicated_price_markup - price_markup)/price_markup_adjustment_delay
price_markup_adjustment_delay = 2

Indicated price markup is computed from price markup and demand–supply effect on price markup and is expressed as:

indicated_price_markup = price_markup * demand_supply_effect

Demand–supply effect on price markup is a function of demand–supply ratio and is expressed graphically as:

demand_supply_effect = GRAPH(demand_supply_ratio)
(0.00, 0.588), (0.5, 0.8), (1.00, 1.00), (1.50, 1.25), (2.00, 1.45), (2.50, 1.65), (3.00, 1.85), (3.50, 2.10), (4.00, 2.38), (4.50, 2.63), (5.00, 2.95)

Demand–supply ratio is simply the ratio of electricity demand to electricity supply and is expressed as:

demand_supply_ratio = electricity_demand_MW/total_electricity_supply_MW

Price of electricity should not be less than the electricity cost, and hence price of electricity is computed as the maximum of either cost of electricity or (1 + price markup) × cost of electricity and is expressed as:

price_of_electricity = MAX(cost_of_electricity,(1 + price_markup) * cost_of_electricity)

Cost of electricity increases with time and is expressed graphically as:

cost_of_electricity = GRAPH(TIME)
(0.00, 32.5), (3.00, 33.0), (6.00, 33.0), (9.00, 33.5), (12.0, 34.0), (15.0, 35.5), (18.0, 36.0), (21.0, 36.5), (24.0, 38.0), (27.0, 39.5), (30.0, 41.0)

Figure 5.13 shows simulated total energy supply, total energy demand and price. The price increases with time mainly due to the increase in cost. But in the later period (after 22–30 years), the demand–supply deficit causes the price to increase at a relatively higher rate.

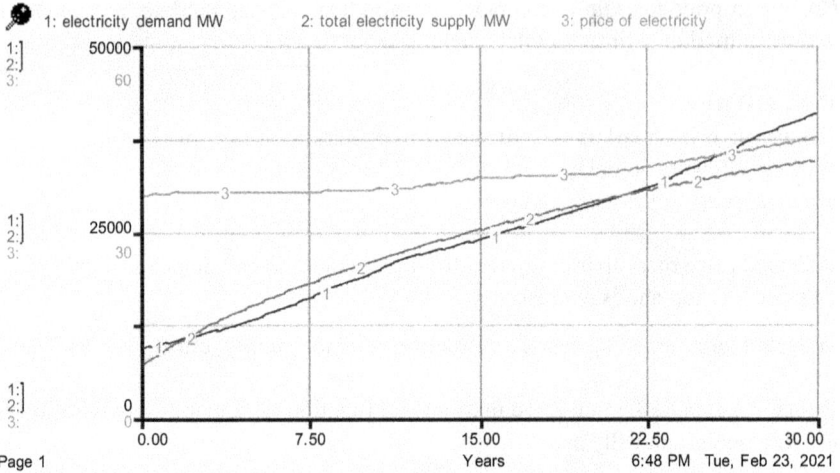

1: electricity demand MW    2: total electricity supply MW    3: price of electricity

**FIGURE 5.13** Simulated total energy supply, total energy demand and price.

## REFERENCES

Abdulla, I., Arshad, F. M., Bala, B. K., Noh, K. M., & Tasrif, M. (2014). Impact of CPO export duties on Malaysian palm oil industry. *American Journal of Applied Sciences*, 11(8): 1301–1309.

Adhiswara, R., Abdullah, A. G., & Mulyadi, Y. (2019). Long-term electrical consumption forecasting using artificial neural network (ANN). *Journal of Physics Conference Series*. 1402 033081.

Akhwanzada, S. A, & Tahar. R. M. (2012). Strategic forecasting of electricity demand using system dynamics approach. *International Journal of Environmental Science and Development*. 3(4): 328–333.

Bala, B. K. (1998). *Energy and Environment: Modeling and Simulation*. Nova Science Publishers, USA.

Bala, B. K. (2006). Computer modelling of energy and of environment for Bangladesh. *International Agricultural Engineering Journal*. 15: 151–160.

Bala, B. K., Alam, M. S., & Debnath, N. (2014). Energy perspective of climate change: The case of Bangladesh. Strategic Planning for Energy and the *Environment*. 33(3): 6–22.

Bala, B. K., Bhuiyan, M. G .K., Alam, M. M., Fatimah, M. A., Sidique, S. F., & Alias, E. F. (2017). Modeling of supply chain of rice in Bangladesh. *International Journal of Systems Science: Operations and Logistics*. 4(2): 181–197.

Bala, B. K., Fatimah, M. A., & Kushairi, M. N. (2017). *System Dynamics: Modelling and Simulation*. Springer.

Ford, A. (1997). System dynamics and the electric power industry. *System Dynamics Review*, 13(1): 57–85.

Hecht-Nielsen, R. (1989). *Theory of the backpropagation network*. Proceedings of International 1989 Joint Conference on Neural Networks. IEEE. Washington DC, pp.593–605

Jäger, T., Schmidt, S., & Karl, U. (2009). A system dynamics model for the German electricity market – model development and application. *European Institute for Energy Research (EIFER)*, Karlsruhe, Germany. 1–45.

Lazarus, M., Hipel, D. V., Hill, D., & Margolis, R. (1995). A guide to environmental analysis for energy planners. *Stockholm Environmental Institute*, Boston.

Meadows, D. L. (1970). *Dynamics of Commodity Production Cycles*. Wright-Allen Press Inc., Cambridge, MA, USA.

Meier, P. (1986). *Energy Planning in Developing Countries: An Introduction to Analytical Methods*. Westview Press, Boulder, Co, USA.

Mutingi, M., Mbohwa, C., & Kommula, V. P. (2017). System dynamics approaches to energy policy modelling and simulation. *Energy Procedia*. 121:530–539.

Munasinghe, M. (1990). *Energy Analysis and Policy*. Butterworths, London, UK.

Munasinghe, M., & Meier, P. (2008). *Energy Policy Analysis and Modelling*. Cambridge University Press, Cambridge, UK.

Nail, R. F. (1992). A system dynamics model for national energy policy planning. *System Dynamics Review*. 8: 1–19.

Nerlove, M. (1958). Adaptive expectations and cobweb phenomena. *Quarterly Journal of Economics*. 72: 227–240.

Panklib, T., Prakasvudhisarn, C., & Khummongkol, D. (2015). Electricity consumption forecasting in Thailand using an artificial neural network and multiple linear regression. *Energy Sources, Part B*, 10:427–434.

Sahara, M., Fatimah, M. A., Bala, B. K., & Abdulla, I. (2015). System dynamics analysis of the determinants of the Malaysian palm oil price. *American Journal of Applied Sciences*, 12(5): 355–362.

Song, H., Chen, Y., Zhou, N., & Chen, G. (2020). *Electricity Consumption Forecasting for Smart Grid using the Multi-Factor Back-Propagation Neural Network*. https://arxiv.org/abs/1902.10823 accessed on 30 December 2020.

Sterman, J. D. (1981). The energy transition and the economy: A system dynamics approach. *PhD thesis. MIT*, USA.

Sterman, J. D. (2000). *Business Dynamics, Systems Thinking and Modeling for a Complex World*. McGraw-Hill Higher Education, Boston, USA.

Teimoury, E., Nedaei, H., Ansari, S., & Sabbaghi, M. (2013). A multi-objective analysis for import quota policy making in a perishable fruit and vegetable supply chain: A system dynamics approach. *Computers and Electronics in Agriculture*. 93: 37–45.

Vogstad, K. (2004). A system dynamics analysis of the Nordic electricity market: The transition from fossil fuelled toward a renewable supply within a liberalised electricity market. *PhD thesis. Norwegian University of Science and Technology*.

Wasserman, P. D. (1989). *Neural Computing: Theory and Practice*. Van Nostrand Reinhold, New York.

Zhang, Q., Yang, S. X., Mittal, G. S., & Yi, S. (2002). Prediction performance indices and optimal parameters of rough rice drying using neural networks. *Biosystems Engineering*. 83(3): 281–290

Zhao, II. X., & Magoulès, F. (2012). A review on the prediction of building energy consumption. *Renewable and Sustainable Energy Reviews* 16(6): 3586–3592

# 6  Energy Use and Environmental Impact

## 6.1  INTRODUCTION

Energy use and production can be major sources of serious environmental impact. This impact, in turn, can threaten the overall social and economic development of energy utilization (Bala, 1998). At regional and global levels, fossil fuel consumption leads to acid rain and, most likely, to global warming, and both of these can disrupt natural systems and economic activity. At a local level, the major share of energy consumption in developing countries is traditional biomass, and this places added stress on woodlands and farmlands, further contributing to soil erosion and habitat loss, and can lead to high levels of indoor pollution.

Energy-environmental issues can be grouped by their relative geographic scale of impact into global, regional and local. The impacts with a global scale originate from local activities and affect global conditions. These issues include global climate change from greenhouse gas emissions, stratospheric ozone depletion and habitat destruction with associated reduction of global biodiversity. Some energy use and production activities result in emissions and lead to damage to conditions ten or hundreds of kilometers away. We refer to these as regional issues. Finally, local issues refer to situations where impacts, in general, occur at or near the site of energy use or production. These include urban and indoor air pollution, ground water contamination and solid waste production.

## 6.2  GLOBAL ISSUES

Some environmental issues have a widespread impact that can be sufficiently far removed from their sources, and these are climate change, sometimes referred to as the greenhouse effect, the depletion of stratospheric ozone and reduction in biodiversity (Bala, 2003; Lazarus et al., 1995). The first two are occasionally confused; some of the same chemical compounds have a role in both processes, but they pose quite distinct threats.

### 6.2.1  Global Climate Change

Global warming, climate change and the greenhouse effect are common expressions used to describe the threats to human and natural systems resulting from continued emissions of greenhouse gases from human activities. These emissions are changing the composition of the atmosphere at an unprecedented rate. While the complexity of the global climate system makes it difficult to accurately predict the impact of these changes, the evidence from modeling studies by the Intergovernmental Panel on Climate Change (IPCC) indicates that the global mean temperature will increase by 1.5 to 4.5° with a doubling of carbon dioxide concentration relative to pre-industrial levels (IPCC, 1992). Given the current trends in the emissions of greenhouse gases, this doubling, with its attendant increase in global temperature, would likely happen in the middle of the 21st century.

### 6.2.1.1  Greenhouse Effect

The greenhouse effect is a relatively well-understood natural phenomenon. The earth receives a relatively constant amount of energy from the sun in the form of incoming solar radiation. The atmosphere and surface of the earth reflect some of this radiation, most of which is in the form of visible light, directly back into space, but absorbs the majority. An amount of energy equal to that in the radiation absorbed is ultimately re-emitted to space as thermal (heat) or outgoing radiation, thereby maintaining an energy balance between incoming and outgoing energy. This balance keeps the earth's temperature at an equilibrium level. Figure 6.1 shows the basic mechanisms of the greenhouse effect.

The greenhouse gases in the atmosphere absorb some of the outgoing radiation on its way to space from the surface of the earth. These are principally water vapor ($H_2O$), carbon dioxide ($CO_2$), methane ($CH_4$), nitrous oxide ($N_2O$), and ozone ($O_3$) and these gases together act as a transparent atmospheric blanket that allows sunlight to warm the earth but keeps infrared radiation (heat) from leaving the earth and radiating out to space. Without this atmospheric blanket of greenhouse gases, the equilibrium surface temperature of the earth would be approximately 33°C cooler than today's levels, averaging -18°C rather than +15°C, and making the earth too cold to be habitable. It is this blanketing effect of the atmosphere that is referred to as the greenhouse effect. A greenhouse is a useful analogy; the atmosphere behaves somewhat like the glass pane of a greenhouse, letting in invisible or shortwave radiation but impeding somewhat the exit of thermal energy, thereby increasing the equilibrium temperature inside the greenhouse.

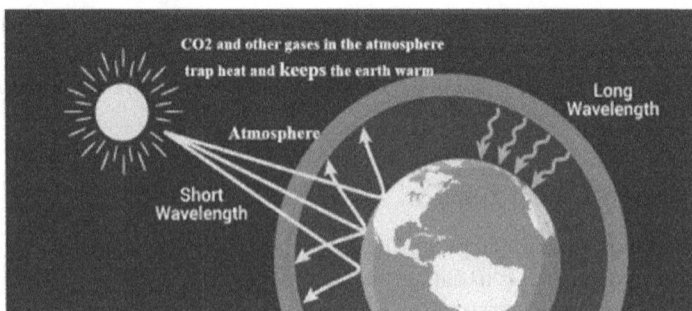

**FIGURE 6.1**   The greenhouse effect (BioNinja, 2021).

The present concern with global warming is not centered on the natural greenhouse effect of the atmosphere on global equilibrium temperature and climate. Rather, it arises from the poles, additional global warming that may occur due to the rapidly increasing concentrations of heat-trapping greenhouse gases. Measurements taken at remote locations around the globe have revealed that current concentrations of greenhouse gases in the atmosphere substantially exceed their pre-industrial levels. The primary human activities that are responsible for this growth in the atmospheric concentrations of these gases are the combustion of fossil fuels and the reduction of carbon stored in biomass through the conversion of forests and other natural land types into settlements, agricultural land, and other uses.

The combustion of all carbon-based fuels, including coal, oil, natural gas and biomass, releases carbon dioxide ($CO_2$) and other greenhouse gases into the atmosphere. Over the past century, emissions of greenhouse gases from a combination of fossil fuel use, deforestation and other sources have increased the effective thickness of the atmospheric blanket by increasing the concentration of greenhouse gases in the troposphere or lower part of the atmosphere. It is this thicker blanket that is thought to be triggering changes in the global climate. Figure 6.2 shows the possible elements of a global temperature increase. Table 6.1 lists the most important greenhouse gases, together with their major sources of current concentrations, and the rate at which they have recently been increasing in the atmosphere, The global warming potential (GWP) values in Table 6.1 are taken from an assessment of the IPCC (IPCC, 1994).

Carbon monoxide (CO), nitrogen oxides ($NO_2$), non-methane hydrocarbons (NMHC) and methane ($CH_4$) are all thought to contribute indirectly to global warming by affecting the atmospheric concentration of other greenhouse gases. Because of an incomplete understanding of the chemical processes involved, these indirect contributions to global warming are more uncertain than the contributions of direct greenhouse gases ($CO_2$, $CH_4$, $N_2O$, CFC).

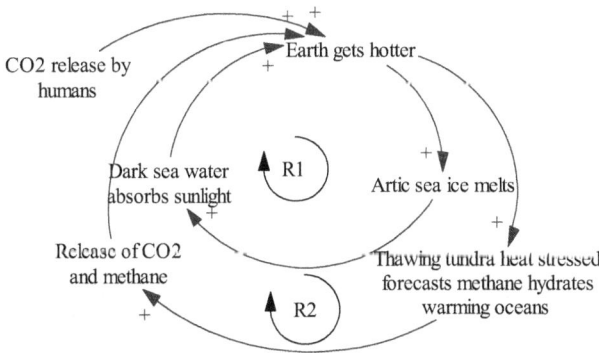

**FIGURE 6.2** Possible elements of a global temperature increase.

**TABLE 6.1**
**Overview of Common Greenhouse Gases**

| Gas | Principal sources | Current rates of emissions and increasing concentration | Global warming potential (100-year time horizon) | Approximate contribution to global warming |
|---|---|---|---|---|
| Carbon dioxide ($CO_2$) | Fossil fuel combustion Deforestation Cement manufacturing | 26,000 Tg/yr emitted 5%/yr increase in concentration | 1 | 84% |
| Methane ($CH_4$) | Natural gas and oil production natural gas pipe lines coal mining fossil fuel combustion (minor) agricultural rice cultivation fermentation and waste disposed | 300 Tg/yr emitted 0.9%/yr increase in concentration | 24.5 | 11% |
| Nitrous oxide ($N_2O$) | Fossil fuel combustion (minor), biomass burning and agriculture (fertilizer use) | 6 Tg/yr emitted and 0.8%/yr increase in concentration | 320 | 5% |
| Chlorofluorocarbons (CFCs, HCFCs) | Industrial uses including refrigerants, foam blowing, solvents and fire retardants | 1 Tg/yr emitted and 4%/yr increase in concentration | 140–12,100 depending on gas | |
| Gases that may have indirect effects on climate change | | | | |
| Carbon monoxide (CO) | Fossil fuel combustion (especially vehicles), biomass burning, including biomass fuel combustion | 200 Tg/yr emitted | | |
| Nitrogen oxide (NOx) | Fossil fuel burning, biomass burning | 66 Tg/yr emitted | | |
| Non-methane hydrocarbons | Fossil fuel combustion solvent use | 20 Tg/yr emitted | | |

Some gases may have cooling effects as well. Recent evidence suggests that the overall role of chlorofluorocarbons (CFCs), once thought to be major contributors to global warming, is no longer clear. By reducing the concentration of stratospheric ozone, CFCs may have a cooling effect, approximately canceling out the direct warming effect of the CFC molecule themselves. The emission of sulfur compounds (such as $SO_2$), which leads to the formation of sulfate aerosols, may have a cooling effect in the Northern Hemisphere. However, this cooling effect, like the sulfate aerosols themselves, is highly localized and relatively short-lived.

The role of water in the atmosphere is complex. Global warming will increase evaporation and thus increase moisture in the atmosphere. Since water vapor acts as a greenhouse gas, this leads to a global warming effect. Increased water vapor will also likely increase cloud formation. However, the net feedback effect of cloud formation is uncertain, depending upon the types of clouds that are formed, since water vapor in clouds can both reflect incoming sunlight and trap outgoing radiation.

Under ideal conditions, the use of biomass fuels will not lead to net $CO_2$ emissions. Plants take up carbon dioxide as they grow to construct organic (carbon-containing) biological molecules that make up the bulk of their (dry) mass. When the biomass in plants is eaten, burned or decomposed, the carbon is released again (in large measure) as $CO_2$ and is restored to the pool of carbon dioxide in the atmosphere. This recycling is part of a natural process called the carbon cycle. If the rate at which the biomass is harvested for fuel is balanced by the rate of biomass growth, then no net $CO_2$ emissions will occur. In cases where biomass is removed but does not (or is not allowed to) grow back, the use of biomass fuels can yield net $CO_2$ emissions. This is the case in instances where fuelwood consumption in a region takes place at a rate faster than forests can grow back or when carbon in the soil is depleted by sub-optimal forestry or agricultural practices.

While there is general (but not total) agreement among atmospheric scientists that increases in atmospheric concentrations of greenhouse gases will result in an average increase in global temperatures, assuming current rates of emissions continue, some uncertainty remains. There are some complicated interactions and feedbacks between the atmosphere, the oceans, the continents and the biosphere. While very powerful and complex climate models now exist for analyzing the interactions between greenhouse gas emissions and the climate, there still remain major scientific uncertainties and modeling challenges yet to be solved.

### 6.2.1.2   Impacts of Global Climate Change

The warming of the earth may, in turn, have numerous secondary effects, some of which have potentially serious impacts on the wellbeing of both humans and the plants and animals with which we share this planet. These effects include a rise in sea levels due to the melting of polar ice, changes in precipitation patterns, and changes in vegetation. The timing and spatial distribution of their effects around the globe are as yet extremely uncertain. The implications of these effects on humans are discussed briefly (Bala, 2003; Lazarus et al., 1995; Parry, 1990).

## Sea Level Rise

Two processes could contribute to sea-level rise: the thermal expansion of the oceans and the melting of polar ice caps, snow cover and glaciers. Many coastal communities, particularly those in large alluvial flood plains and low-lying islands, are particularly vulnerable. Low-lying cities and crop lands may be partially submerged or subject to more frequent flooding by tides and storms, and highly productive coastal ecosystems that humans depend on, particularly estuaries, may suffer a loss in productivity. This loss in productivity is due both to increasing salinity in surface and ground water and to increasing water levels. Furthermore, as highly reflective polar and glacial ice and snow melt to expose less reflective earth or open sea water, the earth becomes a better absorber of solar radiation. This may further increase the rate at which global temperatures rise. Similarly, as the artic tundra warms, methane could be released from methane hydrates and contribute to further warming.

## Increased Climatic Variability and Storm Intensity

Changes in temperature are likely to create changes in wind and precipitation patterns; that is, some places will become wetter, others drier. In addition, the timing of wet and dry seasons may change, and what precipitation does fall may do so in a more concentrated fashion or more gradually. At present, the various climate models are not in full agreement as to which regions will be affected in what way, but given the importance of rainfall patterns on the ability of agriculture to sustain human populations, particularly in heavily populated areas (Southeast Asia, for example), any change in precipitation is of concern. Another prediction of some climate models is that severe storms, including tropical typhoons and hurricanes, will become more frequent or more severe, posing an added threat to coastal and island populations. Finally, changes in precipitation will affect the availability and quality of ground and surface water.

## Changes in Vegetation

Changes in the distribution of plants brought on by changes in temperature and precipitation also have implications for human wellbeing. Some interpretations of climate model results show the grains belts of the northern hemisphere shifting north by hundreds of kilometers. However, it is uncertain whether the full set of contributions to maintain agricultural productivity will remain. Forests, animal habitats and ecosystems as a whole may be unable to tolerate climatic changes and unable to migrate as quickly as climates shift to other areas. At the same time, higher levels of $CO_2$ in the atmosphere increase the rate at which certain plants grow, and higher temperatures would benefit some plant species while being detrimental to others.

### 6.2.2 STRATOSPHERIC OZONE DEPLETION

Encircling the earth is a thin, protective layer of ozone ($O_3$) that screens 99% of the sun's harmful ultraviolet light. The ozone occupies the outer two-thirds of the

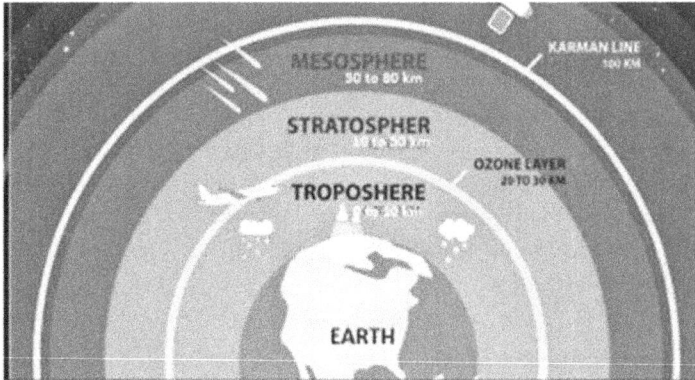

**FIGURE 6.3**   Ozone layer above the earth.

stratosphere, 20 to 50 km above the earth's surface (Figure 6.3). The screening effect of the ozone layer protects all organisms from the damage caused by ultraviolet light.

Depending upon where it occurs, ozone ($O_3$) plays two very different roles in the atmosphere. In the troposphere, where it is produced through the interaction of sunlight, NO, and volatile organic compounds (VOCs), ozone can be a major local and regional pollutant that can cause acute respiratory symptoms and damage to materials, crops and forests. Tropospheric ozone is also a greenhouse gas.

### 6.2.2.1   Extent and Effect of Ozone Layer Depletion

In the stratosphere, however, ozone naturally occurs in much higher concentrations and plays a different beneficial role. The stratospheric ozone layer intercepts much of the ultraviolet (UV) part of the radiation emitted by the sun. Invisible to the human eye, this UV radiation increases the risk of skin cancer, cataracts, and immune system problems in human beings.

In recent years, dramatic reductions in stratospheric ozone concentrations – up to 50% in some polar regions – have been recorded. A smaller reduction of 5–10% has been detected in the middle and upper altitudes, while tropical regions appear to be unaffected thus far. A sustained reduction of 10% in ozone concentrations could lead to a 25% increase in non-melanoma skin cancers and a 7% increase in eye damage from cataracts.

Furthermore, exposure to UV radiation can damage agricultural yields, phytoplankton and terrestrial ecosystems in two general ways. First, UV light causes damage to biological functions in plants and microorganisms, which can result in stunted growth and lower viability. Second, UV radiation can modify the genetic material in plant and animal cells (DNA), resulting in potential damage to cell function and mutations that can influence the viability of seeds, pollen, eggs and sperms. Since some species are more resistant to UV radiation than others, an increase in such radiation could alter the species balance in some ecosystems.

### 6.2.2.2  Activities That May Deplete the Ozone Layer

Several human activities can lead to stratospheric ozone destruction. Chlorofluorocarbons (CFCs) pose the largest and most immediate threat to the ozone layer. Certain types of energy sector equipment, most notably electric refrigerators and air conditioners, contain CFCs, which are also used for purposes such as the cleaning of computer chips, and in foam paste manufacture. Released through leakage or when old appliances are discarded, CFCs rise to the stratosphere where their reaction with sunlight can yield free chlorine, a catalyst for ozone destruction.

### 6.2.3  BIODIVERSITY AND HABITAT LOSS

Biodiversity is perhaps the most discussed ecological issue of recent years. The variety of living organisms may be significantly affected by human activities associated with energy supply and use, particularly those that lead to local or global climatic change or alter land use patterns. The term biodiversity encompasses not only the variety of plant and animal species, including the less visible world of microscopic plants, insects, fungi and bacteria inhabiting a given area, but also the genetic variability inherent among individual organisms within a species, variability that contributes to the ability to adopt to changing circumstances.

Several energy supply options can affect biodiversity. Facility construction and resource extraction can disturb natural land areas and thereby endanger sensitive ecosystems. By their nature, some energy resources – surface-mined coal, biomass, solar, and wind – have particularly large land requirements. However, the extent of actual harm depends on several factors, including the previous land use, the sensitivity of local ecosystems, the reversibility of changes, and the specific practices employed. In some cases, such as degraded land converted into multiple species of biomass plantation, biodiversity might actually be enhanced. Measures such as reclamation can reduce the impact of coal surface mining, but regional flora and fauna may not fully return. Central station solar technologies can cover large desert or other ecosystem areas, while rooftop solar collectors ostensibly sacrifice little in terms of natural habitat. Hydroelectric developments inundate areas behind dams and change the timing and amount of water and sediment flows below them, potentially affecting diversity in aquatic ecosystems and decreasing biodiversity directly or indirectly, adding to the stress on selected plants and animals.

## 6.3  REGIONAL ISSUES

The distinction between global and regional issues is obviously difficult to define precisely. For example, land degradation, a regional issue, is intimately related to a reduction in biodiversity, a global issue. In this section, a group of environmental issues such as land water use and degradation, ocean contamination, mobilization of toxic contaminants, acid deposition and radioactive waste are discussed (Bala, 2003; Lazarus et al., 1995).

## 6.3.1   Land and Water Use and Degradation

Land degradation – soil erosion, deforestation, desertification and so on – is arguably the single most visible and immediate environmental concern in many developing countries. It has many forms and many causes. It is also a controversial issue, where some may view forest clearing as habitat threatening, climate altering and deforestation. Others may see it as an opportunity to exploit natural resources or agricultural productivity. In addition, the question of reversibility is an essential one. Land cleared for charcoal production or other purposes may or may not regenerate, deforestation may or may not be permanent, and soil may or may not be degraded in the process. Land factors such as post-harvest land management and soil and climate characteristics are important determinants of the fate of cleared land. Table 6.2 shows the levels and causes of soil degradation by region. The largest single global cause of soil degradation is overgrazing, accounting for 35% of the global total, as shown in Table 6.2. Fuelwood exploitation, on the other hand, is estimated to account for only 7% of the global degraded area, with greater relative impotence in Africa (13%) and Central America (18%) compared with other regions.

These data concur with the general understanding that fuelwood uses only one, often less important, of many factors that can result in soil or land degradation. Similarly, the collection of subsistence levels of fuelwood is only one of several processes that contribute to the clearing of forested lands; commercial logging and land clearing for agricultural expansion are often major contributors to deforestation. Furthermore, despite indicative aggregate figures, there are no generalizable results or relationships that can be used to estimate the role of fuelwood harvesting on land and soil degradation. While fuelwood scarcity and land degradation problems are often of regional importance, they tend to be dependent on highly localized conditions.

As noted above, another major potential contributor to water and land degradation is water in impoundment for hydroelectric and other purposes. High head hydroelectric dams can flood large areas, forcing resettlement of local human populations and displacing or killing animal and plant populations. By their nature, hydroelectric dams can also change the timing and magnitude of water flows in larger rivers. These can affect downstream and upstream fish populations, the availability of water to sustain aquatic ecosystems, and agriculture and mariculture to recharge ground water and provide for domestic needs. In addition, by changing the quantity of water available, the quality of water can also suffer. Lower water flows can cause an increase in the concentrations of salts, toxic metals, fertilizers, pesticides and herbicides. Hydro projects can also change the amount and timing of sediment flow in rivers, potentially affecting the fertility of farmlands in downstream areas or changing flood patterns. Operation of hydroelectric facilities sometimes involves rapid fluctuations of the amount of water released as the demands for electricity by consumers change over a day, week or year. These fluctuations can cause rapid changes in water levels downstream and in the reservoir behind

**TABLE 6.2**
**Levels and Causes of Soil Degradation by Region**

| Name of continent | Total degraded area (million ha) [% of regretted area] | World fuel exploitation | Land conservation and logging | Overgrazing | Agricultural activity | Industrialization |
|---|---|---|---|---|---|---|
| Africa | 494 [22%] | 13% | 14% | 49% | 24% | |
| Europe | 219 [23%] | | 38% | 23% | 29% | 9% |
| Asia | 747 [20%] | 6% | 40% | 26% | 27% | |
| Oceania | 103 [13%] | | 17% | 80% | 8% | |
| North America | 96 [5%] | | 4% | 30% | 66% | |
| Central America | 63 [25%] | 18% | 22% | 15% | 45% | |
| South America | 243 [14%] | 5% | 41% | 28% | 26% | |
| World | 1964 [17%] | 7% | 30% | 35% | 28% | 1% |

the dam that are disruptive to ecosystems and human activities alike. The large surroundings (lakes) behind hydroelectric dams may also change regional weather patterns by altering the extent, timing and location of evaporation of water.

The large-scale use of biomass energy can also have a regional impact on water and land use. If large tracts of forest lands are cleared to plant fuel crops or for direct use of fuel, erosion can result, affecting ecosystems and human activities downstream. The intensive irrigation of biomass crops may deplete surface and ground water, leaving less water of potentially poorer quality for areas downstream. This may happen in the absence of vegetation if biomass crops that are particularly adept at tapping the water table replace natural vegetation that transpires less water into the atmosphere. Intensive use of fertilizers, herbicides and pesticides in biomass production can also affect downstream water quality, as these will run off or erode soil or leach into ground water.

Other energy sources and technologies also have land water use impacts. Oil and gas production sites can disturb sensitive ecosystems, consume water resources, and produce drilling waste and localized oil spills that can contaminate surface and ground water. Coal mining, particularly surface or strip mining, can degrade large areas. Electric production facilities, particularly those using wind and solar resources, can also require a significant land area.

## 6.3.2 ACID DEPOSITION

Acid deposition results when nitrogen and sulfur oxides ($NO_x$ and $SO_2$) react in the atmosphere with oxygen and water droplets to form nitric and sulfuric acids ($HNO_3$ and $H_2SO_4$). As the water droplets condense, they fall as rain, snow or fog, hence the common name acid rain. We should note that while the acid in rain is the most frequently discussed pathway for these compounds to return to earth, nitrate and sulfate ions ($NO_3$ and $SO_4$) can also combine with positive ions or adhere to the surface of particles in the atmosphere, sometimes falling to earth in a dry form. $SO_x$ and $NO_x$ can also adhere to soil or plants, eventually reacting with water and oxygen to form acids. As a consequence, the terms acid rain and acid precipitation are somewhat incomplete, though a more common term is the broader phenomena of acid deposition (Figure 6.4), the term we use most frequently here.

The standard measure of acidity is the pH scale. pH is equal to the base 10 logarithm of the concentration of hydrogen ions ($H^+$) and is given on a scale of 0 to 14, with low pHs being indicative of highly acid solutions and high pHs being indicative of highly alkaline solutions.

Neutral pH, the pH of distilled water, is 7.0, and physiological pH is the pH most commonly found in plant and animal cells, typically between 6 and 8. In the atmosphere, water reacts with $CO_2$ to form carbonic acid ($H_2CO_3$), a weak acid, and as a consequence, the pH of rain and snow in the absence of all pollutants would be about 5.6. Figure 6.5 shows a pH scale to indicate acid-base level.

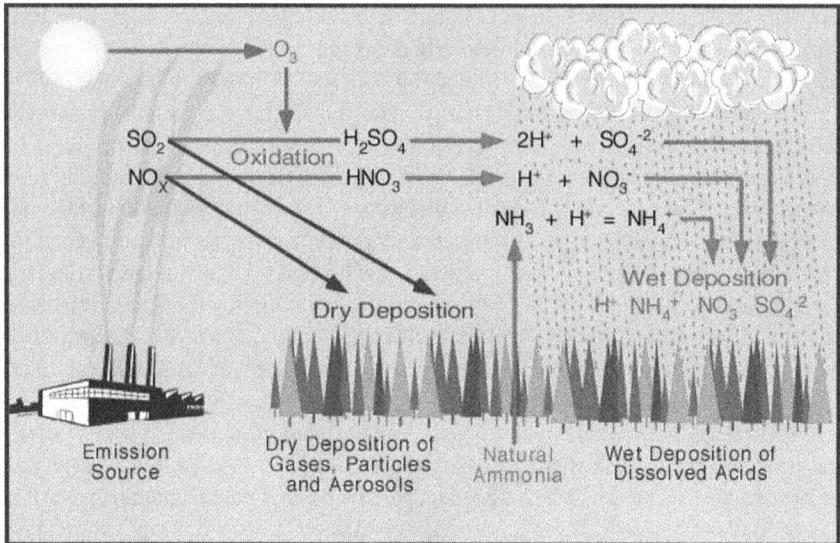

**FIGURE 6.4**  Acid deposition (Pidwirny, 2009).

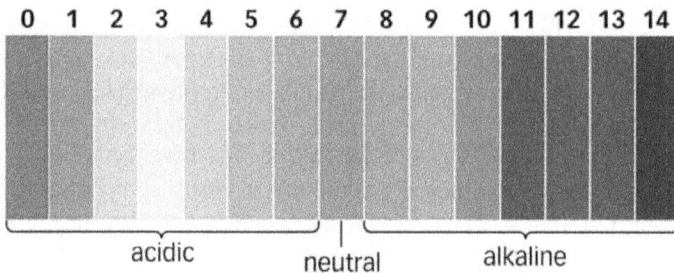

**FIGURE 6.5**  pH scale to indicate the acid-base level (Pidwirny, 2009).

### 6.3.2.1  Impact of Acid Precipitation

The effects of acid rain may vary considerably with the vegetation, soil type and weather condition in a given area. Acid deposition can threaten acid-sensitive plants while encouraging the growth of other acid-loving ones. Under some conditions, the addition of sulfate and nitrate to the soil helps to replace lost nutrients and aids plants growth. In other instances, however, acid deposition can cause lakes and streams to become acid, damage trees and other plants, damage man-made structures and help mobilize toxic compounds naturally present in soil and rocks. Acid rain has been implicated in the death of fish and other aquatic life in otherwise pristine lakes in the northeast United States, southeast Canada, and Scandinavia. Lakes and soils with minimum buffering capacity (the ability to maintain pH in response to the addition of acids), such as many of those found in these areas, are particularly susceptible to acid rain. The lowered pHs in some North American and Scandinavian lakes has resulted in the loss of or shift in species composition of phytoplankton (including algae) that are the base of the aquatic food chain and caused direct and indirect damage to aquatic invertebrates (e.g., insects and small crustaceans), amphibians and fish. The gradual die-off of forests in Germany, Sweden and other areas has also been attributed to the effects of acid deposition. Plants are affected by acid rain in several ways, including direct erosion of cellular structures in leaves; interference with cell processes and the uptake of gases (including $CO_2$) from the atmosphere; alteration of soil chemistry and the activity of bacteria and other microorganisms in soil; interference with plant reproduction; and the weakening of plants susceptible to disease and pests. Buildings and other structures, including many ancient cultural landmarks, are being degraded by acid rain, particularly those structures made of minerals, such as limestone, that are more soluble in more acidic solutions. In soils with limited buffering capacity, the acidified water flowing through the soil can dissolve and mobilize potentially toxic minerals, such as aluminum, leading to elevated concentrations in streams and lakes. A nutrient with small quantities of aluminum can become toxic to fish and other organisms at higher concentrations found in acidified watersheds.

### 6.3.2.2   Sources of Acid Precipitation

While natural sources account for a significant, though uncertain, fraction of the atmospheric sulfur and nitrogen oxides that are the precursors of acid deposition, human sources appear to be the major cause of recent declining trends in the pH of rainfall. While some industrial sources of emissions, particularly the smelting of metals, are important sources of sulfur oxides, the energy sector accounts for a large fraction of these emissions. Sulfur oxides are produced during the combustion of coal, which contains varying amounts (about 0.5 to 5 or more percent) of sulfur, and during the combustion of fuel oil, particularly the heavier grades. These fuels are most commonly used in large industrial facilities and electric power generation. Nitrogen oxides are produced at varying rates by all types of fossil and biomass fuel combustion. The nitrogen in the $NO_x$ produced during combustion is derived both from nitrogen in the fuel and from the molecular nitrogen ($N_2$) that makes up nearly four-fifths of the air we breathe. Gasoline-powered autos and trucks are major emitters of $NO_x$.

Though acid deposition can be a local phenomenon, particularly in urban areas and in areas near large point sources of emissions, the extent to which acid gases are carried by prevailing wind patterns makes acid rain a truly regional issue, one that frequently crosses national boundaries. For example, many of the acidified lakes in eastern North America are hundreds of kilometers from major sources of emissions. Likewise, emissions from as far away as the United Kingdom have contributed to acid rain and forest decline in Scandinavia. Automobile use in southern California is probably a major contributor to low pH rain and snow in the Colorado Rockies, well over 1000 kilometers away.

### 6.3.3   Mobilization of Toxic Contaminants and Bioconcentration

Emissions to the air and water from energy production and use activities can also lead to the mobilization of toxic contaminants, which in turn can have a far-reaching impact. Once introduced into the environment, toxic materials can be transported in a variety of ways (e.g., wind and ground water).

When lead is used as a performance-enhancing additive in automobile fuels, and the leaded fuels are burned, the lead that comes out of the tailpipe can end up blown by the prevailing winds or carried in rivers and in oceans where it can have several effects. Since lead is concentrated in the ocean food chain, the danger of chronic or acute poisoning is acute for large animals (including humans).

The food chain denotes the linkage between predators and prey, producers and consumers. The primary producer takes in solar energy, carbon dioxide, water and nutrients and produces plant biomass. These plants are eaten by animals or microbes called primary consumers, which may be eaten by other animals and so on up to carnivores, the animals at the top of the food chain. Some natural or managed food chains may be very short, e.g., the grass is eaten by deer, and the deer are eaten by tigers. Others may be much longer, as in the ocean, where phytoplankton is eaten by zooplankton, which is in turn eaten by tiny fish, which are eaten by large fish and so on up to the top carnivore, such as a shark or human. No matter what the length of the food chain is, conservation of energy dictates that there is biomass in the order of tenfold less at each level of the chain. This reduction in biomass, which can be several orders of magnitude for a longer food chain, is important because some pollutants are retained in the bodies of the organisms that take them in, then passed on to the organisms higher up the food chain. Thus, a compound that is present at a level too low to cause biological problems, say one part per million (ppm) in a primary producer, may be a thousandfold more concentrated in a top carnivore and may at that level be quite toxic.

The classic example of a bioconcentrated toxin is DDT, which was used extensively as a pesticide in the 1950s and 1960s. This chemical was found at very high concentrations in the blood and eggs of large birds, including brown pelicans and other top carnivores, often causing failure in reproduction or other effects. In the energy sector, other potentially bioconcentrated toxins include pesticides and herbicides sometimes used on biomass crops or to maintain road, powerline or pipeline right of ways, and compounds or metals that are produced, discarded or released during petroleum refining, oil and gas exploration, and geothermal power generation.

### 6.3.4   Ocean Contamination

The most visible and prevalent example of the direct spillage of energy products into oceans is that of oil spills. Crude oil and refined products spill during the routine operation of offshore oil rigs, from oil tanker filling and offloading operations, during the cleaning of tankers, as spillage from other ships that use petroleum fuels and as a result of leakage from undersea pipelines, as well as during less frequent but better-publicized oil tanker accidents and blowouts at offshore oil platforms.

These spills are toxic to many forms of marine life, as well as fouling beaches and affecting other ecosystems and man-made installations along the shoreline. Oil floating on the ocean surface can coat marine birds, making them unable to fly, reducing the insulating properties of their feathers and often eventually killing them. Birds can ingest oil when they try to preen it out of their feathers, and the developing embryos inside eggs can be killed if oil gets on the egg. Oil spills can also disrupt the food chain by killing phytoplankton and zooplankton at or near the ocean surface. Toxic and carcinogenic compounds in oil products can cause death and illness in these organisms or become bioconcentrated in the food chain. Heavier oil products and the heavier fraction of crude oils sink to the bottom where they can coat shellfish, making shellfish and other invertebrates inedible. Damage from oil spills may persist for many years, as compounds stained in oils can remain both in the bodies of organisms and in marine sediments. Oil spills can be spread rapidly by tides, currents and winds, making them a regional as well as local threat.

Hydroelectric projects on major rivers can affect ocean systems by changing the timing and amount of fresh water and sediment flows that reach the ocean. Much of the total production of plant and animal biomass in the ocean takes place near land, and many animals live at least part of their lives in or near fresh water or in estuaries where fresh water from a river mixes with the salt water of the oceans. Changing the timing and amount of fresh water can change environments at the ocean's edge sufficiently as to affect the breeding and growth of organisms there, which can have an impact on the marine ecosystem in the entire region. Too little sediment flow can reduce the fertility of the ocean by reducing the input of needed inputs from the land. Too much sediment flow, on the other hand, can increase the turbidity of the water, resulting in reduced photosynthesis and productivity. Excess sediments can also bury key marine habitats such as shellfish.

### 6.3.5 RADIOACTIVE WASTES

Nuclear energy from fission, and perhaps in the future from fusion, has tremendous theoretical potential as an energy source. As of 1989, 23 countries have used nuclear fission to generate electricity, the combined amount accounting for slightly more than 2% of total global commercial energy production. To date, with the obvious exceptions of nuclear weapon detonations and accidents at the Chernobyl power plant in April 1986, exposers of the public and environment to radioactive pollution have been limited.

Radioactive materials undergo the spontaneous transformation of their atomic nuclei. Nuclei undergoing transformations emit energy (radiation) in the form of alpha particles, beta particles, gamma rays or individual neutrons. Alpha particles are massive and unable to penetrate human skin and are therefore dangerous and capable of causing great damage only if they are emitted inside the human body after ingestion or inhalation. Plutonium-239, for example, is a powerful emitter of alpha particles. Beta particles are much lighter and more capable of penetrating barriers. Like alpha particles, they are most harmful if emitted externally, but externally emitted beta particles can also be dangerous. Gamma rays and neutrons are capable of traveling hundreds of kilometers and penetrating solid walls.

Exposures to radioactivity from the routine operation of a nuclear power station are relatively minor. Releases of tritium and krypton gas during the reprocessing of nuclear fuels are the largest emissions, and these represent only small additions to natural background radiation levels. More serious environmental questions are related to the sustained management of nuclear wastes, the possible proliferation of nuclear weapons and the prevention of catastrophic releases from reactors and reprocessing plants. Radioactive wastes must be safely managed for long periods of time, tens of thousands of years for materials with long half lives. No proposals for the long-term disposal of high-level nuclear wastes have been able to overcome political and technical objections, and the question of what to do with nuclear wastes remains unsolved.

Another consideration in weighing the attributes and problems of nuclear energy is the potential for nuclear weapons; that is, the potential for nuclear materials from power reactors to be diverted for no peaceful uses. All fission reactions produce materials suitable for making nuclear weapons. The amount of fissionable material required to make a weapon depends upon the material's critical mass. For plutonium-239, the critical mass lies in the range of 4 to 8 kilograms, and for uranium-235, the high end of the range is only 25 kilograms. Even the most meticulous national or international accounting and security measures cannot ensure that amounts sufficient for weapon building do not go missing. Even smaller amounts of plutonium-239 could be used, for example, to poison the water supply of a major city.

The promise of nuclear energy has been offset by the threat of potential large-scale environmental disasters. Large-scale radiation releases, eventually resulting in tens, even hundreds or thousands of deaths, widespread property loss and eco-system contamination are improbable but not impossible. More than with other energy technologies, assessing the environmental impact of nuclear energy hinges upon estimating the probabilities and acceptability of rate and catastrophic effects. As a consequence, debate on this issue is likely to continue.

## 6.4   LOCAL ISSUES

Some or all of the issues discussed above could be considered regional as well as local in scale. However, environmental issues whose effects are mostly local are discussed below (Bala, 2003; Lazarus et al., 1995).

### 6.4.1   URBAN AIR POLLUTION

Power plants that burn fossil or biomass fuels emit pollutants that can cause health hazards in the locality of the power plants. These pollutants are carbon monoxide,

$NO_x$, $SO_x$, hydrocarbons and other compounds. Emissions vary with the type of power plant and the fuel used.

The production, distribution and combustion of fossil fuels, particularly combustion in motor vehicles, emit carbon monoxide, $NO_x$, $SO_x$, hydrocarbons and other compounds. In high enough concentrations, the molecules can react with each other, additional compounds already present in the atmosphere and sunlight to yield photochemical smog. If sufficiently severe, smog can be a hazard to human health, livestock and ecosystems, as well as damaging buildings or other structures.

The combustion of wood in wood stoves releases particulate matter, carbon monoxide, hydrocarbons and other pollutants. At most times of the year, this does not cause serious problems in the rural areas where wood is typically burned, especially in developed countries, but under atmospheric conditions known as temperature inversions, which occur relatively frequently in the winter in some temperate localities, wood smoke can be retained within small valleys, allowing pollutants to build up to levels that can exacerbate or cause respiratory problems.

## 6.4.2   INDOOR AIR POLLUTION

Indoor fuel combustion can create elevated concentrations of carbon monoxide, carbon dioxide, sulfur and nitrogen oxides, particulate matter and polycyclic aromatic hydrocarbons. Globally, biomass fuels used for cooking, lighting and heating are the most common source of these emissions. Over half of the world's population depend upon fuelwood, crop residues or dung to meet their daily cooking energy requirements.

The type and level of emissions depend on several factors: fuel type, the design of the end use device, and the nature of the combustion. Smith (1987) notes that while biomass fuels can be burned with minimal smoke production and with the release of few toxic contaminants under normal operating conditions, household-size stoves emit a wide range of pollutants. The resulting indoor pollution concentrations and human exposures depend on additional factors such as dwelling architecture, indoor/outdoor air exchange rate and the direction of an individual's exposures to the emission source. Due to the extensive use of biomass fuels and the close proximity of human receptors, particularly women and children, to the emission sources, resulting doses can be quite high. Smith (1987) calculates that in India, a tonne of particulates emitted by household biomass stoves may produce over 500 times the combined human dose of a tonne of particulates emitted by a coal-fired power plant. In general, biomass fuel use under enclosed conditions can be expected to result in high emissions of and human exposures to carbon monoxide, particulates and hydrocarbons, and there is evidence to validate this expectation.

Much of the evidence on the health effects of indoor air pollution is anecdotal or conflicting, suggesting the need for further research. For two potentially important health categories, birth weight and cardiovascular disease, no studies are available. The conflicting evidence is surprising since the studies of direct health impacts must account for a number of factors such as the delayed effects of chronic exposures and their interaction with other health variables such as smoking and diet.

In countries where biomass fuels are not widely used, indoor air pollution may still be a concern. Emission sources include improperly vented propane or natural gas, cooking and heating appliances, household chemicals, cigarettes and radioactive radon gas. Generally, the most effective approach for controlling these hazards is to identify and minimize or eliminate the pollutant source. In the case of very tightly sealed houses where adequate source control is not possible, the installation of air to air heat exchangers can provide adequate ventilation while minimizing heat loss.

### 6.4.3   Surface and Ground Water Pollution

Water is an essential resource for all life and is required for agriculture, industry and domestic use. It can be polluted by different sources such as agricultural run-off, municipal sewage and energy-related water use. Water pollution may be classified as:

(i)   excess nutrients from sewage and soil erosion, which can cause algae blooms that eventually deplete the oxygen content of water;
(it)   pathogens from sewage that spread disease;
(iii)   heavy metals and synthetic organic compounds from industry, mining and agriculture;
(iv)   thermal pollution, which can alter the chemistry and structure of aquatic ecosystems;
(v)   acidification;
(vi)   suspended and dissolved solids; and
(vii)   radioactive pollution.

Energy production and consumption are most closely associated with the last five types of emissions. The mechanics and consequences of acidification and radioactive pollution are discussed previously in the section regarding regional environmental issues.

At coal mines or coal-burning facilities, water can be polluted through the use of particulate technologies such as coal-washing or from pollutants carried in run-off or as leachates from tailings and storage piles. Mine tailings often release contaminants that were previously bound in impermeable rock formations. The quantity and coal type of contaminants released depends on a number of factors, including the local geology, hydrology and type. Emissions from coal mine tailings commonly include beryllium, cadmium, copper and zinc. Metals released into ground and surface water can be bioconcentrated in food chains or ingested directly by humans and other organisms. The deposition of airborne $SO_2$ and $NO_x$ leads to the acidification of lakes and streams, which has a significant impact on aquatic ecosystems. If mine tailings contain sulfur compounds, acid drainage from piles that are exposed to water can contribute to the acidification problem. Aluminum, cadmium, mercury and lead become more soluble as acidification progresses, resulting in higher levels of mobilized metals being released to ground and surface water.

Ash piles at thermal power plants are another potential source of water pollutants. Leachate and run-off from ash piles commonly contain heavy metals including arsenic, beryllium, cadmium, copper, iron, lead, mercury, nickel and vanadium. The type of contaminants present in each situation depends upon the fuel type, combustion method, and pollution control methods used. Residues from power plant boilers can contaminate water used in maintenance operations with the heavy metals listed above.

Leaching domestic, commercial or industrial fuel storage tanks can contaminate surface or ground water with gasoline, heating oil or petroleum products. Drilling for oil can release brine deposits and damage surrounding fresh water ecosystems. The negative environmental impact of oil spills has been discussed in the section on regional environmental issues. Water withdrawal for mine drainage can cause salt water intrusion into fresh water aquifers. Uranium mines and mills, fuel fabrication plants, nuclear power plants, fuel reprocessing plants, and nuclear waste storage facilities all have the potential to release radioactive materials into ground and surface water.

### 6.4.4   SOLID AND HAZARDOUS WASTES

Energy-related solid wastes are primarily generated by mining and fuel combustion. The largest two sources of such wastes are coal mining and coal-burning power plants, respectively. The mining of copper and other metals used in the transmission and distribution of electricity also produces solid wastes. Other sources of energy-related solid wastes are oil and natural gas drilling, oil-fired power plants and the nuclear fuel cycle, although these are relatively minor sources of solid wastes in comparison to coal technologies.

Solid wastes can also be used for energy. In 15 countries, energy is recovered from the incineration of municipal wastes. The type and quantity of pollutants emitted by waste incineration plants depend on factors such as the composition of the waste materials and the combustion and emission control technologies employed.

### 6.5   POLLUTION AND GLOBAL WARMING

Pollution means either that a poison or destructive agent (such as crude oil) has been added to the environment or that a nutrient or waste product has been added in concentrations at which the impact is destructive. Environmental degradation, on the other hand, means that something useful has been taken away from the environment or destroyed. Erosion of top-quality agricultural soil, deforestation on a watershed or reduction in a species directly in a community are examples.

Pollution may be categorized as air pollution, water pollution, and solid-waste pollution. Emissions of pollutants from energy production and use to air and bodies of water and solid waste emissions are discussed below.

## 6.5.1 AIR EMISSIONS

A polluted atmosphere is generally considered to be an unnatural atmosphere. It may be due to the addition of pollinations from nature or the activities of man. About 90% of air pollution is accounted for with five substances known as primary pollutants: carbon monoxide, oxides of nitrogen, oxides of sulfur, hydrocarbons, and particulate matter. Carbon dioxide, methane, nitrous oxides and CFCs are important greenhouse gases that are considered to affect the climatic change leading to global warming.

## 6.5.2 GLOBAL WARMING

The major greenhouse gases that directly affect global climate are carbon dioxide ($CO_2$), methane ($CH_4$), nitrous oxide ($N_2O$) and chlorofluorocarbons (CFCs). A number of gases may also indirectly affect the global climate. Carbon monoxide (CO), nitrogen oxides ($NO_x$), and non-methane hydrocarbons (NMHC) are all thought to contribute indirectly to global warming by affecting the atmospheric concentration of other greenhouse gases. Some of the major greenhouse gases directly affecting the global climate are discussed below (Bala, 1998; Rao, 1992; Kapoor, 1989; Pandey and Cerney, 1989).

**Carbon dioxide**
Carbon dioxide ($CO_2$) is a major greenhouse gas both in terms of quantity emitted and overall effect on global warming and is released whenever a fuel that contains carbon is combusted or oxidized. It is released in quantities generally proportional to the carbon content of the fuel. $CO_2$ emission factors are principally estimated based on fuel carbon content. Carbon dioxide is not directly toxic to most plants and animals. Thus, its principal environmental impact is on the climate. $CO_2$ emissions from fossil and biomass fuels are categorized separately as non-biogenic and biogenic carbon dioxide, respectively. As $CO_2$ is taken up and emitted by many terrestrial and oceanic sources and sinks, its lifetime in the atmosphere is difficult to specify but may be in the order of 100 years.

**Non-biogenic (fossil fuel) $CO_2$**
Non-biogenic emissions are those derived from the combustion of fossil fuels and other fuels that are essentially non-renewable. Non-bioorganic emissions constitute a net addition of $CO_2$ to the atmospheric pool of the gas, at least on a human timescale.

**Net Biogenic $CO_2$**
Biogenic emissions of $CO_2$, in contrast, result from biomass combustion and do not constitute net additions of $CO_2$ to the atmosphere under sustainable biomass

harvesting conditions. Under these conditions, the $CO_2$ released upon combustion of biomass-derived fuels can be recaptured during photosynthesis in the next biomass growth cycle. The non-sustainable harvesting of biomass, leading to soil and land degradation and, in extremes, deforestation and desertification, will cause the net addition of $CO_2$.

## Methane ($CH_4$)

Methane ($CH_4$) is emitted as a byproduct of fuel combustion through leakage from natural gas, oil, and coal extraction, transmission and distribution facilities, and from other agricultural and natural sources. In general, fuel combustion is a relatively minor contributor to overall $CH_4$ emissions relative to other sources of the gas. Methane is relatively non-toxic to humans and animals, but in high enough concentrations, it can cause suffocation. Methane is, however, a powerful greenhouse gas contributing to global warming directly and through its interactions with tropospheric ozone and stratospheric water vapor.

## Nitrous oxide ($N_2O$)

Of the six or seven oxides of nitrogen, only three nitrous oxides ($N_2O$), nitric oxide (NO), and nitrogen oxide ($NO_2$) are found in any appreciable quantities in the atmosphere. Nitrous oxide is a colorless, odorless gas and a very powerful greenhouse gas. Although the quantities emitted are subject to large uncertainty, they appear to be a small fraction of total nitrogen oxide emissions. The process of $N_2O$ formation during and after combustion is still only partly understood. Unlike other nitrogen oxides, nitrous oxide has a life span in the atmosphere of approximately 150 years

## 6.5.3 AIR POLLUTION

The major air pollutants are oxides of sulfur, carbon monoxides, oxides of nitrogen, hydrocarbons, and particulates. The minor pollutants are ammonia, sulfur and sulfides, chlorine and hydrogen chloride, chlorinated hydrocarbons, bromides, fluorine and fluorides, mercaptans and metals. Some of the major pollutants are discussed below (Bala, 1998; Rao, 1992; Kapoor, 1989; Pandey and Cerney, 1989).

## Carbon monoxide (CO)

CO constitutes the single largest pollutant in the urban atmosphere. It is colorless, odorless and tasteless, and has a boiling point of 192°C. CO is produced in concentrations that vary widely across different types of combustion devices when carbon-based (both fossil and biomass) fuels are burned. It results when combustion of these fuels is incomplete; that is, when the carbon in a fuel is not completely oxidized to carbon dioxide. As a consequence, emissions of carbon monoxide are primarily a function of combustion conditions; inefficient combustion generally increases CO emissions. Motor vehicles tend to be the major source of CO emissions in most areas, with older vehicles being the primary culprits. Carbon monoxide is created in oxygen-starved, fuel-rich combustion conditions, such as by low-speed and idling vehicles in congested urban areas. Household biomass and coal-burning stoves are also significant sources of CO, while industrial boilers and utility power plants, for example, will produce relatively little CO when operated properly. Carbon monoxide is converted in the atmosphere to $CO_2$ and typically remains in the atmosphere for a few months at most. The emission factors for carbon monoxide are derived from many sources.

Carbon monoxide is a local air pollutant with a respiratory impact and contributes both directly and indirectly to the increase in greenhouse gas concentrations in the atmosphere. Its respiratory impact on human and animal health stems primarily from the ability of the CO molecule to bind to hemoglobin, the oxygen-carrying molecule in the blood, which thereby reduces the supply of oxygen to the brain and other tissues. Since carbon monoxide binds more readily to hemoglobin than oxygen, even relatively low concentrations of CO in the air can lead to carbon monoxide poisoning, which is characterized by headaches, dizziness, and nausea in mild cases, and loss of consciousness and death in acute cases.

### Nitrogen oxides ($NO_x$)

Often NO and $NO_2$ are analyzed together in air and referred to as $NO_x$. Nitric oxide (NO) is a colorless gas produced largely by fuel combustion. It is oxidized to $NO_2$ in a polluted atmosphere through photochemical secondary reactions. Nitrogen dioxide is a pungent brown gas with an irritating odor that can be detected at concentrations of about 0.12 ppm. It absorbs sunlight and initiates a series of photochemical reactions. Small concentrations of $NO_2$ have been detected in the lower stratosphere. $NO_2$ is probably produced by the oxidation of NO by ozone. Nitrogen dioxide is of major concern as a pollutant. It is emitted by fuel combustion and nitric acid plants.

Nitrogen oxides ($NO_x$) can contribute to environmental problems in several ways. Short-term exposure to elevated $NO_2$ concentrations (0.2 to 0.5 ppm) can cause respiratory symptoms among asthmatics. Indoor fuel combustion, particularly from gas stoves or traditional fuel use, can lead to elevated indoor levels that have been associated with increased respiratory illness and decreased resistance among children. Nitrogen oxides contribute to the formation of atmospheric ozone and nitrate aerosols. $NO_x$ species may also have a role in global warming, but the extent of this role is still a subject of debate.

Atmospheric emissions of $NO_x$ contribute to the photochemical smog prevalent in many urban areas, and these have a general detrimental effect on the respiratory health of human beings and other animals, as well as on visibility. In high concentrations, $NO_x$ can injure plants, though the required concentrations usually only exist near a large point source of the pollutant. The major hazard to plants from nitrogen oxide emissions may be the effect of $NO_x$ on ozone formation. Atmospheric nitrogen oxides in high concentrations cause respiration system damage in animals and human beings, and even in relatively low concentrations can cause breathing difficulties and increase the likelihood of respiratory infection, especially in asthmatics and other individuals with preexisting respiratory problems. Emission factors for total nitrogen oxides are available for many of the fuel combustion source categories, and most of these are taken or derived from US Environmental Protection Agency (EPA).

### Oxides of sulfur ($SO_x$)

Pollution from oxides of sulfur consists primarily of two colorless gases, $SO_2$ and $SO_3$. The combustion of any material containing sulfur as an impurity will produce these oxides. The amount of $SO_2$ and $SO_3$ formed depends on temperature and the level of $O_2$ in the combustion mixture. More $SO_2$ is oxidized to $SO_3$ at high temperatures, but the concentration of $SO_3$ in an equilibrium mixture is more at low temperatures. $SO_3$ exists in the atmosphere when moisture exists.

Two source categories of sulfur oxides are total sulfur oxide and sulfur dioxide. Total sulfur oxides cover all of the different species of sulfur oxide emissions, while sulfur dioxide covers only $SO_2$, the major $SO_x$ species related to the air by human activities. Quite often, in emission factor literature, total $SO_x$ emission will be expressed as $SO_2$ equivalents. Total $SO_2$ emission factors are available, and most of these factors are drawn from US EPA documents.

Energy-related sulfur dioxide emissions are generally proportional to the fraction of sulfur in fuels such as coal and crude oil. For some fuels, the fraction of sulfur can exceed 10%. The fuels with the most sulfur are the coals and heavy oils used in electric utility and heavy industrial boilers. When these fuels burn, sulfur combines with oxygen in the combustion air to yield $SO_x$. Metal smelters and other industrial processes are also key sources of $SO_x$ emissions.

Sulfur oxides can react with water and oxygen in the atmosphere to yield sulfuric acid, one of the major components of acid rain. $SO_2$ itself can damage plants, with acute exposure to the gas causing death of part or all of a plant and chronic exposure, though the threshold at which plants are affected varies widely among different plant species. In humans, exposure to $SO_2$ at high levels causes respiratory problems in sensitive individuals. In developing countries and other areas where coal is used as home heating or cooking fuels, $SO_2$ can be an important health hazard as an indoor air pollutant.

### Hydrocarbons

Hydrocarbons comprise a group of organic pollutants released in the atmosphere both by natural and man-made processes. Hydrocarbons are emitted from energy sector activities as either products of incomplete combustion of carbon-based fuels or by the evaporation or leakage of fuels and lubricants from fuel production, transport and storage facilities, or from fuel-using devices. As a class, hydrocarbons contribute to the production of photochemical smog and ground-level ozone, which are dangerous to human health due to their effects on the respiratory system. High ozone levels also damage crops, forests and wildlife.

With the exception of methane, hydrocarbons as a class are likely to contribute indirectly to global warming through their effect on tropospheric ozone concentrations. Different hydrocarbon species have different life spans in the atmosphere, with some chemicals having life spans of hours or days, while other less reactive molecules remain in the atmosphere longer. Total hydrocarbon emission factors are available and are derived from the US EPA.

**Particulate matter**

Small solid particles and liquid droplets are collectively called particulate matter and constitute a serious problem. The size of the particles and the chemical composition are both important from the viewpoint of their effects. The particulates are emitted from combustion processes or are carried into from roads, agricultural activities or during the transport or storage of finely divided solid materials such as crushed oil.

Anyone who has traveled down a dusty road can appreciate the effect of particulate emission on the human upper respiratory system (nose, throat), but small particles can also penetrate deep into the lungs, where they can aggravate existing respiratory problems and increase the susceptibility to cold and other diseases. Particulates can also serve as carriers for other substances, including carcinogens and toxic metals, and in so doing can increase the length of time these substances remain in the body. Particulate matter in the air impairs visibility and views, and particulate matter setting on buildings, clothes, and other humans may increase cleaning costs or damage materials. Particulate matter is a notable indoor air pollution in areas when often poorly vented household cooking and heating equipment is used, particularly with smoky fuels such as wet biomass, crop and animal residues, and low-grade coals. Particulate matter can settle on plants, reducing their growth by reducing their uptake of light and carbon dioxide.

The amount of particulate matter emitted during combustion is a function of fuel type, the amount of non-combustible fuel contaminants such as ash present in the fuel, the firing conditions, and the level of pollution control equipment used. Total suspended particle (TSP) emissions cover a wide range of particle sizes, from those that are nearly visible to the naked eye to particles less than one micron in diameter. The size classification of particulate matter is important: the smaller the particle in general, the longer it will remain in the atmosphere, and the farther it can be dispersed from its source, while particles in smaller size ranges are a more serious concern to human health as, unlike larger particles, they are not filtered out by the upper respiratory system. Total particulate emissions give the mass of all particles emitted per unit of fuel consumed or produced and are derived primarily from US EPA.

### 6.5.4 Water Effluents

Emissions of pollutants to bodies of water are covered by a number of emission categories, though at present, there are relatively few emission factors for these effects. The factors that are presently available are principally derived from US Department of Energy and World Health Organization documents. In general, emissions to water are most likely to originate in energy-transforming installations such as oil refineries, petroleum wells, coal mines or ethanol facilities. Water effluents can also be created by aqueous emissions from the cleaning of large fuel combustion facilities such as boilers. Water effluents are typically measured in mass units with the exception of radioactive emissions, which are measured in units of radiation loading. Brief descriptions of the categories of water emissions are provided below.

Emissions of solids are described in two categories: suspended solids and dissolved solids. Suspended solids are materials that are mixed into water but not dissolved, such as slurries of mud, sand, ash or other particles, while dissolved solids include salts and other materials in solution in effluents and in bodies of water receiving the pollutants. Suspended solids reduce the visibility and penetration of sunlight into water, potentially affecting the behavior of fish and other species, and reducing the productivity of marine and aquatic plants. Depending on the nature of the suspended material, suspended solids can also affect water chemistry, as can dissolved solids. Both types of water emissions can affect humans through their impact on drinking water quality and on the quality of water used for recreation and industrial purposes.

Oxygen demand is a measure of the amount of oxygen that would be required by material present in a water sample. The two measures of oxygen demand are biochemical oxygen demand (BOD) and chemical oxygen demand (COD). Energy sector sources of oxygen-demanding effluents include petroleum refineries and fuel-alcohol production facilities. The effect of effluents that raise BOD and COD is to decrease the amount of oxygen available for aquatic animals by increasing the demand for oxygen by microorganisms in the water that degrade organic matter. In extreme circumstances, as in rivers and lakes heavily polluted with sewage or industrial effluents, the amount of oxygen in the water can fall to zero, resulting in the death of most larger aquatic flora and fauna. Oxygen-demanding effluents can also adversely affect water used for drinking, irrigation, industrial process, fishing and recreation.

Several different types of emission of metals into water include cadmium, chromium, copper, iron, mercury and zinc. Energy sector activities that can produce these emissions include oil refining, oil and coal extraction, and coal processing, though other industrial activities probably release large quantities of these effluents to water bodies compared with energy-related sources.

Salts encompass a number of chemicals, including sulfates, chlorides, nitrates and others that are soluble in water; that is, they dissolve to yield anions and cations in solution. Salts affect aquatic ecosystems in different ways depending on the species present and the concentrations. Some species have rather narrow ranges of salt tolerance and are adversely affected when average concentrations rise above a certain level. As salt concentrations increase, water may be treated or may be unfit for human use, such as irrigation or drinking water.

## 6.5.5   SOLID WASTES

Several different categories of solid waste emissions are provided, but, as with water, there are at present relatively few emission coefficients available for these categories. Most emission factors are derived from US Department of Energy documents.

Mining wastes are solid wastes from mining operations, including such energy-sector activities as the mining and processing of coal and oil shale. Scrubber sludge is an effluent of some concern for coal-fired industrial and electricity generation equipment.

## 6.6  EMISSION FACTORS

Emission factors or coefficients describe the quantity of a pollutant that is released per unit of fuel consumed, produced or lost. In Environmental Database (EDB) terminology, emission factors describe the relationship between an energy activity or source category and an effect category (a specified pollutant). A few examples are given in Table 6.3. Emission factors are measured or estimated numbers that relate effects to sources. Emission factors are thus numbers that allow energy/environmental planners to estimate quantities of emissions and other environmental effects or impacts associated with activities such as tonnes of fuels burned using coal stoves or numbers of barrels of oil passing through a refinery.

Emission factors are determined by empirical measurement or various estimation techniques. Emission factors are typically measured or estimated by researchers working for governmental or international agencies, industrial firms, universities or research institutes.

Data on emission factors are available from different documents and databases. There are several major general sources of emission factors, including those collected by the US EPA, the IPCC, the CORINAIR data assembled by researchers in Europe, and the World Health Organization's Rapid assessment of sources of air, water, and land pollution. The emission factors of some common fossil fuels are given in Table 6.4.

## 6.7  MODELING OF POLLUTION AND GLOBAL WARMING

The pollution and global warming causes considered are $CO_2$ emissions from natural gas, oil and coal-based power plants, and radiation pollution and hazards from

**TABLE 6.3**
**Sample Emission Factor**

| Source | Type of pollution/ climate changer | Effect | Unit | Emission factor |
|---|---|---|---|---|
| Coal-fired stove | Climate change | $CO_2$ | kg/tonne | 2555.0 |
| Gasoline engine | Air pollution | CO | g/tonne | 401.4 |
| Diesel engine | Air pollution | CO | g/tonne | 41.0 |
| Ethanol production | Solid waste pollution | Total | g/lit | 38.1 |

**TABLE 6.4**
**Emission Factors for the Generation of Electricity from Different Sources of Fuels**

| Type of power plant | Emission factor Tonnes $CO_2$/GWh | Source |
|---|---|---|
| Coal | 888 | WNA Report (2020) |
| Oil | 735 | WNA Report (2020) |
| Natural gas | 500 | WNA Report (2020) |
| Hydroelectric | 26 | WNA Report (2020) |
| Nuclear | 28 | WNA Report (2020) |
| Solar PV | 85 | WNA Report (2020) |
| Wind | 26 | WNA Report (2020) |
| Biomass | 45 | WNA Report (2020) |

a nuclear power plant. The generation of electricity from the power plants using natural gas, oil and coal emits $CO_2$, and the emissions are computed by multiplying the amount of electricity generated by a fossil fuel by the corresponding emission factor. The radiation pollution level emitted by a nuclear power plant is about 0.035 microsieverts per hour (Chia et al., 2015). The thermal pollution is set to a value of 1 to indicate the expected impact of thermal pollution if there is a power plan, otherwise it is 0. The radiation harmful index is computed from the relation:

$$\text{Radiation harmful index n} = (\text{Radiation pollution} \times 0.001 \times 24 \times 365)/100 \tag{5.1}$$

The modeling of pollution and global warming consists of two sectors: emission sector and radiation sector. Emission sector includes electricity generation from natural gas-based power plant, oil-based power plant and coal-based power plant, and their $CO_2$ emissions, while radiation sector includes electricity generation from nuclear power plant and radiation pollution and radiation hazards.

**Emission Sector**
A stock–flow diagram of electricity generation from natural gas, oil and coal-based power plants and their contributions to pollution and global warming is shown in Figure 6.6.

The fundamental equations that correspond to the major variables shown in Figure 6.6 are as follows:

**Natural gas power plant**
The natural gas power plant is increased by its construction rate and reduced by its capacity discarded.

natural_gas_installed(t) = natural_gas_installed(t - dt) + (natural_construction_rate - natural_gas_discard_rate) * dt
INIT natural_gas_installed = 7500

The construction rate of natural gas power plant has a time delay called construction delay, and the construction rate is computed from natural gas power plant under construction and natural gas power plant construction delay and is computed as:

**INFLOWS**
natural_construction_rate = natural_gas_under_construction/natural_gas_construction_delay
natural_gas_construction_delay = 3.5

The discard rate of natural gas power plant is computed from natural gas power plant installed and discard fraction.

**OUTFLOWS**
natural_gas_discard_rate = natural_gas_installed * natural_gas_discard_factor
natural_gas_discard_factor = 0.001

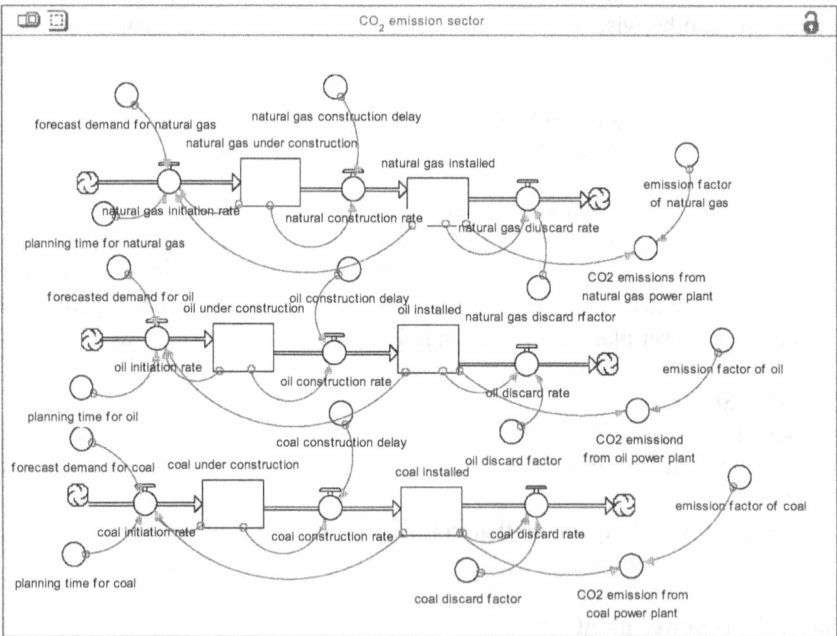

**FIGURE 6.6** Stock–flow diagram of pollution and global warming from natural gas, oil and coal-based power plants.

The natural gas power plant under construction is increased by its construction initiation rate and decreased by construction rate.

natural_gas_under_construction(t) = natural_gas_under_construction(t - dt) + (natural_gas_initiation_rate - natural_construction_rate) * dt
INIT natural_gas_under_construction = 2500

The initiation rate of natural gas power plant depends on forecasted demand, natural gas power plant installed, natural gas power plant under construction and planning time and is computed as:

**INFLOWS**
natural_gas_initiation_rate = (forecast_demand_for_natural_gas - natural_gas_under_construction - natural_gas_installed)/planning_time_for_natural_gas
forecast_demand_for_natural_gas = 25000
planning_time_for_natural_gas = 30

**Oil-based power plant**
The oil-based power plant is increased by its construction rate and reduced by its capacity discarded.

oil_installed(t) = oil_installed(t - dt) + (oil_construction_rate - oil_discard_rate) * dt
INIT oil_installed = 3500

The construction rate of oil-based power plant depends on oil-based power plant under construction and the construction delay and is computed as:

**INFLOWS**
oil_construction_rate = oil_under_construction/oil_construction_delay
oil_construction_delay = 3.5

The discard rate of oil-based power plant is computed from oil-based power plant installed and discard fraction.

**OUTFLOWS**
oil_discard_rate = oil_installed * oil_discard_factor
oil_discard_factor = 0.001

The oil-based power plant under construction is increased by its construction initiation rate and decreased by construction rate.

oil_under_construction(t) = oil_under_construction(t - dt) + (oil_initiation_rate - oil_construction_rate) * dt
INIT oil_under_construction = 1500

The initiation rate of oil-based power plant depends on forecasted demand, oil-based power plant installed, oil-based power plant under construction and planning time and it is computed as:

**INFLOWS**
oil_initiation_rate = (forecasted_demand_for_oil - oil_under_construction - oil_installed)/planning_time_for_oil
forecasted_demand_for_oil = 10000
planning_time_for_oil = 30

### Coal-based power plant
The coal-based power plant is increased by its construction rate and reduced by its capacity discarded.

coal_installed(t) = coal_installed(t - dt) + (coal_construction_rate - coal_discard_rate) * dt
INIT coal_installed = 2000

The construction rate of coal-based power plant depends on coal-based power plant under construction and the construction delay and is computed as:

**INFLOWS**
coal_construction_rate = coal_under_construction/coal_construction_delay
coal_construction_delay = 3.5

The discard rate of coal-based power plant is computed from coal-based power plant installed and discard fraction.

**OUTFLOWS**
coal_discard_rate = coal_installed * coal_discard_factor
coal_discard_factor = 0.001

The coal-based power plant under construction is increased by its construction initiation rate and decreased by construction rate.
coal_under_construction(t) = coal_under_construction(t - dt) + (coal_initiation_rate - coal_construction_rate) * dt
INIT coal_under_construction = 1000

The initiation rate of coal-based power plant depends on forecasted demand, coal-based power plant installed, coal-based power plant under construction and planning time and is computed as:

**INFLOWS**
coal_initiation_rate = (forecast_demand_for_coal - coal_under_construction - coal_installed)/planning_time_for_coal

forecast_demand_for_coal = 500
planning_time_for_coal = 30

## $CO_2$ emissions

$CO_2$ emissions from oil power plant is computed from oil-based power plant capacity and emission factor for oil as:

CO2_emissions_from_oil_power_plant = oil_installed * emission_factor_of_oil
emission_factor_of_oil = 735

$CO_2$ emissions from natural gas power plant is computed from natural gas-based power plant capacity and emission factor for natural gas as:

CO2_emissions_from_natural_gas_power_plant   =   natural_gas_installed   *
emission_factor_of_natural_gas
emission_factor_of_natural_gas = 500

$CO_2$ emissions from coal power plant is computed from coal-based power plant capacity and emission factor for coal as:

CO2_emission_from_coal_power_plant = coal_installed * emission_factor_of_coal
emission_factor_of_coal = 888

Figure 6.7 shows the emissions of $CO_2$ from natural gas, oil and coal-based power plants, and for all power plants, the emissions of $CO_2$ increase with time since the capacity of each power plant increases with time. The increase in the emissions of $CO_2$ with the increase of energy production and use is a globally observed

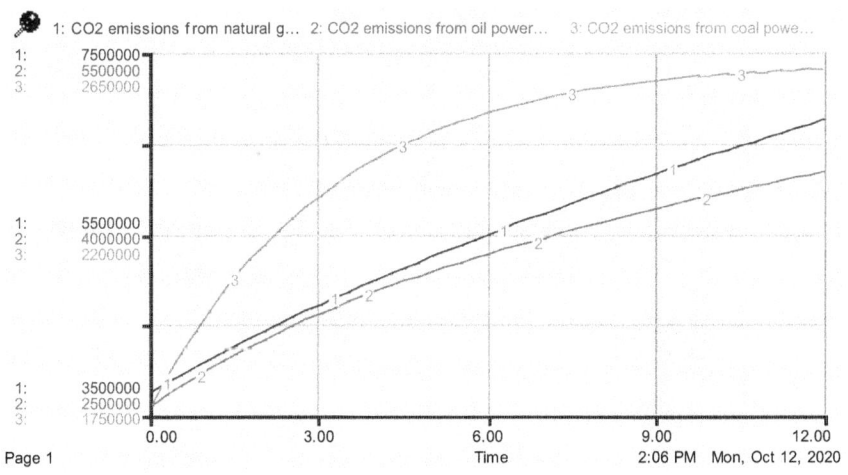

**FIGURE 6.7**   Emissions of $CO_2$ from natural gas, oil and coal-based power plants.

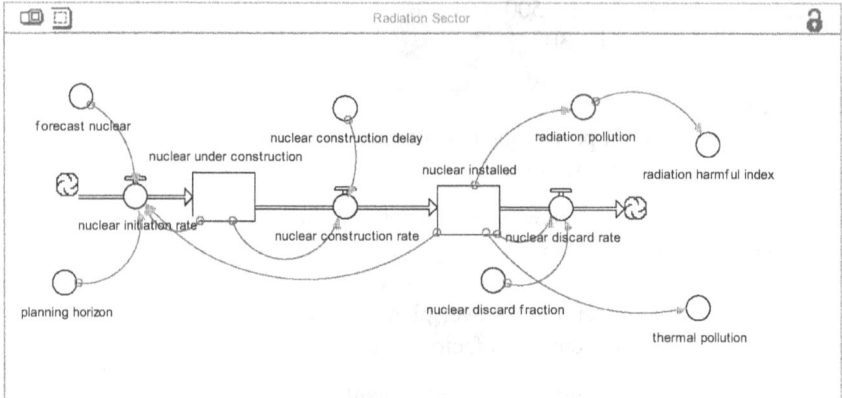

**FIGURE 6.8** Stock–flow diagram of radiation pollution and hazard from nuclear power plants.

environmental impact of energy production and use for development and economic growth (Bala et al., 2014; Bala, 2006).

### Radiation Sector
A stock–flow diagram of electricity generation from a nuclear power plant and its contribution to nuclear pollution and hazard is shown in Figure 6.8.

The fundamental equations that correspond to the major variables shown in Figure 6.8 are as follows:

### Nuclear power plant
The nuclear generation of electricity is increased by its construction rate and reduced by its capacity discard.

nuclear_installed(t) = nuclear_installed(t - dt) + (nuclear_construction_rate - nuclear_discard_rate) * dt
INIT nuclear_installed = 0

The construction of a nuclear power plant takes time called construction delay, and the construction rate is computed from nuclear power plant under construction and nuclear power plant construction delay.

### INFLOWS
nuclear_construction_rate = nuclear_under_construction/nuclear_construction_delay
nuclear_construction_delay = 5

The discard rate of nuclear generation of electricity is computed from nuclear power plant installed and discard fraction.

## OUTFLOWS
nuclear_discard_rate = nuclear_installed * nuclear_discard_fraction
nuclear_discard_fraction = 0.001

The nuclear power plant under construction increases by its initiation and decreases by its construction rate.

nuclear_under_construction(t) = nuclear_under_construction(t - dt) + (nuclear_ initiation_rate - nuclear_construction_rate) * dt
INIT nuclear_under_construction = 2500

The initiation rate of nuclear power plant depends on forecasted nuclear power and planning horizon and is computed as:

## INFLOWS
nuclear_initiation_rate = (forecast_nuclear - nuclear_under_construction - nuclear_ installed)/planning_horizon
forecast_nuclear = 2500
planning_horizon = 30

### Radiation pollution and hazards
The radiation pollution level emitted by a nuclear power plant is about 0.035 microsieverts per hour, and in STELLA is expressed as:

radiation_pollution = IF(nuclear_installed > 0)THEN(0.035)ELSE(0)
The thermal pollution is set to a value of 1 to indicate the expected impact of thermal pollution if there is a power plant, otherwise it is 0. In STELLA, it is expressed as:

thermal_pollution = IF(nuclear_installed > 0)THEN(1)ELSE(0)

The radiation harmful index is computed from the relation

$$\text{Radiation harmful index} = (\text{Radiation pollution} \times 0.001 \times 24 \times 365)/100$$

In STELLA it is expressed as:

radiation_harmful_index = (radiation_pollution * 0.001 * 24 * 365)/100

Figure 6.9 shows simulated nuclear power installed, radiation pollution, radiation harmful index and thermal pollution. A nuclear power plant produces no carbon emissions, but safety and thermal pollution pose challenges.

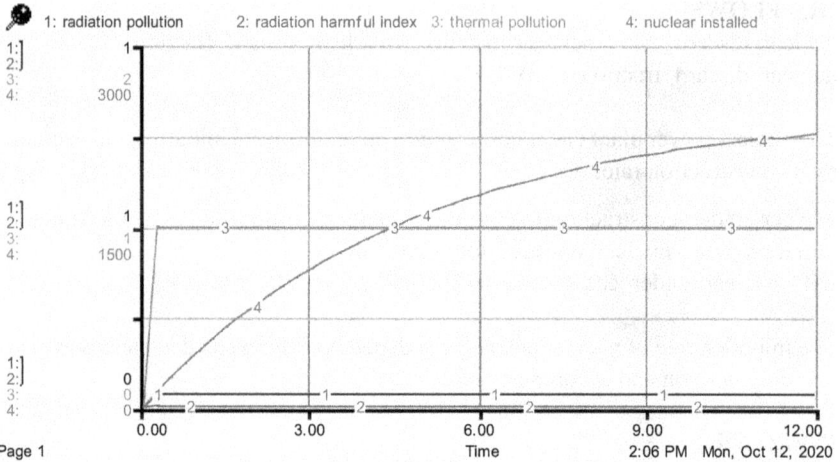

**FIGURE 6.9**   Pollution and hazard from nuclear power plant.

## Listing of STELLA program

### $CO_2$ emission sector

coal_installed(t) = coal_installed(t - dt) + (coal_construction_rate - coal_discard_rate) * dt

INIT coal_installed = 2000

### INFLOWS

coal_construction_rate = coal_under_construction/coal_construction_delay

### OUTFLOWS

coal_discard_rate = coal_installed * coal_discard_factor

coal_under_construction(t) = coal_under_construction(t - dt) + (coal_initiation_rate - coal_construction_rate) * dt

INIT coal_under_construction = 1000

### INFLOWS

coal_initiation_rate = (forecast_demand_for_coal - coal_under_construction - coal_installed)/planning_time_for_coal

### OUTFLOWS

coal_construction_rate = coal_under_construction/coal_construction_delay

natural_gas_installed(t) = natural_gas_installed(t - dt) + (natural_construction_rate - natural_gas_discard_rate) * dt

INIT natural_gas_installed = 7500

**INFLOWS**
natural_construction_rate = natural_gas_under_construction/natural_gas_con-struction_delay

**OUTFLOWS**
natural_gas_discard_rate = natural_gas_installed * natural_gas_discard_factor
natural_gas_under_construction(t) = natural_gas_under_construction(t - dt) + (natural_gas_initiation_rate - natural_construction_rate) * dt
INIT natural_gas_under_construction = 2500

**INFLOWS**
natural_gas_initiation_rate = (forecast_demand_for_natural_gas - natural_gas_under_construction - natural_gas_installed)/planning_time_for_natural_gas

**OUTFLOWS**
natural_construction_rate = natural_gas_under_construction/natural_gas_con-struction_delay
oil_installed(t) = oil_installed(t - dt) + (oil_construction_rate - oil_discard_rate) * dt
INIT oil_installed = 3500

**INFLOWS**
oil_construction_rate = oil_under_construction/oil_construction_delay

**OUTFLOWS**
oil_discard_rate = oil_installed * oil_discard_factor
oil_under_construction(t) = oil_under_construction(t - dt) + (oil_initiation_rate - oil_construction_rate) * dt
INIT oil_under_construction = 1500

**INFLOWS**
oil_initiation_rate = (forecasted_demand_for_oil - oil_under_construction - oil_installed)/planning_time_for_oil

**OUTFLOWS**
oil_construction_rate = oil_under_construction/oil_construction_delay
CO2_emissions_from_oil_power_plant = oil_installed * emission_factor_of_oil
CO2_emissions_from_natural_gas_power_plant = natural_gas_installed * emission_factor_of_natural_gas
CO2_emissions_from_coal_power_plant = coal_installed * emission_factor_of_coal
coal_construction_delay = 3.5
coal_discard_factor = 0.001

emission_factor_of_coal = 888
emission_factor_of_oil = 735
emission_factor_of_natural_gas = 500
forecast_demand_for_oil = 10000
forecast_demand_for_coal = 500
forecast_demand_for_natural_gas = 25000
natural_gas_construction_delay = 3.5
natural_gas_discard_factor = 0.001
oil_construction_delay = 3.5
oil_discard_factor = 0.001
planning_time_for_coal = 30
planning_time_for_natural_gas = 30
planning_time_for_oil = 30

**Radiation sector**
nuclear_installed(t) = nuclear_installed(t - dt) + (nuclear_construction_rate - nuclear_discard_rate) * dt
INIT nuclear_installed = 0

**INFLOWS**
nuclear_construction_rate = nuclear_under_construction/nuclear_construction_delay

**OUTFLOWS**
nuclear_discard_rate = nuclear_installed * nuclear_discard_fraction
nuclear_under_construction(t) = nuclear_under_construction(t - dt) + (nuclear_initiation_rate - nuclear_construction_rate) * dt
INIT nuclear_under_construction = 2500

**INFLOWS**
nuclear_initiation_rate = (forecast_nuclear - nuclear_under_construction - nuclear_installed)/planning_horizon

**OUTFLOWS**
nuclear_construction_rate = nuclear_under_construction/nuclear_construction_delay
forecast_nuclear = 2500
nuclear_construction_delay = 5
nuclear_discard_fraction = 0.001
planning_horizon = 30
radiation_harmful_index = (radiation_pollution * 0.001 * 24 * 365)/100
radiation_pollution = IF(nuclear_installed > 0)THEN(0.035)ELSE(0)
thermal_pollution = IF(nuclear_installed > 0)THEN(1)ELSE(0)

## 6.8   MODELING OF ACID RAINFALL

The largest source of acid deposition in many parts of the world is sulfur dioxide ($SO_2$). Sulfur dioxide is formed from many sources, but the predominant anthropogenic source is coal burning in power plants. Once $SO_2$ is emitted into the atmosphere, it undergoes a series of chemical reactions that transform it into either sulfurous acid ($H_2SO_3$) or sulfuric acid ($H_2SO_4$). Some of the emitted $SO_2$ is oxidized to $SO_3$, and this conversion can be approximated by a first-order rate constant $k_1$.

$$k_1 = 0.1 \ hr^{-1} \tag{6.2}$$

Some of the emitted $SO_2$ combines with water ($H_2O$) to form $H_2SO_3$, and this conversion can be approximated by a first-order rate constant $k_2$.

$$k_2 = 0.03 \ hr^{-1} \tag{6.3}$$

Finally, some of the emitted $SO_3$ combines with water ($H_2O$) to form $H_2SO_4$, and this conversion and deposition process can be approximated by a first-order rate constant $k_3$.

$$k_3 = 0.03 \ hr^{-1} \tag{6.4}$$

To calculate the acid rainfall at a site far from the source, we must calculate the emissions of $SO_2$, the dispersion of the emission, the transformation of $SO_2$ emissions into acid, the travel time, and the deposition rate of the acid. We are interested in determining the levels of depositions of $SO_2$ and acid deposition at various distances from the source. The assumptions made in the computation are:

(1) The site is directly downwind of the source.
(2) Wind direction and velocity during the simulation period are constant.
(3) Emission rate of $SO_2$ from the source is constant.
(4) Natural inflow of $SO_2$ into the $SO_2$ reservoir is constant.
(5) Assume the one-time inflow of $SO_2$ from the coal-based power is oblique.

We can start the problem by considering a stock of $SO_2$ in the atmosphere, and the primary contributor to this stock is a one-time influx of $SO_2$ from the source. We can consider this stock as a parcel of air and also that the parcel of air travels downward from the source (Kabashi et al., 2012). The stock–flow diagram of the acid rainfall model is shown in Figure 6.10.

The fundamental equations that correspond to the major variables shown in Figure 6.10 are as follows:

$SO_2$ stock is increased by $SO_2$ emission rate and natural $SO_2$ emission rate and decreased by $SO_3$ emission rate and acid deposition 1.

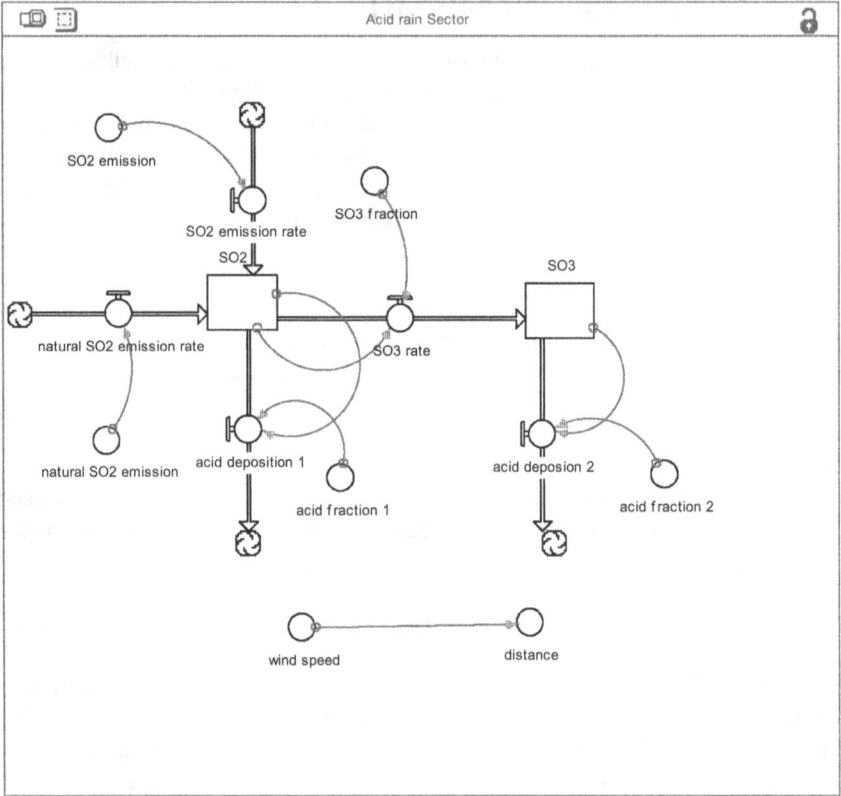

**FIGURE 6.10** Stock–flow diagram of the acid rainfall model.

$SO2(t) = SO2(t - dt) + (SO2\_emission\_rate + natural\_SO2\_emission\_rate - acid\_deposition\_1 - SO3\_rate) * dt$
INIT SO2 = 550

$SO_2$ emission rate is simply the $SO_2$ emission and is expressed as:

**INFLOWS**
SO2_emission_rate = SO2_emission
SO2_emission = 0
Natural $SO_2$ emission rate is simply the natural $SO_2$ emission and is expressed as:
natural_SO2_emission_rate = natural_SO2_emission
natural_SO2_emission = 42

Acid deposition 1 is computed from $SO_2$ and its acid fraction 1 and is expressed as:

**OUTFLOWS**
acid_deposition_1 = SO2 * acid_fraction_1
acid_fraction_1 = 0.03
$SO_3$ emission rate is computed from $SO_2$ and its $SO_3$ fraction and is expressed as:
SO3_rate = SO2 * SO3_fraction
SO3_fraction = 0.1

$SO_3$ stock is increased by $SO_3$ rate into $SO_3$ stock and decreased by acid deposition 2:

SO3(t) = SO3(t - dt) + (SO3_rate - acid_deposition_2) * dt
INIT SO3 = 0

Acid deposition 2 is computed from $SO_3$ and its acid fraction 2 and is expressed as:

acid_deposition_2 = SO3 * acid_fraction_2
acid_fraction_2 = 0.03

Distance from the source is computed from time elapsed and wind speed and is expressed as:

distance = TIME * wind_speed
wind_speed = 4

Finally, we should recognize that these processes occur over time in a parcel of air that is being transported due to wind speed. Thus, we can translate time into distance based on the velocity of the wind. In this way, we can track the concentration, transformation and deposition of $SO_2$ over time and distance. Figure 6.11 shows the concentration of $SO_2$ and acid deposition 1 over distance. For this example, the acid rainfall is high within the first 20 km from the source and may cause damage to vegetation within this distance.

**Listing of STELLA program**

**Acid rain sector**
SO2(t) = SO2(t - dt) + (SO2_emission_rate + natural_SO2_emission_rate - acid_deposition_1 - SO3_rate) * dt
INIT SO2 = 550

**INFLOWS**
SO2_emission_rate = SO2_emission
natural_SO2_emission_rate = natural_SO2_emission

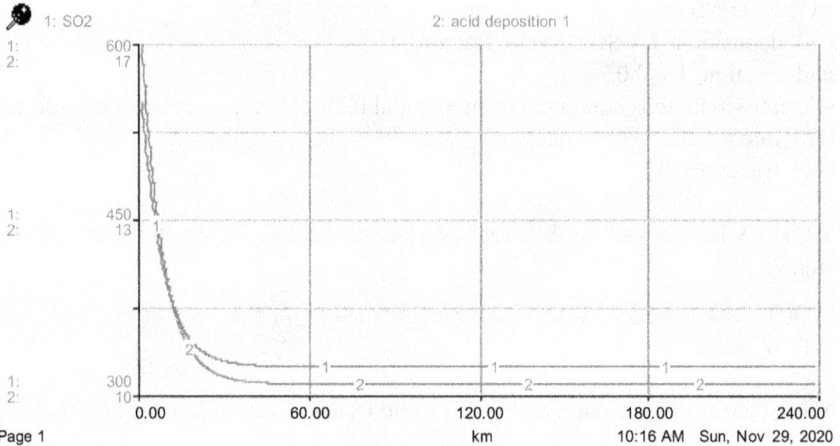

**FIGURE 6.11**   Concentration of $SO_2$ and acid deposition over distance.

## OUTFLOWS
acid_deposition_1 = SO2 * acid_fraction_1
SO3_rate = SO2 * SO3_fraction
SO3(t) = SO3(t - dt) + (SO3_rate - acid_deposition_2) * dt
INIT SO3 = 0

## INFLOWS
SO3_rate = SO2 * SO3_fraction

## OUTFLOWS
acid_deposition_2 = SO3 * acid_fraction_2
acid_fraction_1 = 0.03
acid_fraction_2 = 0.03
distance = TIME * wind_speed
natural_SO2_emission = 42
SO2_emission = 0
SO3_fraction = 0.1
wind_speed = 4

## REFERENCES

Bala, B. K. (1998). *Energy and Environment: Modelling and Simulation.* Nova Science
    Publishers, Inc., New York.
Bala, B. K. (2003). *Renewable Energy.* Agrotech. Publishing Academy, Udaipur-313002, India.
Bala, B. K. (2006). Computer modelling of energy and environment for Bangladesh.
    *Agricultural Engineering International.* 15(4): 151–160.

Bala, B. K., Alam, M, S., & Debnath, N. (2014). Energy perspective of climate change: The case of Bangladesh. *Strategic Planning for Energy and the Environment*. 33(3): 6–22.

BioNinja. (2021). Greenhouse effect. Published online. https://ib.bioninja.com.au/_Media/greenhouse-effect_med.jpeg

Chia, E. S., Lim, C. K., Ng, A., & Nguyen, N. H. L. (2015). The system dynamics of nuclear energy in Singapore. *International Journal of Green Energy*. 12(1): 73–86.

IPCC. (1992). Scientific Assessment of Climate Change. 1992 Supplement, *Intergovernmental Panel on Climate Change*. United Nations Environment Programme and the World Meteorological Organization.

IPCC. (1994). Radioactive Forcing of Climate Change, *the 1994 Report of Scientific Assessment Working Group of the IPCC*. Summary of Policymakers, Intergovernmental Panel on Climate Change, United Nations Environment Programme and the World Meteorological Organization.

Kapoor, B. S. (1989). *Environmental Engineering: An Overview*. Khanna Publishers, Delhi.

Kabashi, S., Bekteshi, S., Ahmetaj, S., Kabashi, G., Podrimqaku, K., Veliu, V., Jonuzaj, A., & Zidanšek, A. (2012). Dynamic modelling of air pollution and acid rain from energy system and transport in Kosovo. *Open Journal of Air Pollution*. 1: 82–96.

Lazarus, M., Hipel, D. V., Hill, D., & Margolis, R. (1995). *A Guide to Environmental Analysis for Energy Planners*. Stockholm Environmental Institute, Boston.

Pandey, G. N., & Cerney, G. C. (1989). *Environmental Engineering*. Tata McGraw-Hill Publishing Company Limited. New Delhi.

Pidwirny, P. (2009). *Fundamentals of Physical Geography* (e-book). Published online.

Parry, M. L. (1990). *Climate Change and World Agriculture*. Earthscan Publications Limited, London.

Rao, C. S. (1992). *Environmental Pollution Control Engineering*. Wiley Eastern Limited, New Delhi.

Smith, K. R. (1987). *Biofuels, Air Pollution, and Health: A Global Review*. Plenum Press, New York.

World Nuclear Association. (2020). Comparison of lifecycle, greenhouse gas emissions of various electricity generation sources. *WNA Report*. Page 1– 6.

# 7 Modeling of Integrated Energy Systems

## 7.1 INTRODUCTION

Integrated energy systems can be classified as integrated rural energy systems and integrated national energy systems. Integrated rural energy systems are dominated by animate energy sources and biomass or related energy sources, and integrated national energy systems are dominated by conventional energy sources such as oil and gas and electricity with the integration of renewable energy sources such as solar and wind.

The source and the sink (consumption) of rural energy systems in developing countries are interrelated. Animate energy, biomass or related energy, as well as commercial energy sources, are the sources of energy supply. However, activities involved in the net flow of energy in the rural sector include household, crop production, livestock, agro-industry, trade, health, housing, education and recreation. The physical quality of life (PQLI) of rural people is determined by these activities. Huq (1975) initiated energy modeling in Bangladesh and proposed a qualitative model for integrated energy flow for rural energy systems, as shown in Figure 7.1.

## 7.2 QUALITATIVE INTEGRATED ENERGY MODEL FOR RURAL FARMING SYSTEMS

The farming system elements in developing countries are crops, cattle, poultry birds, fish and forests. The farming systems may be considered as a set. Crops, cattle, poultry birds, fish and forests are the elements (set variables) of the set. Therefore, farming systems may be defined by the following equation:

$$\text{Farming system} = (\text{crop, cattle, poultry birds, fish, forest}) \qquad (7.1)$$

Each element may be considered a subset. The subsets may be defined as:

$$\text{Crop} = (\text{cash crop, food crop, vegetable}) \qquad (7.2)$$

$$\text{Cattle} = (\text{bullock, cow, buffalo, goat}) \qquad (7.3)$$

$$\text{Poultry bird} = (\text{fowl, duck, pigeon}) \qquad (7.4)$$

$$\text{Fish} = (\text{different species of fish}) \qquad (7.5)$$

$$\text{Forest} = (\text{fruit and timber, firewood, palm, bamboo}) \qquad (7.6)$$

DOI: 10.1201/9781003218401-7

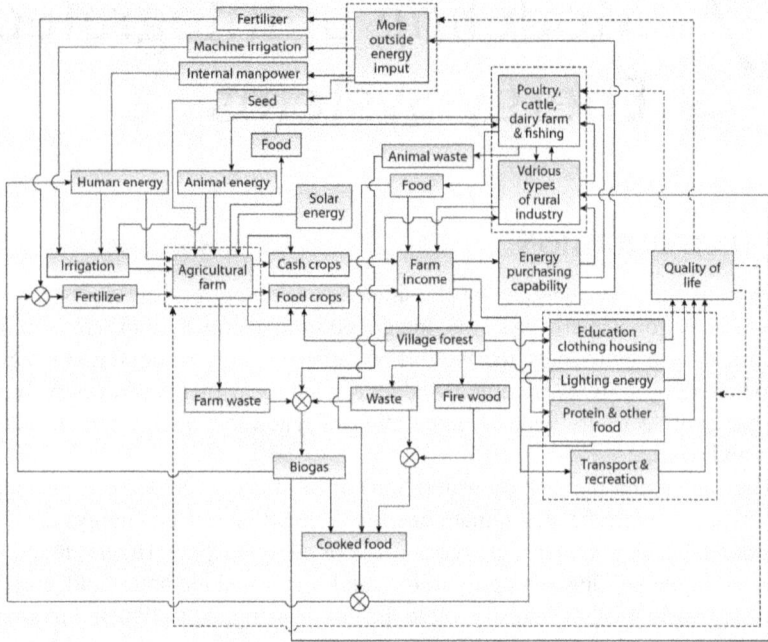

**FIGURE 7.1**   Huq's model for an integrated rural energy system (Revised).

All the elements of the set and the subsets will be considered dynamic variables.

In developing countries, biomass fuels (i.e., agricultural residues, fuelwood and dung) provide over 80% of total energy consumption. The rural farming sector is the source of biomass fuel as well as an energy-consuming sector (as inputs of different energetics) of the rural area. The qualitative integrated energy model can be modified for a rural farming system having the above-mentioned dynamic variables (Equation (7.1) to Equation (7.6)), including other energy-consuming variables. It can also be justifiably assumed that an integrated energy model for farming in a developing country may be in effect treated as the model for an integrated rural energy system.

## 7.2.1   ENERGY DEMAND FOR THE FARMING SECTOR

The various types of energy sources directly used for farming are:

(a) Human and animal muscle power, and diesel for land preparation, planting, harvesting, transportation of crops and livestock maintenance.
(b) Human muscle power, diesel fuel and electricity for irrigation.
(c) The indirect use of energy for facilitating agricultural production through chemical fertilizer and pesticide.

(d) Crop residues, cowdung, and tree biomass used as fuels for domestic cooking.

(e) Energy in the form of heat produced by the combustion of biomass fuels or other fuels, and some amount of mechanical power necessary for the post-harvest processing of crops. Some of this processing may take place at the household level and also in rural industries.

### 7.2.1.1   Categories of Energy in the Rural Farm

Energy flow in rural farming systems can be best described by Huq's model, but the energy use patterns for crop production depend on economic, technological and social constraints (Bala et al., 1989). However, it is possible to break up all the various activities that need energy in the farm into two categories:

(i) Energy that is applied for obtaining the finished products of the farm, namely, food crops and cash crops.

(ii) Energy that is consumed for making life possible in the farming sector.

### 7.2.1.2   Other Forms of Energy in the Farm

It is important to note that in the setting up of an overall energy model for a farming system, a few other types of energy directly or indirectly flowing into farming activities must also be taken into account. These are solar energy (providing heat and chemical energy); chemical energy provided by animal waste, which in turn is a partial output from the food provided to farm animals; chemical energy provided by artificial fertilizers, which are, in general, energy-intensive products of industry; rainwater with stored potential energy derived ultimately from solar energy; and irrigation water provided either by muscle power or machine power (from fossil fuel or electricity), or a combination of these. The existence of energy feedback loops at various stages can be identified in this case.

### 7.2.2   ENERGY MODEL WITH SOME FEEDBACK LOOPS

Figure 7.1 represents a qualitative model showing the forward and backward flow of energy. An attempt has been made to make the model as complete as possible with factors known about Bangladesh (Alam, 1991). While setting up such a model, one must remember that energy, including all losses, must balance. Many facts are either not known or cannot be evaluated in terms of numerical quantities. It may be stated that the aim of the farming sector is to produce edible, saleable or combustible outputs so as to secure a basic standard of living as a minimum goal. This standard of living is commonly measured in terms of income per capita of the human beings subsisting on a particular sector. Another way of looking at the output of the sector is to determine how much energy could have been bought (from sources outside of this sector) from the net produce of the farm after all the energy in its various forms inside the farm have been accounted for. As is also known, energy consumption per capita is an index of the standard of living in a community. If some of the energy purchasing capability of a farming sector can be reinvested in the farm for the purposes of artificial irrigation by power pumps or chemical fertilizers for crops, or even better food for animals, etc., in an ideal situation, the farm income per capita (or energy purchasing capability) should increase at a faster rate than the rate of reinvestment. Saturation, however, will be reached at some stage, but as pointed out earlier, the usefulness of the model would lie, on the one hand, in deciding the most suitable point of injection of energy and on the other hand, in raising the point (or standard of living) where saturation is reached. However, it is not difficult to realize that quality of life is not the same as "standard of living", although the farmer, to a large extent, is dependent on the latter. In order to represent the performance of the model in characterizing the dependence of quality of life on energy consumption, a mathematical model for physical quality of life as a function of energy consumption (PQLI = f(kWh/cap)) is now available (Alam et al., 1991).

### 7.2.3 ENERGY MODEL: FORECASTING, POLICY ANALYSIS AND OPTIMIZATION

The ultimate aim in analyzing the efficiency of any industrial production system, whether with the help of a computer model or any other means, is to secure: optimization of the entire process so as to achieve, in terms of energy, the maximum energy purchasing capability with the minimum application of energy; status projection; or analysis of the alternative courses of the policy.

## 7.3   OPTIMIZATION OF INTEGRATED ENERGY SYSTEMS FOR RURAL FARMING SYSTEMS

Systems analysis is needed to construct scenarios for long-term energy development in selected rural areas. Subhash and Satsangi (1990) developed a cost-optimized energy plan for the Fatehpur village cluster in India in 2001 using a generally applicable methodology of input-output analysis. Alam et al., (1990) further developed Huq's model based on the system dynamics methodology of Forrester for energy planning. They also discussed the results of different energy policies. Parikh (1985) and Parikh and Kromer (1985) developed a linear programming model for rural energy systems to optimize the revenues from crop and energy production subject to the constraints of limited resources. Biomass allocation for food, fodder, fuel and fertilizer is the most crucial question in the case of Bangladesh, and they applied this model for the rural energy system in Bangladesh. This model in the base case provides insights into the present behavior of different income groups as regards the choice of fuels and allocation of biomass for various purposes. It is shown that, due to the high needs and prices of fuels, the biomass for fuels takes priority over feed and fertilizers. In fact, landless farmers and small farmers burn 80% of animal dung rather than use it for fertilizers. The model also shows that unless substantial amounts of fertilizers are used, the small and middle farmer would have feed and fuel shortages on adopting high-yielding varieties (HYV) that minimize the straw–grain ratio. Similarly, when the population increases further, middle farmers also become vulnerable in meeting their feed, fuel and fertilizer requirements. To mitigate these effects, improved stoves and other measures would be necessary to increase biomass use efficiencies considerably. Singh and Subharayan (1986) reported a study on the optimization of net returns and net energy through the optimal allocation of inputs on different size groups of farms in the Meerut district of Uttar Pradesh, India. The study reveals that the farmers have to forgo very little in terms of net returns (monetary) if they have to adopt energy-maximizing plans. The energy maximizing plans are very much consistent with net returns (in monetary terms) with maximizing plans. So, the planners can give emphasis to energy maximizing plans that will be readily acceptable to the farmers since these plans can give higher monetary returns than those under the existing plan and are almost consistent with the net returns optimization plan. Devadas (1997) reported a linear programming model for a rural energy system in India with the objective function of maximizing the revenue of the rural energy system subject to a number of energy and non-energy related constraints. The model was used to quantify the major yields as well as the by-products of the different sectors, and the model was a modified version of the linear programming model developed by Parikh (1985) for Bangladesh.

The purpose of this section is to develop a linear programming model of an integrated rural energy system to maximize the net returns through the optimal use of available resources for different income groups.

### 7.3.1 VERBAL DESCRIPTION OF THE MODEL

A linear programming model for rural farming systems can be constructed for optimizing net revenues and net energy separately on different sized groups of farms from crop and energy production. The model takes into consideration the following crop commodities and cropping patterns, activities of energy production and purchase, activities of irrigation methods, activities of fertilizer provision, activities of draft power, and availability of crop residues; requirements of food and energy by class, and availability of land and other resources. In addition, the model has the flexibility of introducing several land classes or subregions. Energy demands for cooking are considered in competition with energy demands for agriculture. The model is applicable to low-income developing countries.

### 7.3.2 LINEAR PROGRAMMING MODEL

#### 7.3.2.1 Objective Function

To address the problem of the optimization of an integrated rural energy system, the problem may be stated as a linear programming model with the goal of maximizing profit as the objective function to a set of constraints. The objective function here maximizes the net revenues subject to energy resources available at different sizes of farms. The net revenue here refers to gross revenue from crops, crop residue, fuelwood, leaves and twigs, and dung minus the cost of purchasing fertilizers, commercial energy, feed and hired laborers, as well as the cost of crop residue, fuelwood, leaves and twigs, and dung. The crops are selected according to the agro-climatic conditions, and their initial pattern is the one that presently exists. Livestock is assumed to be given as present, and its maintenance is imperative. The objective function for the rural area is to maximize:

$$
\begin{aligned}
Z = \sum_j &(\text{yield} \times \text{area} \times \text{price revenue from crops and byproducts} + \text{yield} \\
&\times \text{forest area} \times \text{price revenue from fuelwood, and leaves and twigs}) \\
-\sum_j &(\text{cost of crop residue used as feed, fuel, and other uses} + \text{cost of} \\
&\text{cowdung} + \text{cost of fuelwood, and leaves and twigs} + \text{cost of} \\
&\text{bought nutrients} + \text{cost of bought feed} + \text{cost of hired labors})
\end{aligned}
\tag{7.7}
$$

where j is the income class index.

The maximization is subject to the constraints of resource availability, individually or collectively. For example, each income class has private assets such as land, livestock, etc., as well as access to the collective resources such as wood resources or unused biomass resources from other income classes, such as dung and crop residues, which are exchanged on cost.

### 7.3.2.2  Crop Production and Crop Residue

Each income class has a fixed amount of land and broad allocations of crop production. The fertilizer yield responses are assumed. The crop residue coefficients for each selected crop are given approximately. Thus, on the basis of yield, land allocation and crop coefficients, crop production and residues are generated separately for each income class. The residues have the following uses: cattle feed, household cooking and fertilizer, and other purposes such as construction, mats, furniture stuffing, etc., to be given exogenously. All residues from different crops are added for a given income class, which allocates them to the above uses depending on requirements and other opportunities.

### 7.3.2.3  Livestock Sector: Maintenance and Services

Cattle and buffaloes are considered in the model because they have high feed requirements, highly volatile dung production and provide services. The number of animals and their distribution between various income classes are considered to be exogenously given. Two types of animals are considered: working and non-working animals. Their feed requirements, dung coefficients and collection coefficients are different. When the cultivation is fully mechanized, i.e., the land is plowed fully by tractors, and there are no working animals.

The exogenously given share of the feed required for cattle could be obtained from pastures, and the remaining need is to be used from crop residues. When that is not sufficient, the feed could be bought. The calorie and protein content of the available feed must be greater than or equal to the calorie and protein levels

**TABLE 7.1**
**Nutrient Content of Organic Fertilizers (on Dry Matter Basis)**

| Source of organic fertilizers | Unit | P | N | K |
|---|---|---|---|---|
| Crop residue | kg/tonne | 2.5 | 0.8 | 0.7 |
| Dung | kg/tonne | 10.0 | 5.0 | 12.0 |
| Biogas sludge | kg/1000 m³ | 16.0 | 14.5 | 10.0 |

required by animals. In addition to the feed, the human labor required to maintain the animals is considered.

### 7.3.2.4   Fertilizer Sector

The levels of fertilizer application in terms of kg/ha for N, P and K are exogenously given corresponding to the yield level devised by each income class. There are four ways of obtaining fertilizers: using crop residue (i.e., burning or plowing back straw) on the ground, using dung, using biogas sludge and purchasing chemical fertilizers. The first three groups are organic fertilizers. The nutrient contents of organic fertilizers are given in Table 7.1.

The plowing is mainly done by tractors and, to a lesser extent, by animals, and each requires different amounts of human labor days, animal labor days and diesel consumption per hectare of plowing.

### 7.3.2.5   Energy from Agriculture

The energy supplies that are used in households and agriculture are classified into three categories:

(a) Non-commercial energy that is gathered or produced within the agricultural system. These are dung, crop residues and three types of fuelwood: fuelwood 1 includes leaves and twigs; fuelwood 2 is obtained employing human labor from natural forests; and fuelwood 3 is harvested from plantations that grow it commercially, requiring investment.

(b) One type of secondary energy is biogas. This biofuel is processed at conversion facilities to obtain a more efficient and high-value secondary energy form. A biogas plant produces methane from anaerobic digestion and sludge that could be used as fertilizer. This energy form requires an initial investment, but in the static model, it is considered after deriving its annual cost, assuming a certain rate of return. Biogas plants that produce methane from anaerobic digestion and sludge that can be used as fertilizer have an initial cost of US$500.00 for a 2 m³/day production of methane.

(c) Three types of commercial energy are considered that are purchased with cash: kerosene, diesel and electricity. These fuels are purchased in rural areas and from urban areas, and only in the absence of other fuels.

#### 7.3.2.6 Energy Demand Sector

*Household Sector*

This sector includes all households with different income classes in the rural areas. The energy used by rural households is assumed to be mainly for cooking and lighting. Only two sources are considered for lighting: kerosene and electricity. The values taken for these two sources are 25 l for kerosene and 160 kWh of electricity annually per household.

*Agricultural Sector*

This sector includes the energy use of tractors and irrigation pumps, and competition of each use for activities such as plowing and irrigation are considered with other methods such as animals and humans.

#### 7.3.2.7 Objective Function and Constraints

The objective function is to:

Maximize

$$Z = \sum_c \sum_s \sum_j (Y_{cj} \times L_{csj} \times P_c) + \sum_c \sum_s \sum_j (Y_{rc} \times L_{csj} \times P_r)$$

yield of crop × area × price + yield of crop residue × area × price

$$+ (Y_{fw} \times A'_{fw} \times P_{fw}) + (Y_{lt} \times A_{fw} \times P_{lt})$$

+ yield of fuelwood × area × price + yield of leaves and twigs × area × price

$$+ (A_h \times c_b^d \times P_d) - \sum_c \sum_s \sum_j (h_e \times f_e \times L_{csj} \times P_e)$$

head of animal × dung × price − electricity for irrigation × price

$$- \sum_s \sum_j (h_d \times f_d \times L_{cjs})$$

− diesel for irrigation × price

$$- \sum \sum \sum (P_r \times (F'_{csj} + N'_{csj} + Q'_{csj} + Q'_{csj})$$

− price × use from residue for feed + nutrient + cooking + other

$$- \sum_j (H_j^e \times u_j^{le} \times P_{el}) - \sum_j (H_j^k \times u_j^{lk} \times P_{lk})$$

− electricity for household × price − diesel for household × price

$$- \sum_n P_n \times B_{nj}$$

− price × bought nutrient

$$- \sum_j (P_d \times (N_j^d + e_b^d \times Q_j^{db} + Q_j^d) - \sum_n P_n \times B_{nj}$$

− price × use from dung for manure + biogas + cooking − price × bought nutrient

$$-p\sum_{j} P_f \times B_{fj} - \sum_{j} P_k \times B_{kj} - \sum_{j} P_{fw} \times Q_j^{fw}$$

– price × bought feed – price × energy bought – price × fuel wood

$$-\sum_{s} P_{lt} \times Q_j^{lt} - \sum P_a \times L_{sa} \qquad (7.8)$$

– price of leaves and twigs – cost of plowing by animal

$$-\sum_{s} P_t \times L_{st} - w \times H_s^d$$

– cost of plowing by tractor – cost × hired labor

where

$L_{csj}$ = land under crop
P = price
Y = yield
$A_{fw}$ = forest area
$A_h$ = animal heads
B = bought feed/nutrient or energy
$H^e$ = number of electrified households
$H_s$ = hired labor
N = amount of dung or crop residue used
Q = amount of dung, crop residue, biogas and wood energy available
f = fraction
h = energy required
w = wage rate

c s j indicates crop, season and income class, respectively.

**The constraints are**

1. **Crop land area availability constraint**

$$\sum_{c} \sum_{s} \sum_{j} L_{csj} \quad \leq \quad \alpha L$$

$\qquad$ land under $\qquad\qquad$ cropping intensity ×
$\qquad$ crop $\qquad\qquad\qquad$ land area

$\qquad$ (7.9)

where

$L_{csj}$ = land under crop
L = land area
A = cropping intensity

## 2. Cropping pattern constraint
### (a) Crop constraint

$$\sum_j L_{psj} \leq L_p$$
$$\quad\text{land under}\quad\text{land suitable for}$$
$$\quad\text{pth crop}\qquad\text{pth crop}$$
(7.10)

where
$L_{psj}$ = land under crop
$L_p$ = land suitable for pth group of crops

### (b) Crop diversification constraint

$$L_{csj} \geq L_c$$
$$\text{land under}\quad\text{minimum area}$$
$$\text{crop c}\qquad\text{for crop c}$$
(7.11)

where
$L_c$ = minimum crop area

## 3. Crop residue availability constraint

$$-L_{csj} \times Y_{rc} + F_{csj}{}^r + N_{csj}{}^r + Q_{csj}{}^{ro} \leq 0$$
$$-\text{land used}+\text{residues}+\text{feed}+\text{nutrient}+\text{other uses}\leq 0$$
$$\times\text{yield}\qquad\qquad\text{from crop residues}$$
(7.12)

where
$F_{cs}$ = crop residues
$Y_{rc}$ = yield of crop residues
$L_{csj}$ = land used
$N_{csj}{}^r$ = crop residue as a nutrient in fields
$Q_{csj}{}^o$ = crop residue used for other purposes

## 4. Animal feed availability constraints
### (a) Calorie balance

$$-(\text{cal})^{pat} - (\text{cal})^b \times B^f - (\text{cal}) \times F^r + f_b{}^{cal} \times A_h \leq 0$$
$$-\text{grazing}-\text{bought feed}-\text{crop residues}+\text{calorie need}\times \leq 0$$
$$\text{calorie}\quad\text{calorie}\qquad\text{calorie}\qquad\text{animal head}$$
(7.13)

where
$(\text{cal})^{past}$ = calorie from grazing
$(\text{cal})^b$ = calorie from bought feed
$B^f$ = bought feed
$(\text{cal})^r$ = calorie from residue

$F^r$ = crop residue
$f_b^{cal}$ = calorie requirement per animal
$A_h$ = animal heads

## (b) Protein balance

$$-(prot)^{pat} - (prot)^b \times B^f - (prot)^r \times F^r + f_b^{prot} \times A_h \leq 0$$
$$\begin{array}{cccc} - \text{grazing} & - \text{bought} & \text{feed-crop residues} + \text{protein need} \times \leq 0 \\ \text{protein} & \text{protein} & \text{protein} & \text{animal head} \end{array} \quad (7.14)$$

where
$(prot)^{past}$ = protein from grazing
$(prot)^b$ = protein from bought feed
$(prot)^r$ = protein in residue
$f_b^{prot}$ = protein requirement per animal

## 5. Animal dung availability constraints

$$N_j^d + e_g^d \times Q_j^{db} + Q_j^{dr} + e_b^d \times A_{hj} \leq 0$$
$$\begin{array}{ccccc} \text{dung as} + \text{fraction} \times \text{dung} + \text{dung} + \text{dung yield} \times \leq 0 \\ \text{manure} & \text{gathered} & \text{cooking} & \text{animal head} \end{array} \quad (7.15)$$

where
$N_j^d$ = dung used as manure
$c_b^d$ = fraction of dung
$Q_j^{db}$ = biogas produced from dung
$Q_j^d$ = dung used for cooking
$e_b^d$ = dung required for biogas for 1000 m³ of biogas
$A_{hj}$ = animal head

## 6. Fertilizer nutrient availability constraint

$$\sum_c F_{ncj} - B_j^n - (nut)^{dn} \times N_{nj}^d - (nut)^m \times N_{nj}^r - N_{nj}^b \times (nut)^{bn} \leq 0$$
$$\begin{array}{cccccc} \text{Fertilizers} - \text{fertilizers} - \text{nutrient} \times \text{dung} - \text{nutrient} \times \text{residue} - \text{nutrient} \times \text{slurry} \leq 0 \\ \text{applied} & \text{bought} & \text{used} & \text{used} & \text{used} \end{array}$$
$$(7.16)$$

The above equation is for each type of nutrients N, P and K, represented by n = 1, 2 and 3, respectively.

where
$F_{ncj}$ = application of fertilizers
$B_j^n$ = purchased chemicals
$(nut)^{dn}$ = nutrient of n type from dung

$N_{nj}^{d}$ = dung used as nutrient type n

$(nut)^{rn}$ = nutrient of n type from residue

$N_{nj}^{r}$ = residue used as nutrient type n

$(nut)^{bn}$ = nutrient of n type from biogas slurry

$N_{nj}^{b}$ = biogas slurry used as nutrient type n

## 7. Energy use constraints

### (a) Household cooking

$$-u_{j}^{cb} \times Q_{j}^{db} \quad - \quad u_{j}^{cd} \times Q_{j}^{d} \quad - \quad u_{j}^{cr} \times Q_{j}^{r} \quad - \quad u_{j}^{cf} \times Q_{j}^{fw}$$

biogas × dung − dung used − residue used − fuelwood used
used           for cooking   for cooking    for cooking     (7.17)

$$- u_{j}^{ct} \times Q_{j}^{lt} + Q_{j}^{ce} \leq \quad 0$$

− leaves wigs + cooking energy ≤ 0
for cooking    requirements

where

$Q_{j}^{ce}$ = -cooking energy requirement

$u_{j}^{cb}$ = useful energy from biogas

$Q_{j}^{db}$ = -biogas energy from dung

$u_{j}^{cd}$ = useful energy from dung

$Q_{j}^{d}$ = dung available

$u_{j}^{cr}$ = useful energy from crop residues

$Q_{j}^{r}$ = residue available

$u_{j}^{cf}$ = useful energy from fuelwood

$Q_{j}^{fw}$ = fuelwood

$u_{j}^{ct}$ = useful energy from leaves and twigs

$Q_{j}^{lt}$ = leaves and twigs available

### (b) Household lighting

$$-u_{j}^{le} \times H_{j}^{e} \quad - \quad u_{j}^{lk} \times H_{j}^{k} \quad + Q_{j}^{l} \leq \quad 0$$

electricity ×  − kerosene    + lighting energy ≤ 0
household     for lighting   requirements        (7.18)

where

$Q_{j}^{l}$ = -lighting requirement

$u_{j}^{lk}$ = kerosene consumption

$H_{j}^{k}$ = -number of non-electrified house

$u_{j}^{le}$ = electricity consumption

$H_{j}^{e}$ = -number of electrified house

8. **Energy supply constraints**
   (a) **Fuelwood constraint**

$$\sum_j Q_{kj}^{\,w} \quad - \quad Y_{fw} \times A_f \quad \leq \quad 0$$

$$\text{wood energy} \quad - \text{fuel yield} \times \text{forest}$$
$$\text{used} \qquad\qquad\qquad\qquad \text{area}$$

(7.19)

where

$Q_{kj}^{\,wt} = $ -fuelwood
$Y_{fw} = $ -yield of fuelwood
$A_f = $ -forest area

(b) **Leaves and Twigs constraint**

$$\sum_j Q_j^{\,lt} \quad - \quad Y_{lt} \times A_f \quad \leq \quad 0$$

$$\text{leaves twigs} \quad - \text{fuel yield} \times \text{forest}$$
$$\text{used} \qquad\qquad\qquad \text{area}$$

(7.20)

where

$Q_j^{\,lt} = $ -leaves and twigs
$Y_{lt} = $ -yield of leaves and twigs
$A_f = $ forest area

(c) **Biogas conversion constraint**

$$Q_j^{\,db} \quad = \quad Q_{bj}$$

$$\text{biogas} \quad = \quad \text{biogas}$$
$$\text{used} \qquad\quad \text{supplied}$$

(7.21)

where

$Q_{bj} = $ -biogas supply

9. **Irrigation constraints**
   (a) **Electricity balance**

$$\sum_c \sum_s \sum_j f_e \times h_e \times L_{csj} \quad - \quad Q_{elec} \quad \leq 0$$

$$\text{electricity} \times \text{fraction} \times \text{land under} \quad - \text{electricity} \leq 0$$
$$\text{for irrigation} \qquad\quad \text{crop} \qquad\qquad \text{for irrigation}$$

(7.22)

where

$h_e = $ -electricity required
$f_e = $ -fraction of farm irrigated
$Q_{elec} = $ -electricity for irrigation

### (b) Diesel balance

$$\sum_{c}\sum_{s}\sum_{j} f_d \times h_d \times L_{csj} \quad - \quad Q_{diesel} \quad \leq 0$$

$$\text{diesel for} \times \text{fraction} \times \text{land under} - \text{diesel for} \leq 0$$
$$\text{irrigation} \qquad\qquad \text{crop} \qquad\qquad \text{irrigation}$$

(7.23)

where
$h_d$ = diesel required
$f_d$ = fraction of farm irrigated
$Q_{diesel}$ = diesel for irrigation

## 10. Animal and tractor power requirements (season basis)

$$-\sum_{j} L_{csj} + \quad L_{sa} \quad + \quad L_{st} = 0$$

land under    + land plowed  + land plowed = 0
crop            by animal        by tractor

(7.24)

$$c_{sa} \times L_{sa} \qquad\qquad = \qquad A_{an}$$

animal × land plowed  =  available
days      by animal        animal days

(7.25)

where
$c_{sa}$ = animal days per ha
$L_{sa}$ = land plowed by animal
$L_{st}$ = land plowed by tractor
$A_{sn}$ = animal days available
$L_{csj}$ = land under crop

Where land is plowed fully by tractor, and land plowed by animals is zero.

## 11. Human labor constraint (season basis)

$$\sum_{j} h_{cs}^{\ 1} \times L_{csj} \leq (H_s^{\ sl} + H_h^{\ 1})$$

labor        ≤  (seasonal + hired)
requirements    labor      labor

(7.26)

where
$H_s^{\ sl}$ = -seasonal labor
$H_h^{\ 1}$ = -hired labor
$h_{cs}^{\ 1}$ = labor requirement
$L_{csj}$ = land under crop

### 7.3.3 MODEL VALIDATION AND POLICY ANALYSIS

The linear programming model presented here can be used to allocate resources with the objective of maximizing the total income from a rural energy system subject to the constraints set by equality and inequality constraints. The model quantifies the yields of major crops as well as the by-products, along with dung and fuelwood produced. To build up confidence in the model and to use it for policy planning, the model must be validated (Bala et al., 2017). The model should be run using the base year data, and the results of the model should be compared with the actual data (Parikh, 1997). The validated model should be used for policy planning, such as what will happen if the efficiencies of non-conventional fuels are increased by 50% or if an afforestation program is launched to increase rural forest cover by 50%?

## 7.4 OPTIMIZATION OF SOLAR HYBRID ENERGY SYSTEMS USING GENETIC ALGORITHMS

In most developing countries, much of the population living in remote and isolated locations do not have access to the electrical grid line. Yet, there are alternatives. Renewable energy can offer an ideal source of electricity for an island or other isolated places far from the national grid. Schmid (2003) explained such a new technical model for mini-grids that could offer a power supply of high quality.

Energy, information and communication are the fundamental needs for economic and social development in rural regions. While these are features of most industrialized countries and urban areas, the provision of these basics is still important in the rural areas of developing countries. While small-scale renewable power supplies, particularly solar home systems, have much to offer (mainly for lighting, radio or TV, possibly refrigeration or mobile phone recharging), these cannot provide enough electric power for significant economic development to take place.

Energy is needed for economic and social development (Bala, 1998; Bala, 2003). The utilization of renewable energy is the key to sustainable integrated coastal development and sustainable development of remote and isolated areas. These can be tidal power (Salequzzaman, 2003), a solar-diesel hybrid mini-grid (Houqc et al., 1989; Islam, 2004; Bala, 2006a and 2006b; Bala et al., 2006) and an optimal mix mini-grid consisting of solar, wind and diesel (Houqe et al., 1989).

Hybrid systems are recognized as viable alternatives to reticulated grid supplies or conventional fuel-based remote area power supplies (Wichert, 1997). Developing countries with abundant solar radiation are suitable candidates for rural electrification using solar PV–diesel hybrid systems for remote areas where grid connections are not economically feasible. PV–diesel systems have greater reliability in comparison with diesel-only systems. Furthermore, PV–diesel hybrid systems are much more economical for the rural electrification of remote areas and produce less pollution. In order to supply electricity from a hybrid system such as

PV–diesel, its design must be optimal in terms of system configuration and operational control.

### 7.4.1 SOLAR PV–DIESEL HYBRID SYSTEM

A hybrid system is a combination of one or more resources of renewable energy such as solar, wind, mini/micro hydropower and biomass with other technologies such as batteries and diesel generators. To obtain electricity economically and reliably from a hybrid system, its design must be optimized in terms of renewable energy units, battery systems, diesel generator systems and operation. The architecture of a typical PV–diesel hybrid system is shown in Figure 7.2 and consists of a solar module, storage battery, inverter, charge controller and diesel generator. The mode of operation of a hybrid system is as follows. In normal operation, solar–PV supplies the load demand. The excess energy from the PV above the hourly demand is stored in the battery until the battery is fully charged. The main purpose of the storage of electricity into the battery system is to import or export energy depending upon the situation. When the output from the PV system exceeds the load demand and the state of charge in the battery is maximum, the excess energy is fed/dissipated to some dump load or remains unused. A diesel backup system is operated at times when the PV system fails to satisfy the load and when the battery storage is depleted.

The criterion for selecting the best hybrid energy system combination for a proposed site is based on the trade-off between reliability, cost and minimum use of diesel generator sets. For different renewable combinations, the output of the optimal sizing and operation is a set of component sizes for a given application together with recommendations for system operation. The component sizes are restricted to that available in the markets. From the cost comparisons of different combinations, the most economical system is selected that ensures power supply

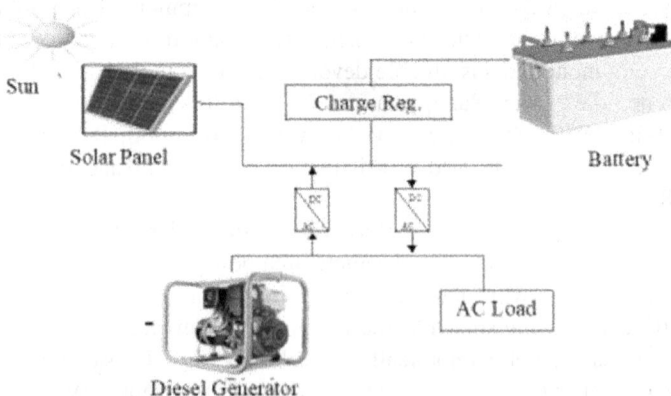

**FIGURE 7.2**   PV–diesel hybrid system.

continuity. The basic power modules of hybrid energy systems considered here are solar PV with battery and diesel backup.

## 7.4.2 GENETIC ALGORITHM

A PV–diesel hybrid system is more reliable in terms of electricity production and less costly than systems that use a single source of energy. When designing a hybrid system, both the sizing of the elements and the most adequate control strategy must be achieved. Achieving a good control strategy is essential since the performance of a PV hybrid system can be significantly affected by relatively small changes made in the control strategy. The optimal design of such a system can not be achieved easily using classical optimization methods. Genetic algorithms appear to be an adequate search technique for such complex problems.

A genetic algorithm is a global optimization method based on the principle of survival of the fittest (Darwin's hypothesis of evolution). The basic principles of the genetic algorithm are attributed to Holland (1975) and were further developed for engineering applications by Goldberg (1989) and Michalewicz (1996).

Genetic algorithms simulate the phenomena of reproduction, selection, crossing and mutation that are observed in nature using a computer program. A population of candidate solutions to a problem may be manipulated by the genetic algorithm. The candidate solutions are typically binary strings, but any representation may be used. At every generation, some of the candidate solutions are paired, and parts of each individual are mixed to form two new solutions. This is called a crossover. A uniform crossover exchanges individual bits, whereas a multipoint crossover exchanges whole substrings. Additionally, every individual is subjected to random change or mutation. The next generation is produced by selecting individuals from the current one on the basis of their fitness, which is a measure of how good each candidate solution is. Eventually, the population should become saturated with individuals of very high fitness.

The salient features of the genetic algorithm used here for the optimal design of a solar–diesel hybrid system for the electrification of the isolated island Sandwip in Bangladesh are described, and the details are given, in Dufo-Lopez and Bernal-Agustin (2005) and Bernal-Agustin et al. (2005). The PV–diesel system is studied using an hourly time step for one hour. Every hour the following inputs are estimated: the current from the PV, which depends on solar insolation; the AC load current, which depends on predicted load; and the battery state of charge (SOC). This information is needed for system simulation.

The genetic algorithm (GA) used here is divided into two parts: a main algorithm and a secondary algorithm. The main one searches for the optimal configuration of solar PV panels, batteries and diesel generators, minimizing the total net present cost of the system ($C_{TOT}$), which includes all the costs throughout the useful lifetime of the system, and these are converted into present values using an effective discount rate. This cost is given by:

$$C_{TOT} = C_{SEC} + C_{ACQ\_PV} + C_{ACQ\_B} + C_{ACQ\_BCH} + C_{ACQ\_GEN} \qquad (7.27)$$
$$+ C_{REP\_BCH} + C_{O\&M\_PV} + C_{O\&M\_B}$$

where

$C_{ACQ\_B}$ = acquisition costs of the battery
$C_{ACQ\_BCH}$ = acquisition costs of the battery charger
$C_{ACQ\_GEN}$ = acquisition costs of the diesel generator
$C_{ACQ\_PV}$ = acquisition costs of the PV panels
$C_{O\&M\_B}$ = costs of maintenance of the batteries throughout the life of the system
$C_{O\&M\_PV}$ = costs of maintenance of the PV panels throughout the life of the system
$C_{REP\_BCH}$ = costs of replacing the battery charger throughout the life of the system
$C_{SEC}$ = costs that depend on the optimal strategy
$C_{TOT}$ = total cost throughout the life of the system
$C_{TOT}$ = must be calculated for each combination represented by one of the $N_m$ vectors that constitute the population

The fitness function of the combination i of the main algorithm is assigned according to its rank in the population: rank 1 for the best individual and rank $N_m$ for the worst solution. The fitness function is expressed as:

$$fitness_{MAIN_i} = \frac{(N_m + 1) - i}{\sum_j \left[ (N_m + 1) - j \right]}, \quad j = 1 ..........N_m \qquad (7.28)$$

where
$N_m$ population of the GA

The secondary algorithm works with a Boolean vector with the dispatch strategy. For each vector of the main algorithm, the optimal strategy is obtained by minimizing the non-initial cost, including operation and maintenance costs, by the secondary algorithm. $C_{SEC}$, the costs that depend on the optimal strategy, is expressed as:

$$C_{SEC} = C_{ACQ\_INV} + C_{ACQ\_REG} + C_{REP\_B} + C_{REP\_INV} + C_{REP\_REG} \qquad (7.29)$$
$$+ C_{REP\_GEN} + C_{O\&M\_GEN} + C_{FUEL}$$

where
$C_{ACQ\_INV}$ acquisition costs of the inverter
$C_{ACQ\_REG}$ acquisition costs of the charge regulator
$C_{FUEL}$ cost of the fuel consumed throughout the life of the system
$C_{O\&M\_GEN}$ costs of operation and maintenance of the diesel generator throughout the life of the system
$C_{REP\_B}$ costs of replacing the batteries throughout the life of the system
$C_{REP\_GEN}$ costs of replacing the diesel generator throughout the life of the system

$C_{REP\_INV}$ costs of replacing the inverter throughout the life of the system
$C_{REP\_REG}$ costs of replacing the charge regulator throughout the life of the system

The fitness function of the combination i of the secondary algorithm is:

$$\text{fitness}_{SEC_i} = \frac{(N_{sec}+1)-i}{\sum_j \left[(N_{sec}+1)-j\right]}, \quad j=1\ldots\ldots\ldots N_{sec} \qquad (7.30)$$

where
$N_{sec}$ = number of vectors of secondary algorithm

The first step of optimal sizing consists of a system simulation in order to examine whether the system configuration fulfills the load power supply requirements. The data used in this study are the daily solar radiation and the consumer's power requirements. This is performed in the secondary algorithm, which minimizes the non-initial costs. The second step searches for the optimal system configuration, which minimizes the total costs and pollutant emissions. The details of the computation are given in Dufo-Lopez and Bernal-Agustin (2005) and Bernal-Agustin et al. (2005).

Practically, a genetic algorithm simulates the phenomenon of reproduction, selection, crossing and mutation using a computer program. The solutions to the problem are considered as the individuals of a species. In our design problem, the species are the configuration and control of a hybrid system, and the individuals are the combinations of the variables (genes) that we are trying to optimize.

Individuals are defined by integer codes, and the genetic algorithm searches for possible designs modifying the integer values, and the integer codes representing the individuals correspond to chromosomes. Individuals crossover to give birth to a new offspring in the next generation, and the crossing is to provide a mechanism for the exchange of chromosomes between mated parents. The next generation is produced by selecting individuals from the current one on the basis of their fitness, which is a measure of how good each candidate solution is. The higher the fitness of the individual, the higher the probability of crossover there is. With new generations, the algorithm finds better individuals whose fitness functions are close to optimal. Some individuals perform mutations resulting in some changes in genes, and mutation is an operator that produces spontaneous random changes in various chromosomes.

The size of the population is the number of individuals, and the number of generations is the number of iterations in the solution of the genetic algorithm problem, and these, including crossover and mutation, are the parameters of the genetic algorithm.

The genetic algorithm searches for PV panels, AC generators and inverters that minimize the cost of the system. The main algorithm finds the optimal combinations of the elements minimizing the net present cost, and the net present cost must be calculated for each combination. The fitness function of any combination of the main algorithm is estimated according to its rank in the population, and rank 1 is for the best individual.

Next, the genetic algorithm finds the control strategy minimizing the net present cost of each combination supplied by the main algorithm. The secondary algorithm tries to find the best control for each individual of the main algorithm. The fitness function of any combination of the secondary algorithm is evaluated by assigning rank 1. The stopping criterion should be such that enough numbers of generations are allowed so as to obtain near-optimal solutions.

### 7.4.2.1   Main Requirements for Solar–Diesel Hybrid System Design

The main requirements for a solar–diesel hybrid system design are as follows:

- **(i)** Site information (environmental data), such as solar intensity, ambient temperature, relative humidity and cloudiness, should be collected.
- **(ii)** Electrical load information, such as the load type and time of use of electrical appliances, should be identified.
- **(iii)** The specifications and cost information of the solar panel, battery, inverter, charge regulator and diesel AC generator should be set.

### 7.4.3   CASE STUDY: SANDWIP, AN ISOLATED ISLAND

Sandwip is a small island in Bangladesh that is totally disconnected from the mainland. It is situated in a very remote area and has great potential for developing electricity from solar energy. This case study examines how a decentralized, small-scale, community-based electric power generation system – a hybrid PV–diesel system – can be designed and managed optimally using genetic algorithms for this island (Bala and Siddique, 2009).

Sandwip is a deltaic island in the Bay of Bengal region of Bangladesh, adjacent to Chittagong and a mere 15 km from the mainland. The population is 330,000 and the area is 240 km². The entire island is a mudflat created from the Ganges.

Households in rural Bangladesh are simple and do not require large quantities of electrical energy for lighting and electrical appliances. Salequzzaman (2003) estimated that 15 kWh is required for the electrification of a fishing community on Sandwip. This load is based on two lights (compact florescent bulbs, 15 W), one fan (a ceiling fan, 230 V, 40 W) and one television (230 V, 80 W) for each family in a rural setting on the island. The main use of electricity in this fishing community is for lighting purposes, and the daily demand for a household that uses electricity only for lighting is less than 0.20 kWh/day. The monthly average daily load profile is shown in Figure 7.3.

Information on the average daily solar energy input is shown in Figure 7.4. The maximum average daily solar radiation is 6.0 kWh/m² in the months of March and April, and the minimum solar radiation is 4.0 kWh/m² in the months of July.

The system was designed and optimized using the HOGA software, developed by Dufo-Lopez and Bernal-Agustin (2005) and Bernal-Agustin et al. (2005).

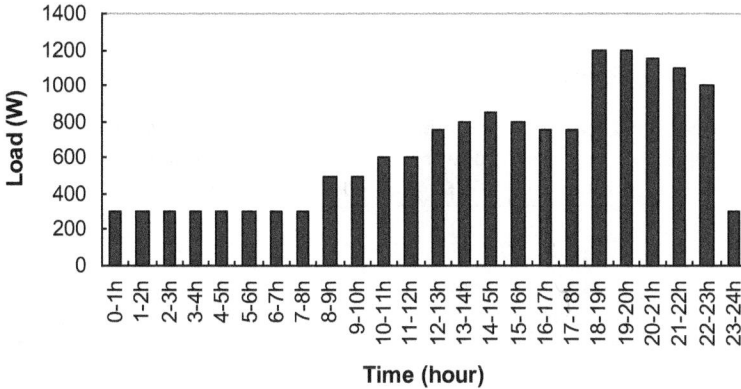

**FIGURE 7.3** Monthly average daily load profile.

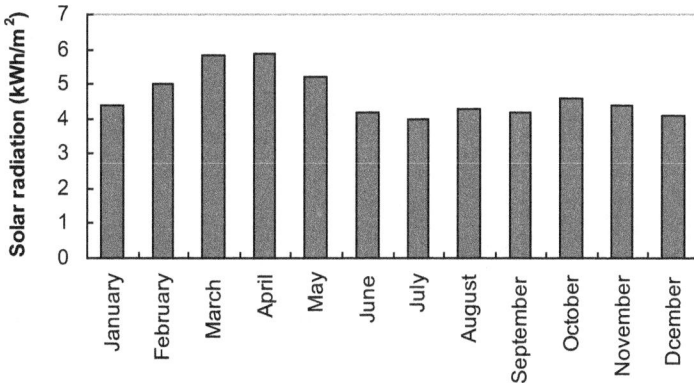

**FIGURE 7.4** Average daily solar insolation.

The load profile considered is shown in Figure 7.3. The daily load profiles are represented by a series of powers, each considered as a constant over a time step of 1 hour. The parameters considered are the crossover rate, 0.9, and the mutation rate, 0.01. The specifications and costs of different PV panels, batteries, inverters and AC generators are shown in Tables 7.2, 7.3, 7.4 and 7.5, respectively.

The optimized system configurations are 4s × 12p 50 W for the PV panels; 4s × 3p 200 Ah for the batteries; 230 V 1.9 kVA for the AC generator; and 3300 VA for the inverter. The system supplies 48 V DC and 230 V AC.

In terms of control strategies, if the power produced by the renewables is higher than the load, the batteries are charged with spare power from the renewables. If the power produced by the renewables is lower, the power not supplied to meet the load is supplied by the batteries (if they cannot supply the whole, the rest will be

**TABLE 7.2**
**PV Modules Specifications and Costs (Euros)**

| Type | Nominal voltage (V) | Shortcut current (A) | Nominal power (Wp) | Acquisition cost (€) | Maintenance cost per year (€/year) | Life span (years) |
|------|------|------|------|------|------|------|
| Panel 0 | 12 | 0 | 0 | 0 | 0 | 25 |
| Panel 1 | 12 | 3.17 | 50 | 387 | 0 | 25 |
| Panel 2 | 12 | 4.80 | 80 | 564 | 0 | 25 |
| Panel 3 | 12 | 7.54 | 125 | 892 | 0 | 25 |

**TABLE 7.3**
**Batteries Specifications and Costs (Euros)**

| Type | Nominal capacity (A h) | Voltage (V) | Acquisition cost (€) | Maintenance cost per year (€/year) | Float life (years) |
|------|------|------|------|------|------|
| Battery 0 | 0 | 12 | 0 | 0 | 50 |
| Battery 1 | 43 | 12 | 155 | 0 | 12 |
| Battery 2 | 96 | 12 | 258 | 0 | 12 |
| Battery 3 | 200 | 12 | 555 | 0 | 12 |
| Battery 4 | 462 | 12 | 1017 | 0 | 12 |

**TABLE 7.4**
**Inverter Specifications and Costs (Euros)**

| Type | Power (VA) | Lifetime (year) | Acquisition cost (€) |
|------|------|------|------|
| Inverter 0 | 0 | 50 | 0 |
| Inverter 1 | 3300 | 10 | 3608 |
| Inverter 2 | 4500 | 10 | 4138 |
| Inverter 3 | 10000 | 10 | 16048 |
| Inverter 4 | 13500 | 10 | 17786 |

supplied by the AC generator). The state of charge (SOC) of the batteries and the minimum power of the AC generator is 40% and 836 W, respectively.

The results were obtained with the following values: main algorithm generations 50; population 20 and secondary generations 25; and population 10. Figure 7.5 shows the changes towards the best total net present cost as a function of the main algorithm generations in an optimization where the number of generations in the main algorithm is 35 and the net present cost is €48,823. The emission is 37 kg $CO_2$.

**TABLE 7.5**
**Generator Specifications and Costs (Euro)**

| Name | Rated power (kVA) | Acquisition cost (€) | Maintenance cost per year (€/year) | Life span (hour) | P.min (% of $P_n$) | Fuel type |
|---|---|---|---|---|---|---|
| Gen0 | 0 | 0 | 0 | 100000 | 30 | Diesel |
| Gen1 | 1.9 | 1269 | 0.2 | 7000 | 30 | Diesel |
| Gen2 | 3.0 | 1514 | 0.2 | 7000 | 30 | Diesel |
| Gen3 | 5.5 | 2314 | 0.2 | 7000 | 30 | Diesel |
| Gen4 | 13.5 | 7200 | 0.2 | 7000 | 30 | Diesel |

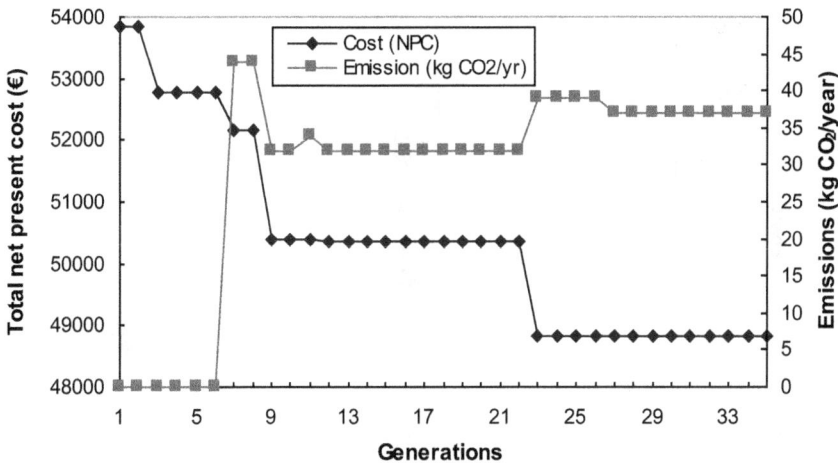

**FIGURE 7.5** Total net present cost and emissions as a function of the main algorithm generations.

The initial selection of individuals from the population, i.e., the components of the hybrid system, and the control strategies selected based on fitness function for the first two generations were poor. However, after 11 generations, the selection of the individuals through crossover and mutation of the codes of the individuals was good, but still not the best. After 26 generations of the crossing over, mutation and selection of the individuals of the candidates of the population, the solution becomes optimum and continues, i.e., the individuals of high fitness are achieved, which remain constant for rest of the period. The details of the individuals of the optimum solution from 26 generations to 35 generations are the same, and these are shown in Figure 7.6 and Figure 7.7. The costs of the different elements of the hybrid PV–diesel system as a percentage of the total net present cost is shown in Figure 7.8.

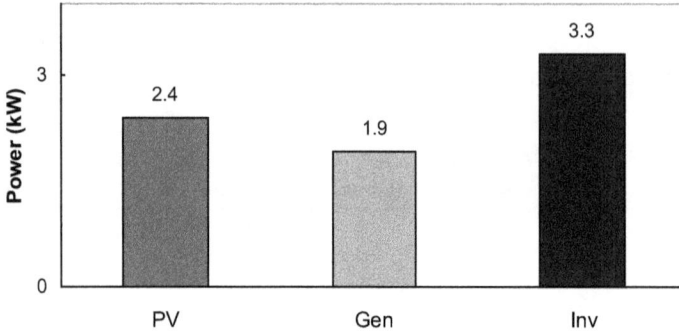

**FIGURE 7.6**   Power generation capacity of a PV generator, AC generator and inverter.

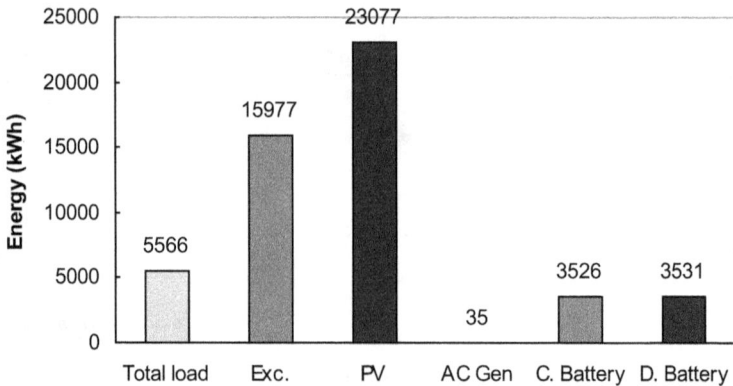

**FIGURE 7.7**   Annual energy balance of the hybrid system for optimal design.

Figure 7.6 shows the power generation capacity of the PV generator, AC generator and inverter. The component capacity of the PV generator, AC generator and inverter are 2.4 kW, 1.9 kVA and 3.30 kVA, respectively. The PV generator is the single largest unit responsible for supplying the electricity.

Figure 7.7 shows the annual energy balance of the hybrid PV–diesel system. The total load and energy charged by the battery are supplied by the solar PV system and the AC generator. The major share of the energy comes from solar PV (23,077 kWh), while the contribution from diesel generation is very small (35 kWh).

Figure 7.8 shows the costs of the different elements of the hybrid PV–diesel system as a percentage of the total net present cost. Batteries cost 33.74% of the total net present cost. The other items have the least share at 4.8% and the PV panels have the largest share at 39.05%. Thus, batteries and PV panels are the most costly items, and the costs of the batteries and the PV panels are the most important leverage points where action programs are essential for the reduction of the production costs of batteries and solar panels.

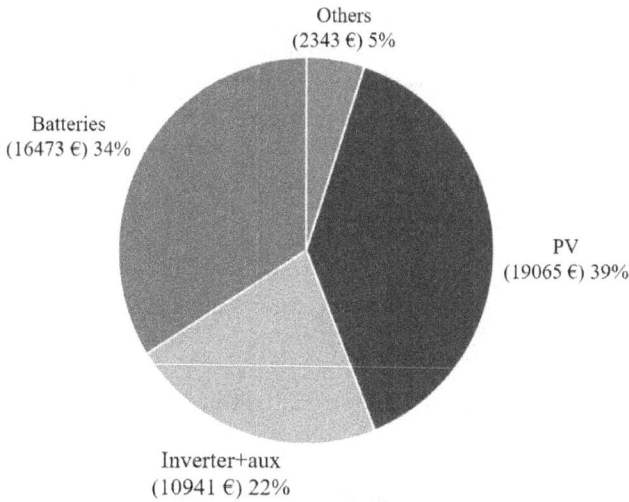

**FIGURE 7.8**   Costs of the different elements of the hybrid system for optimal design.

This study is a typical application of genetic algorithms for the optimal design of component configurations and operational control strategies for a mini-grid for a fixed-load profile. However, a detailed study is required with new data for a year-round load profile using a better solar radiation model. Before installing a solar–diesel hybrid mini-grid in any location, a digital computer model of the system could be extremely useful in examining the optimum size of the solar panels and the battery, and the rating of a generator set; the best operation and control logic; and the likely system performance. It is clear that human life can be sustained on Earth in the future only if, within a short period, we are able to replace conventional energy sources with an alternative source of energy. Solar energy in general, and solar PV in particular, constitutes such a potential alternative.

## 7.5   MODELING AND SIMULATION OF NATIONAL INTEGRATED ENERGY SYSTEMS

### 7.5.1   INTEGRATED ENERGY SYSTEMS

An integrated national energy system is highly complex, containing technological, environmental, socio-economic and political components. Modeling and simulating such a complex system for rational policy planning is a formidable challenge. Coordinated integrated energy planning requires detailed analysis of the interrelations between the different sectors and subsectors of the entire national energy system using a systems approach and their potential energy requirements on the one hand, and energy supply capabilities on the other hand, with a gradual transition to renewable energy resources.

Integrated national energy planning is motivated by a flexible and continuously updated energy strategy to promote the best use of available energy resources for socio-economic development and improvement of quality of life. The specific tasks of integrated national energy planning are: determination of energy needs, selecting an energy mix to meet the energy requirements with minimum cost, conservation of energy, preservation of the environment and so on.

The integrated energy sector consists of interactions of many components such as fossil fuels, renewable energy sources (e.g., biomass, solar, wind, etc.), and the pricing and management of the energy systems. The modeling and simulation of integrated energy systems for national energy planning should be carried out in an integrated manner considering a systems approach. The integrated energy systems must be simulated to understand the systems clearly before their implementation. Since energy systems contain technological, environmental and socio-economic components, they are highly complex. System dynamics, the computer simulation of complex socio-economic systems based on a systems approach, is the most appropriate technique for modeling and simulating such a complex system (Bala et al., 2017; Bala, 1999; Forester, 1968).

An energy system comprises a supply sector, demand sector and emission sector. The supply sector of a typical developing country is categorized as commercial energy and renewables, mainly biomass fuels and, to a limited extent, solar and wind. Commercial energy comprises natural gas, oil and electricity, while biomass fuels include agricultural wastes, animal wastes and fuelwood. In the demand sector, energy consumers make decisions to utilize gas, oil and electricity based on fuel price and availability, whereas biomass fuel consumption is mainly based on the availability of biomass. In the demand sector, the major energy-consuming sectors are residential, industrial and transportation. The supply sectors (commercial and biomass) include fuel supplies and imports. The imports are equal to the shortfalls in domestic supply. Figure 7.9 shows the structure of integrated energy systems of Bangladesh as a typical case study.

Energy modeling for policy planning in Bangladesh was initiated by Huq (1975). Since then, several studies have been conducted on the modeling of integrated energy systems for both micro and macro-level policy analysis and planning in Bangladesh (Bala, 1997a and 1997b; Bala et al., 1999: Bala and Khan, 2003; Bala, 2006a; Bala et al., 2014).

### 7.5.2 System Dynamics and LEAP Model of Integrated Energy Systems

The system dynamics model in combination with the LEAP (Long-range Energy Alternatives Planning) software developed by Bala (2006a; Bala et al., 2014) is considered here as an illustrative example. The system dynamics model is used to project population, cowdung, and agricultural wastes, and these data are fed into the LEAP model (Bala, 1997b and 2006a; Bala et al., 2014). LEAP is a tool that models energy and environmental scenarios and is used to project energy demand situations. The methodologies of system dynamics and LEAP are discussed in Chapter 2 and Chapter 5, respectively.

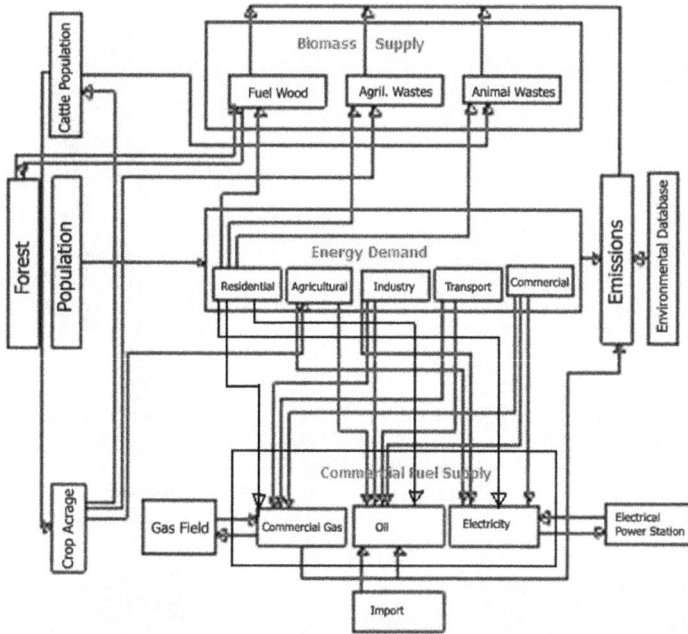

**FIGURE 7.9** Structure of integrated energy systems of Bangladesh.

The principal analytical tool used here is the LEAP model for the Bangladesh energy system (Bala, 1998 and Bala, 1997a and 1997b; Bala et al., 2014), which computes the energy demand and environmental emissions. The overall balances of the energy system in Bangladesh are compiled using a bottom-up approach. This approach starts by looking at various levels of useful energy for functions like cooking, lighting and land preparation using tractors and draft animals. This procedure is followed up to the point where the total energy requirement of the country can be portrayed.

Input data for the base year were collected from secondary sources. Energy demand is projected from activity level and energy intensity, which are based on either interpolation or growth using the LEAP software (Bala, 1998). A system dynamics model was used to provide the input data on crop area, crop production and cattle population to project agricultural wastes and animal wastes (Bala, 1997a and 1997b), and these were used as input into LEAP to project crop and crop wastes, and animal wastes.

## 7.5.3 SIMULATED RESULTS

Simulated electric energy demand projections by Bala et al. (2014) for Bangladesh are shown in Figure 7.10 and Figure 7.11. Figure 7.10 shows the energy demand projection considering 1990 as the base year and shows that the energy demand for

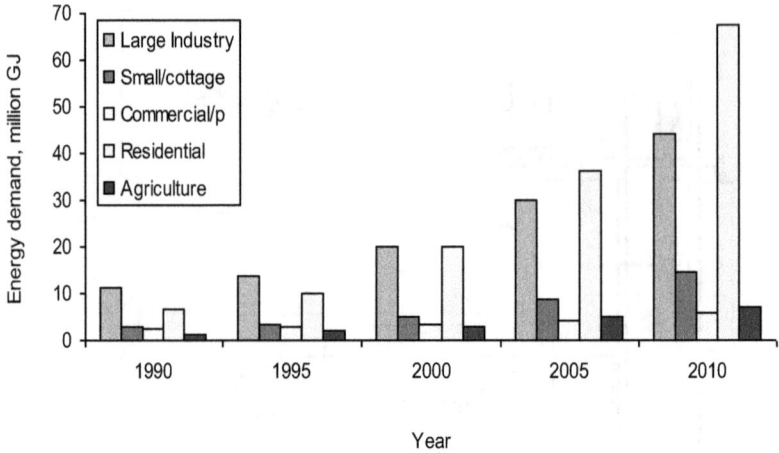

**FIGURE 7.10**   Sector-wise electric energy demand in Bangladesh with 1990 as the base year (Bala, 2006a).

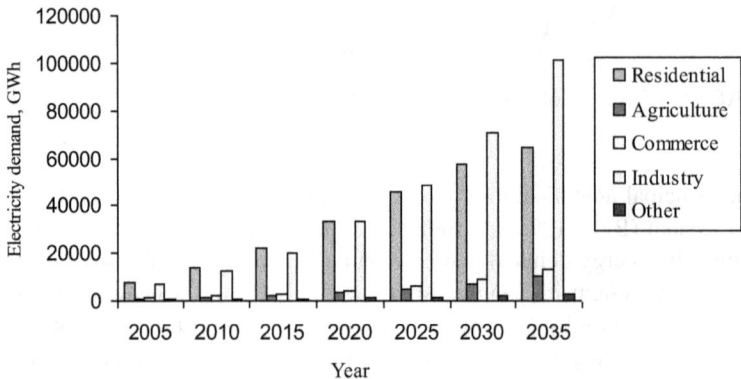

**FIGURE 7.11**   Sector-wise electric energy demand in Bangladesh with 2005 as the base year (Bala, 2006a).

the residential sector increases at a faster rate in comparison to the energy demand for large industry, until 2010. This is mainly due to higher population growth and relatively poor industrial development, as predicted by the trend from the base year. However, the simulated projection of energy demand considering 2005 as the base year shows that energy demand for industry increases rapidly because of the rapid growth of garments and agro-based industries (Figure 7.11).

## 7.5.4   Environmental Effects

Taking 1990 as the base year, emissions of carbon dioxide for energy consumption in Bangladesh are shown in Figure 7.12. The emissions of these gases are increasing.

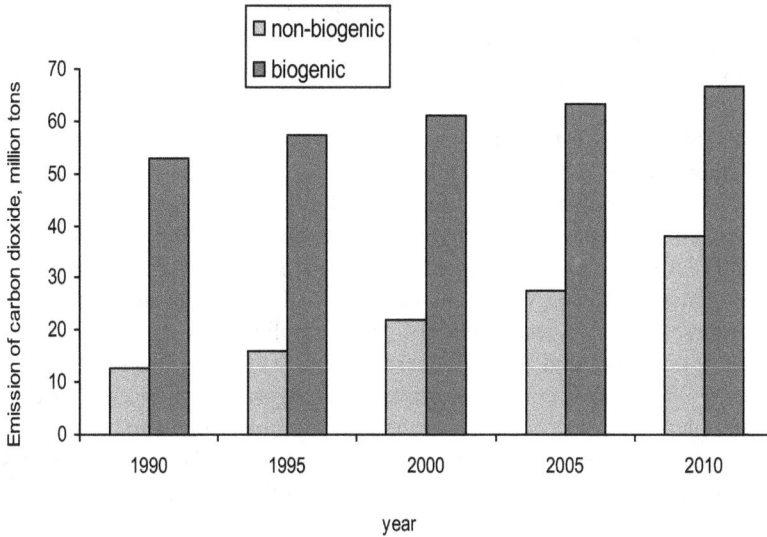

**FIGURE 7.12**  Emissions of carbon dioxide for energy consumption in Bangladesh (Bala, 2006a).

Environmental emissions of non-biogenic $CO_2$ increased from 12.19 million tonnes in 1990 to 36.06 million tonnes in 2010, while biogenic $CO_2$ increased from 49.11 million tonnes in 1990 to 60.36 million tonnes in 2010. To get an idea of how the expected power supply proposed in 2010 affects the $CO_2$ emissions, the model was simulated for $CO_2$ emissions with 2010 as the base year. $CO_2$ emission for the expected power supply of 5936 MW in 2010 to 15880 MW in 2015 and increased from 19.76 million tonnes in 2010 to 58.70 million tonnes in 2015. China is the largest contributor to global warming. In 2017, China's emissions alone reached 9.8 gigatonnes of $CO_2$, which accounted for 27.2% of the global total. Even if the emission intensity is reduced by 60% to 65%, China's total carbon emission will still reach 11.3 gigatonnes by 2030 (Chen et al., 2020). However, the contribution of Bangladesh to global warming is small, and Bangladesh is responsible for a small fraction of the total anthropogenic contribution of $CO_2$ that is only 5% of China's contribution but could be seriously affected by climatic change. However, emissions can be controlled through the application of a suitable carbon tax, and high tax levels would result in a substantial penetration of renewable energy technologies, for example, solar energy, in Bangladesh.

## 7.5.5  Policy Analysis

Basic energy policy is concentrated as follows:

- To provide energy for sustainable growth.
- To meet the energy needs of all the people.

- To ensure the optimum use of renewable energy resources.
- To ensure environmentally sound sustainable programs.
- To ensure public and private participation.

The main policy issue is the supply of electrical energy for sustainable development with a gradual transition to renewable energy resources to ensure economic development and improved quality of life. A broad group of energy policies promotes renewable energy in the electric energy sector today. The simplest schemes consist of both direct subsidies to energy production and indirect mechanisms. A system dynamics model of energy systems can be simulated to develop scenarios that address these policy issues. When designing a policy, it is equally important to consider both supply and demand characteristics and the relevance of each barrier to entry. The introduction of improved cooking stoves to conserve energy, and afforestation programs to enhance carbon sink and supply fuelwood in the biomass energy sector are also advocated. A system dynamics model of energy systems can also be simulated to develop scenarios that address these policy issues.

## REFERENCES

Alam, M. S. (1991). *Integrated modelling of a rural energy system: A system dynamic approach.* PhD Thesis, Bangladesh University of Engineering and Technology, Dhaka.

Alam, M. S., Bala, B. K., Huq, A. M. Z., & Matin, M. A. (1991). A model for the quality of life as a function of electrical energy consumption. *Energy.* 15(4): 739–745.

Alam, M. S., Huq, A. M. Z., & Bala, B. K. (1990). An integrated energy model for a village in Bangladesh. *Energy.* 15: 131–140.

Bala, B. K. (1997a). Computer modeling of the rural energy system and of $CO_2$ emissions for Bangladesh. *Energy.* 22: 999–1003.

Bala, B. K. (1997b). *Computer modelling of energy and environment: The case of Bangladesh.* Proceedings of 15th International System Dynamics conference, Istanbul, Turkey, August 19–22, 1997.

Bala, B. K. (1998). *Energy and Environment: Modelling and Simulation.* Nova Science Publishers, Inc., New York, USA.

Bala, B. K. (1999). *Principles of System Dynamics* (1st ed.). Agrotech Publishing Academy, Udaipur, India.

Bala, B. K. (2003). *Renewable Energy.* Agrotech Publishing Academy, Udaipur-313002, India.

Bala, B. K. (2006a). Computer modelling of energy and environment for Bangladesh. *Agricultural Engineering International.* 15(4): 151—160.

Bala, B. K., (2006b). *Mini-grid for an isolated island – Sandwip in Bangladesh.* Proceedings of the 3rd European Conference on PV-Hybrid and Mini-Grid, Centre de CongrèsCongres, Aix -en -Provence, France, held on May 11—12.

Bala, B. K., Alam, M, S., & Denath, N. (2014). Energy perspective of climate change: The case of Bangladesh. *Strategic Planning for Energy and the Environment, Development,* 33(3): 6–22.

Bala B. K., Bhuiya, S. H., & Biswas, B. K. (1999). Simulation of electric power requirements and supply strategies. *Energy and Environment.* 1: 85–92.

Bala, B. K., Fatimah, M. A., & Kushairi, M. N. (2017). *System Dynamics: Modelling and Simulation.* Springer.

Bala, B. K; Islam, M. M, & Sufian, M. A. (2006). Mini-grid for sustainable integrated coastal development. *Agricultural Engineering Division Journal, IEB*, 32, pp. 55–62.

Bala, B. K., Karim, M. M., & Dutta, D. P. (1989). Energy-use pattern of an electrified village in Bangladesh. *Energy*. 14(2): 61–65.

Bala, B. K., & Siddique, S. A. (2009). Optimal design of a PV-diesel hybrid system for electrification of an isolated island—Sandwip in Bangladesh using genetic algorithm. *Energy for Sustainable Development*. 13: 137–142.

Bernal-Agustín, J. L, Dufo-López, R., & Rivas-Ascaso, D. M. (2005). Design of isolated hybrid systems minimizing costs and pollutant emissions. *Renewable Energy*, 31(14), pp. 2227–2244.

Chen, H., Yang, L., Chen, W., (2020). Modelling national, provincial and city-level low-carbon energy transformation pathways. *Energy Policy* 137, 111096. https://doi.org/10.1016/j.enpol.2019.111096

Devadas, V. (1997). Linear programming model for optimum resource allocation in rural systems. *Energy Sources*. 19: 613–619.

Dufo-López, R., & Bernal-Agustín, J. L. (2005). Design and control strategies of PV-diesel systems using genetic algorithms. *Solar Energy*, 79, pp. 33–46.

Forrester, J. W. (1968). *Principles of Systems.* Allen Wright Press, Cambridge, Massachusetts, USA.

Goldberg, D. E. (1989). *Genetic Algorithms in Search, Optimization and Machine Learning.* Addison-Wesley, Reading, MA, USA.

Holland, J. H. (1975). *Adaptation in Natural and Artificial Systems*, University of Michigan Press, MIT, USA.

Houqe, A., Huq, A. M. Z., & Rahman, S. F. (1989). Possibility of small-scale hybrid electricity generation system from renewable sources in the remote and isolated areas of Bangladesh. *Electrical Engineering Research Bulletin*, 5(1): 45–53.

Huq, A. Z. M. (1975). *Energy modelling for agriculture units in Bangladesh.* Paper presented at the National Seminar on Integrated Rural Development, Dhaka, 1975.

Islam, M. M. (2004). *Design of a Solar-Diesel Hybrid Mini Grid System of Sandwip Island.* M S Thesis, Dept. of Farm Power & Machinery, Bangladesh Agricultural University, Mymensingh, Bangladesh.

Michalewicz, Z. (1996). *Genetic Algorithms + Data Structures = Evolution Programs* (3rd ed.). Springer-Verlag, New York, USA.

Parikh, J. (1985). Modeling energy and agriculture interactions – I: A rural energy systems model. *Energy*. 10(7): 793–804.

Parikh, J. (1997). *Energy Models for 2000 and Beyond.* Tata McGraw Hill, New Delhi, India.

Parikh, J., & Krömer, G. (1985). Modeling energy and agriculture interactions – II: Food–fodder–fuel–fertilizer relationships for biomass in Bangladesh. *Energy*. 10(7): 804–817.

Salequezzaman, M. (2003). *Can Tidal Power Promote Sustainable Integrated Coastal Development in Bangladesh?.* Ph. D. Thesis, Institute for Sustainability and Technology Policy (ISTP), Murdoch University, Western Australia.

Schmid, J. (2003). *Minigrids for rural development and economic growth.* ISET, University of Kassel, Germany.

Singh, I., & Subbarayan, M. (1986). Optimal use of energy inputs on different size-groups of farms in Meerut district of Uttar Pradesh. *Indian Journal of Agricultural Economics*. 41(1): 59–70.

Subhash, C., & Satsangi, P. S. (1990). An integrated planning and implementation-strategy for rural energy systems. *Energy*. 15(10): 913–920.

Wichert, B. (1997). PV-diesel hybrid energy systems for remote area power generation – A review of current practice and future developments. *Renewable and Sustainable Energy Reviews*, 1(3), pp. 209–228.

# 8 Modeling of Rural Energy Systems

## 8.1 INTRODUCTION

The rural energy system in developing countries consists mainly of animal waste, such as cowdung, and agricultural wastes, such as crop residues and fuelwood mainly used for cooking, and, to a lesser extent, diesel for cultivation and electricity for irrigation and rural use in houses. Rural energy applications include cooking, plowing, irrigation and lighting. The major share of rural energy used for cooking is more than 70%, and it is used mostly inefficiently (efficiency less than 10%). Improved stoves can be used for cooking purposes with considerable savings of biomass fuels such as agricultural wastes, briquettes, and fuelwood. Digesters can be used for the production of biogas using animal wastes, and as a bonus, treated slurry can be used as fertilizer. The consumption of fuelwood creates stresses on rural forests resulting in deforestation and reduction in forest cover, which can be avoided by afforestation programs. The patterns of energy use reflect the energy requirements. These issues can be best analyzed within the framework of integrated rural energy systems, and policies can be derived for rural development within the framework of a national energy planning strategy. Thus, a rural energy system plays an important role in the development and growth of rural areas and requires modeling and, hence, planning for sustainable development.

Huq (1975) initiated energy modeling and planning in Bangladesh and proposed a more complete and realistic qualitative model for integrated rural energy utilization. Bala and Satter (1986) developed a simulation model that considers integrated energy use for food production. Parikh and Kromer (1985, a and b) developed a general linear programming model for energy and agricultural interaction in the rural areas of developing countries, and this model was applied to Bangladesh. Alam et al. (1990) developed a system dynamics model for an integrated rural energy system in Bangladesh, and the model has great relevance with historical behavior and is a very useful tool for policy analysis and planning. Bala (1997 and 1998) presented a critical analysis of integrated energy planning and analysis and projections of energy supply and demand and addressed the contribution to global warming of rural energy systems in Bangladesh. Several studies have also been reported on the modeling of rural energy systems (Bala, 1997and 1998; Bala and Satter, 1986 a and b; Alam et al.,1990).

DOI: 10.1201/9781003218401-8

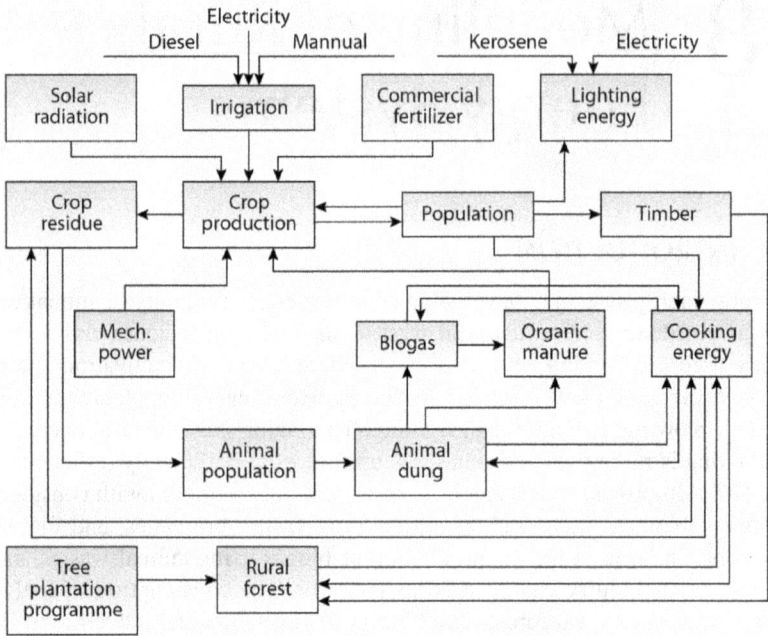

**FIGURE 8.1**   Major components of rural energy systems.

## 8.2   VERBAL DESCRIPTION

The major components of a rural energy system are shown in Figure 8.1 and include population, crop production, beef and dairy cattle production, fuel consumption and rural forests. Crop production depends on the availability of fertilizers, irrigation facilities, mechanical power, and labor. The crop residue derived from crop production is used for animal feed and as cooking fuel by rural households. Beef cattle provide meat, and dairy cattle provide milk. The cowdung produced is used for cooking energy and fertilizers and in recently introduced biogas digesters. Cooking is done most inefficiently by traditional cooking stoves. However, improved cooking stoves are becoming popular. Cooking energy demand depends on family size, and this requirement is supplied by fuelwood, crop residue and animal dung. An afforestation program is in operation to avoid the overcutting of rural forests, which creates an ecological imbalance. A rural electrification program is also in operation to raise the standard of quality of life for rural people.

## 8.3   DYNAMIC HYPOTHESIS

The dynamic hypothesis is a conceptual model typically consisting of a causal loop diagram, stock–flow diagram, or combination thereof. The dynamic hypothesis seeks to define the critical feedback loops that drive the system's behavior. When the model based on the feedback concept is simulated, the endogenous structure of the model should generate the reference mode behavior of the system, and thus the endogenous structure causes changes in the dynamic behavior of the system (Sterman, 2000). Rural integrated energy systems can be represented by causal loop diagrams and stock–flow diagrams, and the simulation model based on these diagrams can generate the dynamic behavior of rural energy systems. Rural energy systems in the form of causal loop diagrams and stock–flow diagrams are hypothesized to generate the observed behavior of rural energy production systems in the reference mode.

## 8.4   SYSTEM DYNAMICS MODELING OF INTEGRATED RURAL ENERGY SYSTEMS

Successful integration of energy systems requires detailed information on dynamic behavior. By substituting models for real systems, designers of integrated energy systems can perform tests to find alternative courses of action and trade-offs with models that are accurate representations of real systems. Integrated rural energy systems are highly complex, containing technological, environmental, socio-economic and political components, and modeling of such complex systems is a formidable challenge. System dynamics methodology is the most appropriate technique for handling such complex systems (Bala et al., 2017; Bala, 1999; Forester, 1968; Sterman, 2000; Mohapatra et al., 1994; Maani and Cavana, 2000).

Rural energy systems in developing countries such as Bangladesh are mainly comprised of biomass, commercial fertilizer and, to a lesser extent, diesel and electricity. Energy is used to grow and prepare food, provide shelter and protect health, and thereby improve the quality of life. The integrated rural energy model described here consists of six sub-models: population, cattle population, crop production, fuel consumption, rural electrification and quality of life.

### 8.4.1   POPULATION SUB-MODEL

The population is divided into sex, and each sex is divided into three age groups, children (0–14), adults (15–55) and old (above 55), to account for the varying effects on births, deaths and labor force participation. The causal loop diagram of the population model is shown in Figure 8.2. Female adult, birth rate and female child form the re-enforcing feedback loop R1. The population of each age group of both sexes forms a balancing feedback loop, with the corresponding death rate (B2–B6). Thus, there are seven feedback loops that simulate population dynamics. Quality of life, food per capita, and population density link population non-linearly. These factors act as the mechanism of population control.

The population is categorized into male population and female population. The stock–flow diagram of the population sub-model is shown in Figure 8.3.

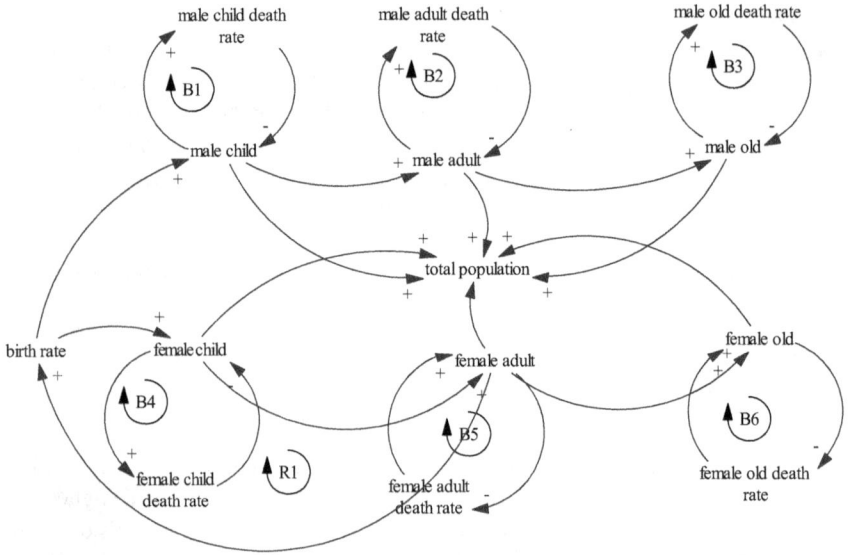

**FIGURE 8.2**  Causal loop diagram of the population model.

The fundamental equations that correspond to the major variables shown in Figure 8.3 are as follows:

**Population stock**
The number of people in each age group for both sexes is represented by stocks or integral equations, which are affected by relevant flow rates of births, growths and deaths.

**Male children**
male_child(t) = male_child(t − dt) + (birth_rate_of_male_child − death_rate_of_male_child − growing_rate_from_male_child) * dt
INIT male_child = 24523000

**Male adult**
male_adult(t) = male_adult(t − dt) + (growing_rate_from_male_child − growing_rate_from_male_adult − death_rate_of_male_adult) * dt
INIT male_adult = 25030000

**Male old**
male_old(t) = male_old(t − dt) + (growing_rate_from_male_adult − death_of_male_old) * dt
INIT male_old = 4762000

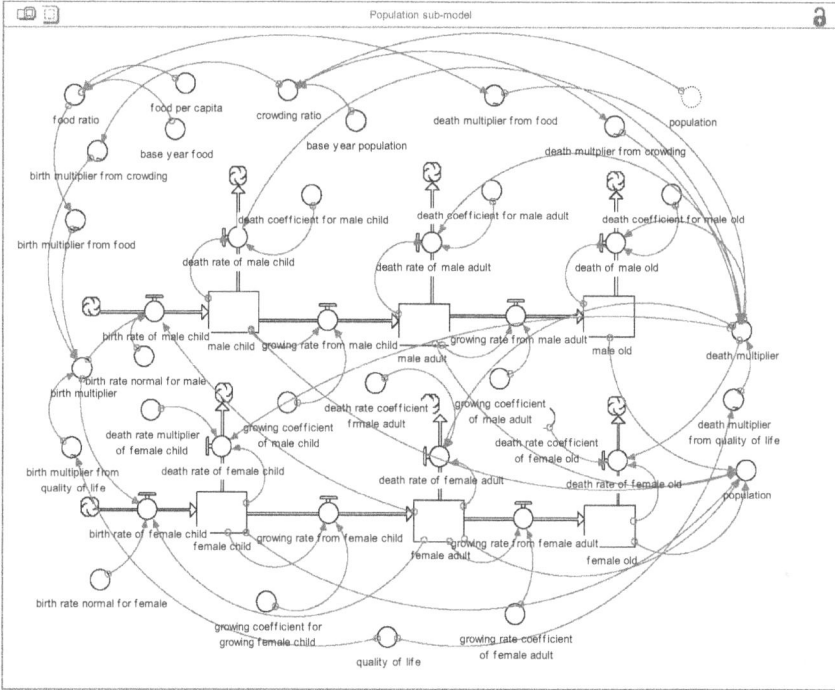

**FIGURE 8.3**   Stock–flow diagram of the population sub-model.

## Female children
female_child(t) = female_child(t – dt) + (birth_rate_of_female_child – growing_rate_from_female_child – death_rate_of_female_child) * dt
INIT female_child = 23391000

## Female adult
female_adult(t) = female_adult(t – dt) + (growing_rate_from_female_child – death_rate_of_female_adult – growing_rate_from_female_adult) * dt
INIT female  adult = 22882000

## Female old
female_old(t) = female_old(t – dt) + (growing_rate_from_female_adult – death_rate_of_female_old) * dt
INIT female_old = 5344000

## Birth rate
Basic birth rate depends on adult female and the coefficient of birth rate per year as a fraction of adult female. It is the rate at which children are being added to the stocks of male children and female children. The basic birth rate depends on and

is altered by so many real-world situations. Hence, in this model, three multiplier effects have been considered: birth rate multiplier from crowding, birth rate multiplier from food and birth rate multiplier from quality of life.

**Birth rate for male children**

**INFLOWS**
birth_rate_of_male_child = female_adult * birth_multiplier * birth_rate_normal_
for_male
birth_rate_normal_for_male = 0.117

**Birth rate for female children**

**INFLOWS**
birth_rate_of_female_child = female_adult * birth_multiplier * birth_rate_
normal_for_female
birth_rate_normal_for_female = 0.096

**Death rate**
Like birth rate, basic death rate depends on respective group of population and the respective coefficient (death rate normal). It is the rate at which each age group is being depleted from the respective population stocks. The basic death rate has been assumed to be affected by food, crowding and quality of life. The death rate multipliers are death multiplier from crowding, death rate multiplier from food and death rate multiplier from quality of life. The death rate multiplier is calculated by multiplication of death multiplier from crowding, death rate multiplier from food and death rate multiplier from quality of life.

**Death rate for male children**

**OUTFLOWS**
death_rate_of_male_child = male_child * death_coefficient_for_male_child *
death_multiplier
death_coefficient_for_male_child = 0.0166

**Death rate for male adult**

**OUTFLOWS**
death_rate_of_male_adult = male_adult * death_coefficient_for_male_adult *
death_multiplier
death_coefficient_for_male_adult = 0.026

**Death rate for male old**

**OUTFLOWS**
death_of_male_old = male_old * death_coefficient_for_male_old * death_
multiplier
death_coefficient_for_male_old = 0.07

**Death rate for female children**

**OUTFLOWS**
death_rate_of_female_child = female_child * death_multiplier * death_rate_
multiplier_of_female_child
death_rate_multiplier_of_female_child = 0.016

**Death rate for female adult**

**OUTFLOWS**
death_rate_of_female_adult = female_adult * death_multiplier * death_rate_coef-
ficient_female_adult
death_rate_coefficient_female_adult = 0.025

**Death rate for female old**

**OUTFLOWS**
death_rate_of_female_old = female_old * death_multiplier * death_rate_coeffi-
cient_of_female_old
death_rate_coefficient_of_female_old = 0.068

**Growing rate**
Growing rate depends on respective population stock and coefficients. It is the rate
at which the respective population migrate out from one stock and migrate into
another.

**Growing rate for male children**

**OUTFLOWS**
growing_rate_from_male_child = male_child * growing_coefficient_of_male_
child
growing_coefficient_of_male_child = 0.072

**Growing rate for male adult**

**OUTFLOWS**
growing_rate_from_male_adult = male_adult * growing_coefficient_of_male_ adult
growing_coefficient_of_male_adult = 0.0185

**Growing rate for female children**

**OUTFLOWS**
growing_rate_from_female_child = female_child * growing_coefficient_for_ growing_female_child
growing_coefficient_for_growing_female_child = 0.072

**Growing rate for female adult**

**OUTFLOWS**
growing_rate_from_female_adult = female_adult * growing_rate_coefficient_of_ female_adult
growing_rate_coefficient_of_female_adult = 0.020

**Birth rate multiplier**
In this model, three multiplier effects have been considered: birth rate multiplier from crowding, birth rate multiplier from food and birth rate multiplier from quality of life.

Birth_multiplier = birth_multiplier_from_crowding * birth_multiplier_from_food * birth_multiplier_from_quality_of_life

**Birth rate multiplier from crowding**
As the crowding ratio remains below normal, there is little effect on birth rate. But when it goes up, birth rate starts declining due to psychological pressure, social stress and conflict. The relationship between birth rate multiplier from crowding and crowding ratio is expressed graphically as:

birth_multiplier_from_crowding = GRAPH(crowding_ratio)
(0.00, 0.994), (0.5, 0.979), (1.00, 0.954), (1.50, 0.936), (2.00, 0.916), (2.50, 0.89), (3.00, 0.868), (3.50, 0.82), (4.00, 0.781), (4.50, 0.745), (5.00, 0.7)

**Birth rate multiplier from food**
The availability of food has a pronounced effect on the birth rate. Food availability alone can regulate both birth and death rates in such a manner that population becomes stationary depending on the availability of food. At zero food, life becomes impossible, and zero birth rate is indicated. At the other extreme, an abundance of food is assumed to raise the birth rate. The relationship between birth rate multiplier from food and food ratio is expressed graphically as:

birth_multiplier_from_food = GRAPH(food_ratio)
(0.00, 1.00), (0.5, 1.01), (1.00, 1.02), (1.50, 1.04), (2.00, 1.06), (2.50, 1.09), (3.00, 1.12), (3.50, 1.14), (4.00, 1.19), (4.50, 1.25), (5.00, 1.29)

## Birth rate multiplier from quality of life
Birth rate is inversely proportional to the quality of life. The effect of quality of life on birth rate is to control population growth through birth control. The relationship between birth rate multiplier from quality of life and quality of life is expressed graphically as:

birth_multiplier_from_quality_of_life = GRAPH(quality_of_life)
(0.00, 1.29), (0.5, 1.14), (1.00, 1.00), (1.50, 0.88), (2.00, 0.788), (2.50, 0.724), (3.00, 0.656), (3.50, 0.616), (4.00, 0.584), (4.50, 0.548), (5.00, 0.512)
quality_of_life = 0.95

## Death rate multiplier
Basic death rate has been assumed to be affected by food, crowding and quality of life. Death rate multipliers are death rate multiplier from food, death rate multiplier from crowding and death rate multiplier from quality of life.

Death_multiplier = death_multiplier_from_food * death_multiplier_from_quality_of_life * death_multplier_from_crowding

## Death rate multiplier from food
Food essentially is the most powerful regulator of population. As food availability approaches zero, death rate approaches infinity. With increasing food availability, death rate still will not be reduced to the desirable extent because of the effect of pressure level of education, health and living condition. The relationship between death rate multiplier from food and food ratio is expressed graphically as:

death_multiplier_from_food = GRAPH(food_ratio)
(0.00, 0.999), (0.5, 0.976), (1.00, 0.944), (1.50, 0.917), (2.00, 0.896), (2.50, 0.872), (3.00, 0.858), (3.50, 0.844), (4.00, 0.832), (4.50, 0.817), (5.00, 0.802)

## Death rate multiplier from crowding
If other influences were to intervene, excessive crowding would eventually limit population. As Forester (1971) suggests, crowding can give rise to pressure, which includes psychological effects, social stresses that cause crime, and conflict. The relationship between death rate multiplier from crowding and crowding ratio is graphically expressed as:

death_multplier_from_crowding = GRAPH(crowding_ratio)
(0.00, 1.02), (0.5, 1.05), (1.00, 1.07), (1.50, 1.09), (2.00, 1.11), (2.50, 1.13), (3.00, 1.15), (3.50, 1.18), (4.00, 1.21), (4.50, 1.25), (5.00, 1.30)

**Death rate multiplier from quality of life**
As the crowding multiplier of death, quality of life multiplier regulates the death rate effectively. The relationship between death rate multiplier from quality of life and quality of life is expressed graphically as:

death_multiplier_from_quality_of_life = GRAPH(quality_of_life)
(0.00, 0.998), (0.5, 0.943), (1.00, 0.892), (1.50, 0.849), (2.00, 0.816), (2.50, 0.788), (3.00, 0.769), (3.50, 0.752), (4.00, 0.739), (4.50, 0.725), (5.00, 0.712)
quality_of_life = 0.95

**Total population**
Population here is total of population of both sexes of different age groups and is expressed as:

population = female_adult + female_child + female_old + male_adult + male_child + male_old

**Food ratio**
Food ratio is the ratio of food per capita to base year food per capita and is expressed as:

food_ratio = food_per_capita/base_year_food
base_year_food = 284
food_per_capita = 284

**Crowding ratio**
Crowding ratio is the ratio of population to base year population and is expressed as:

crowding_ratio = population/base_year_population
base_year_population = 105932000

## 8.4.2   CROP PRODUCTION SUB-MODEL

The crop production considered here is rice production, which provides food and straw. The straw is primarily used as cattle feed and the crop residue is used as a cooking fuel in developing countries. Crop production can be increased by increasing land and irrigated land and increasing yield. However, the land available and land suitable for irrigation have their upper limits. The potential yield can be increased up to a certain limit through research and development. However, the actual yield is affected by regeneration of the yield fallowing the land and degeneration of the yield by cropping intensity. Figure 8.4 shows a causal loop diagram of the crop production model. There are five negative feedback loops and one positive feedback loop. Irrigated land forms one negative feedback loop B1 with increase in irrigated land and also forms another negative feedback loop B5 with discard rate. The yield potential and yield potential increase rate form the

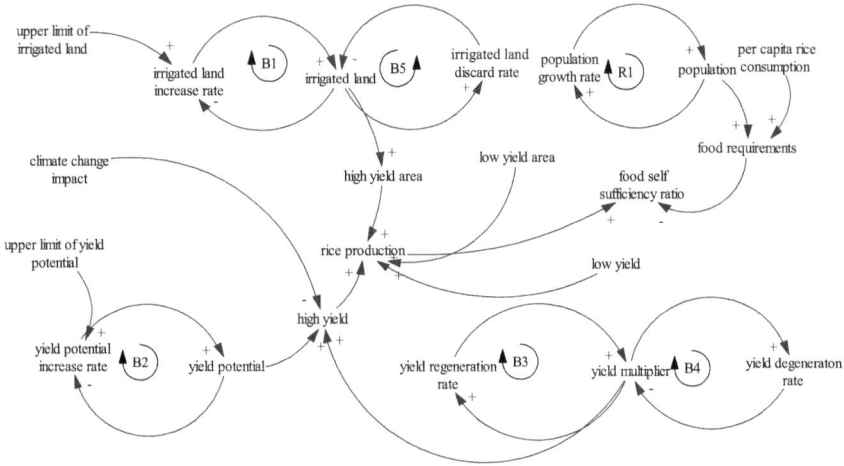

**FIGURE 8.4** Causal loop diagram of the crop production model.

negative feedback loop B2. Yield multiplier forms the negative feedback loop B3 with yield regeneration rate and also forms the negative feedback loop B4 with yield degeneration, respectively. Population and population growth form re-enforcing feedback loop R1. These loops simulate the dynamic behavior of the crop production system.

The stock–flow diagram of a crop production system consists of irrigated land, population, yield potential and yield multiplier. The stock–flow diagram of the crop production system is shown in Figure 8.5. The fundamental equations that correspond to the major variables shown in Figure 8.5 are as follows:

### 8.4.2.1 Irrigated Land

Irrigated land is a stock variable and is increased by increasing rate of irrigated land and decreased by decreasing rate of irrigated land. It is represented by an integral equation.

Irrigated_land(t) = irrigated_land(t − dt) + (increasing_rate − decreasing_rate) * dt
INIT irrigated_land = 3.42E+6

### Increasing rate of irrigated land

Increasing rate of irrigated land is computed using logical IF, which is a built-in function in STELLA. If the high-yield area is less than the upper limit, then the increasing rate of irrigated area will be equal to irrigated area expansion rate, which is calculated by the multiplication of irrigated land by its expansion coefficient. When high yield area is greater than or equal to the upper limit, then the increasing rate of irrigated area will be equal to zero.

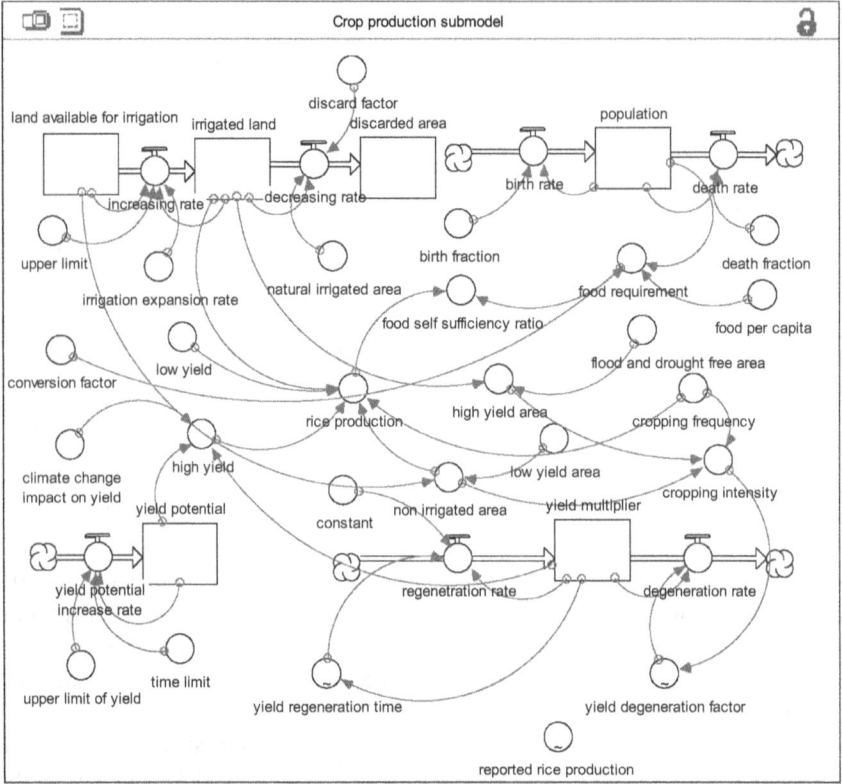

**FIGURE 8.5**    Stock–flow diagram of the crop production sub-model.

**INFLOWS**
increasing_rate = IF(irrigated_land – (0.8E+6)) >= (upper_limit)THEN(0)
ELSE(MIN((land_available_for_irrigation),(irrigation_expansion_rate    *
irrigated_land)))
irrigation_expansion_rate = 0.085
upper_limit = 7.0E+6

**Decreasing rate of irrigated land**
Decreasing rate of irrigated land is computed by multiplying irrigated area by discard factor.

**OUTFLOWS**
decreasing_rate = (irrigated_land – natural_irrigated_area) * discard_rate
natural_irrigated_area = .8E+6
discard_rate = .0085

### Land available for irrigation

Land available for irrigation decreases with the increase in irrigated land and is computed as:

land_available_for_irrigation(t) = land_available_for_irrigation(t – dt) + (-increasing_rate) * dt
INIT land_available_for_irrigation = 4.76E+6

### Land discarded

Land discarded is the accumulation of the discarded irrigated area and increases by the decreasing irrigated area.

Discarded_area(t) = discarded_area(t – dt) + (decreasing_rate) * dt
INIT discarded_area = 0

### Yield multiplier

Yield multiplier is increased by fallowing the land and decreased by cropping intensity and is expressed as:

yield_multiplier(t) = yield_multiplier(t – dt) + (regeneration_rate – degeneration_rate) * dt
INIT yield_multiplier = 1.0

### Regeneration rate

Yield regeneration depends on yield multiplier and yield regeneration time and is expressed as

### INFLOWS

regeneration_rate = (constant – yield_multiplier)/yield_regeneration_time
constant = 1

### Yield regeneration time

Yield regeneration time decreases with the increase in yield multiplier, and the relationship between yield regeneration time and yield multiplier is expressed graphically as:

yield_regeneration_time = GRAPH(yield_multiplier)
(0.00, 120), (0.25, 80.0), (0.5, 30.0), (0.75, 15.0), (1.00, 1.00)

### Degeneration rate

Yield degeneration depends on cropping intensity and yield degeneration factor and is expressed as:

### OUTFLOWS

degeneration_rate = yield_multiplier * yield_degeneration_factor

## Yield degeneration factor

Yield degeneration factor increases with the increase in cropping intensity. The relationship between yield degeneration and cropping intensity is expressed graphically as:

yield_degeneration_factor = GRAPH(cropping_intensity)
(0.00, 0.00), (0.5, 0.005), (1.00, 0.01), (1.50, 0.015), (2.00, 0.025), (2.50, 0.05), (3.00, 0.1)

## Yield potential

Yield potential is increased by the yield potential increase rate and is expressed as:

yield_potential(t) = yield_potential(t – dt) + (yield_potential_increase_rate) * dt
INIT yield_potential = 2.81

## Yield potential increase rate

Yield potential increase rate depends on yield gap and time limit and is expressed as:

## INFLOWS

yield_potential_increase_rate = (upper_limit_of_yield – yield_potential)/time_limit
time_limit = 50
upper_limit_of_yield = 2.81

## High yield area

High yield area is obtained by summation of irrigated area and flood and drought-free area.
High_yield_area = irrigated_land + flood_and_drought_free_area
flood_and_drought_free_area = 0

## Non-irrigated area

Non-irrigated area is the land available for irrigation and low yield area.

Non_irrigated_area = land_available_for_irrigation + low_yield_area
low_yield_area = 0.2E+6

## Cropping intensity

Cropping intensity is calculated by the summation of the product of high yield area and the frequency of cropping and low yield area and then dividing by the total area.

Cropping_intensity =
(high_yield_area * cropping_frequency + non_irrigated_area)/(high_yield_area + non_irrigated_area)
cropping_frequency = 2

## Rice production

Total rice production is the sum of rice production from high yield area and low yield area and is computed as:

rice_production = (irrigated_land * high_yield * cropping_frequency + non_irrigated_area * low_yield)

## High yield

High yield is computed from yield multiplier, yield potential and climate change impact on yield and is computed as:

high_yield = yield_multiplier * yield_potential * (1 – climate_change_impact_on_yield)
low_yield = 2.81
climate_change_impact_on_yield = 0

## Population

Population is increased by birth rate and decreased by death rate and is computed as

population(t) = population(t – dt) + (birth_rate – death_rate) * dt
INIT population = 151.41E+6

## Birth rate

Birth rate is computed from population and birth fraction as:

## INFLOWS

birth_rate = population * birth_fraction
birth_fraction = 0.0205

## Death rate

Death rate is computed from population and death fraction as:

## OUTFLOWS

death_rate = population * death_fraction
death_fraction = 0.0058

## Food requirement

Food requirement is computed from population, food per capita and conversion factor and is expressed as:

food_requirement = population * food_per_capita * conversion_factor
conversion_factor = 1.00
food_per_capita = (188.4/1000)

**Food self- sufficiency ratio**

Food self-sufficiency ratio is the ratio of rice production to food requirement and is expressed as:

food_self_sufficiency_ratio = rice_production/food_requirement

### 8.4.3   CATTLE POPULATION SUB-MODEL

The cattle population sub-model is based on the seminal publication of Meadows (1970) on commodity production cycles. It is divided into two sectors: a beef cattle sector and a dairy cattle sector. In the beef cattle sector, the price links beef cattle population non-linearly through beef inventory, beef production and beef consumption (Bala, 1998), while in the dairy cattle sector, the price links dairy cattle population non-linearly through milk inventory, milk production and milk consumption (Conrad, 2004).

#### 8.4.3.1   Beef Cattle Population Sector

Figure 8.6 shows the causal loop diagram of the beef cattle sector of the cattle population sub-model. There are four negative feedback loops and one positive feedback loop. Desired breeding stock, beef calf and stockers form the negative feedback loop B1 and expected beef price and change in expected beef price rate form the negative feedback loop B2. There are two dominant feedback loops in the beef cattle sector, and these are negative feedback loop 3 and negative feedback loop 4. Beef inventory, price and production form the negative feedback loop B3, while beef inventory, price and consumption form the negative feedback loop B4. Population and population growth form re-enforcing feedback loop R1. These loops simulate the dynamic behavior of the beef production system.

Beef cattle are categorized into three age groups: calf, stocker and mature beef cattle. In addition, there is another stock of beef cattle called breeding stock that initiates beef cattle production. Figure 8.7 shows the stock–flow diagram of the beef cattle sector of the cattle population sub-model.

The fundamental equations that correspond to the major variables shown in Figure 8.7 are as follows:

**Beef calf**

Beef calf is increased by beef calf breeding rate and decreased by weaning into stocker and is expressed as:

beef_calf(t) = beef_calf(t - dt) + (beef_calf_breeding - beef_weaning_rate) * dt
INIT beef_calf = 5000000

**Beef calf breeding rate**

Beef calf breeding depends on the breeding stock and beef cattle fertility and is expressed as:

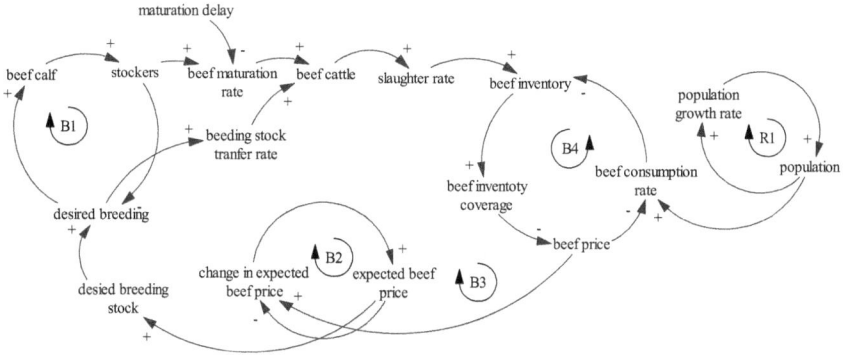

**FIGURE 8.6** Causal loop diagram of the beef cattle sector of cattle population sub-model.

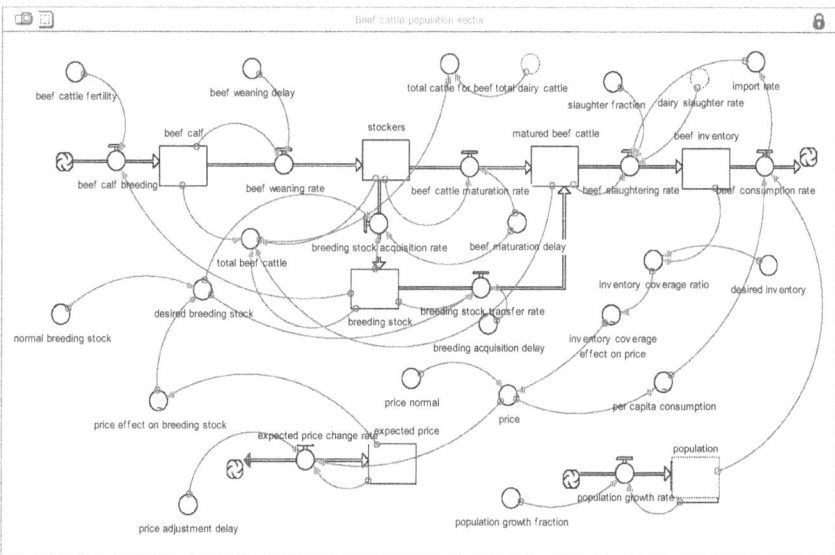

**FIGURE 8.7** Stock–flow diagram of the beef cattle sector of cattle population sub-model.

## INFLOWS
beef_calf_breeding = breeding_stock * beef_cattle_fertility
beef_cattle_fertility = 0.1

## Beef calf weaning rate
Beef calf weaning rate depends on the beef calf and beef calf weaning delay and is expressed as:

**OUTFLOWS**
beef_weaning_rate = beef_calf/beef_weaning_delay
beef_weaning_delay = 0.25

## Stockers
Stockers are increased by beef calf weaning rate into stockers and decreased by maturation into beef cattle and acquisition for breeding stock, and is expressed as:

stockers(t) = stockers(t - dt) + (beef_weaning_rate - beef_cattle_maturation_rate - breeding_stock_aquisition_rate) * dt
INIT stockers = 15000000

## Beef cattle maturation rate
There is a maturation delay between the production initiation and production realization of matured beef, and thus beef cattle maturation rate depends on the stock of stockers and beef cattle maturation delay. The matured cattle production rate is obtained using the third-order smoothing function and is expressed as:

**OUTFLOWS**
beef_cattle_maturation_rate = SMTH3(stockers,beef_maturation_delay)
beef_maturation_delay = 2.25

## Breeding stock acquisition rate
Breeding stock acquisition rate is essentially the acquisition rate of stockers for breeding, and it depends on the breeding stock gap and beef cattle maturation delay. The breeding stock production rate is obtained using the third-order smoothing function and is expressed as:

breeding_stock_aquisition_rate = SMTH3((desired_breeding_stock-breeding_stock),beef_maturation_delay)

## Desired breeding stock
Desired breeding stock is computed from normal breeding stock and price effect on breeding stock and is expressed as:

desired_breeding_stock = normal_breeding_stock * price_effect_on_breeding_stock
normal_breeding_stock = 640000

## Price effect on breeding stock
Price effect on breeding stock is the effect of expected price on the breeding stock. The relationship between breeding stock and expected price is expressed graphically as:

price_effect_on_breeding_stock = GRAPH(expected_price)
(0.00, 0.892), (10000, 1.10), (20000, 1.19), (30000, 1.25), (40000, 1.31), (50000, 1.34), (60000, 1.36), (70000, 1.38), (80000, 1.42), (90000, 1.43), (100000, 1.45)

**Breeding stock**
Breeding stock is increased by beeding stock acquisition rate into breeding stock and decreased by breeding stock transfer rate and is expressed as:

breeding_stock(t) = breeding_stock(t - dt) + (breeding_stock_aquisition_rate - breeding_stock_transfer_rate) * dt
INIT breeding_stock = 3310000

**Breeding stock transfer rate**
Breeding stock transfer rate depends on the breeding stock gap and breeding stock acquisition delay and is expressed as:

**OUTFLOWS**
breeding_stock_transfer_rate = (desired_breeding_stock - breeding_stock)/ breeding_aquisition_delay
breeding_aquisition_delay = 0.5

**Matured beef cattle**
Matured beef cattle is increased by beef cattle maturation rate plus breeding stock transfer rate and decreased by beef cattle slaughtering rate and is expressed as:

matured_beef_cattle(t) = matured_beef_cattle(t - dt) + (beef_cattle_maturation_ rate + breeding_stock_transfer_rate - beef_slaughtering_rate) * dt
INIT matured_beef_cattle = 10500000

**Beef slaughtering rate**
Beef slaughtering rate is computed from the matured beef cattle and slaughter fraction of mature beef cattle, dairy cattle slaughter rate, and import of beef cattle and is expressed as:

**OUTFLOWS**
beef_slaughtering_rate = matured_beef_cattle * slaughter_fraction + dairy_ slaughter_rate + import_rate
slaughter_fraction = 0.5

**Import rate**
Beef production is not sufficient; hence 5% of the consumption of beef is imported, and this is expressed as:

import_rate = 0.05 * beef_consumption_rate

**Beef inventory**
Beef inventory is increased by beef slaughtering rate and decreased by beef con-sumption rate and is expressed as:

beef_inventory(t) = beef_inventory(t - dt) + (beef_slaughtering_rate - beef_con-sumption_rate) * dt
INIT beef_inventory = 10500000

**Beef consumption rate**
Beef consumption rate depends on population and per capita consumption of beef and is expressed as:

**OUTFLOWS**
beef_consumption_rate = population * per_capita_consumption/120

**Per capita beef consumption rate**
Per capita consumption of beef depends on price of beef. The relationship between per capita consumption of beef and price of beef is expressed graphically as:

per_capita_consumption = GRAPH(price)
(0.00, 7.50), (10000, 6.30), (20000, 4.40), (30000, 3.50), (40000, 2.50), (50000, 1.95), (60000, 1.35), (70000, 1.30), (80000, 0.8), (90000, 0.9), (100000, 0.8)

**Expected price of beef**
Expected price is a stock variable and it is represented by an integral equation. Expected price of beef is obtained by averaging the difference between the current market price and the previous expected price over a period of two years and is expressed as:

expected_price(t) = expected_price(t - dt) + (expected_price_change_rate) * dt
INIT expected_price = 70000

**Change in expected price**
Change in expected price of beef depends on the difference between price and expected price, and price adjustment delay, and is expressed as:

**INFLOWS**
expected_price_change_rate = (price - expected_price)/price_adjustment_delay
price_adjustment_delay = 2

**Price of beef**
Price of beef is computed from price normal and the effect of inventory coverage on price and is expressed as:

price = price_normal * inventory_coverage_effect_on_price
price_normal = 60000

**Inventory coverage effect on price**
The price of beef depends on the inventory coverage. The relationship between price and inventory coverage of beef is expressed graphically as:

inventory_coverage_effect_on_price = GRAPH(inventory_coverage_ratio)
(0.00, 1.45), (100, 1.07), (200, 0.922), (300, 0.78), (400, 0.69), (500, 0.6), (600, 0.547), (700, 0.502), (800, 0.443), (900, 0.375), (1000, 0.3)

**Inventory coverage ratio**

Inventory coverage ratio is the ratio of beef inventory to desired inventory and is expressed as:

inventory_coverage_ratio = beef_inventory/desired_inventory
desired_inventory = 450000

**Population**

Population is increased by population growth and is expressed as:

population(t) = population(t - dt) + (population_growth_rate) * dt
INIT population = 118000000

**Population growth rate**

Population growth is computed from population and its growth fraction.

**INFLOWS**

population_growth_rate = population * population_growth_fraction
population_growth_fraction = 0.02

**Total beef cattle**

Total beef cattle is the sum of matured beef cattle, stockers, breeding stock and beef calf.

total_beef_cattle = matured_beef_cattle + stockers + breeding_stock + beef_calf

**Total cattle for beef**

Total cattle for beef is the sum of total dairy cattle and total beef cattle.

total_cattle_for_beef = total_dairy_cattle + total_beef_cattle

### 8.4.3.2 Dairy Cattle Population Sector

Figure 8.8 shows the causal loop diagram of the dairy cattle population sector. There are three negative feedback loops and three positive feedback loops. Heifer and dairy breeding form the re-enforcing positive feedback loop R1 and milk cow, dairy calf and heifer form the second re-enforcing positive feedback loop R2. Expected milk price and change in expected milk price form the negative feedback loop B1. There are two dominant feedback loops in the dairy cattle population sector, and these are negative feedback loop B2 and negative feedback loop B3. Milk inventory, milk price, and production form the negative feedback loop B2, while milk inventory, milk price and consumption form the negative feedback loop B3. The population and the population growth forms re-enforcing feedback loop R3. These loops simulate the dynamic behavior of the milk production system.

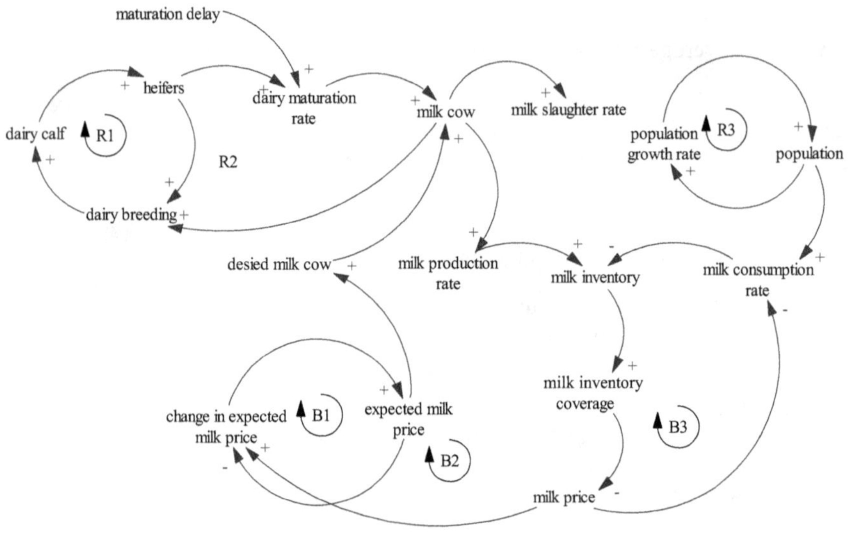

**FIGURE 8.8**  Causal loop diagram of the dairy cattle sector of cattle population sub-model.

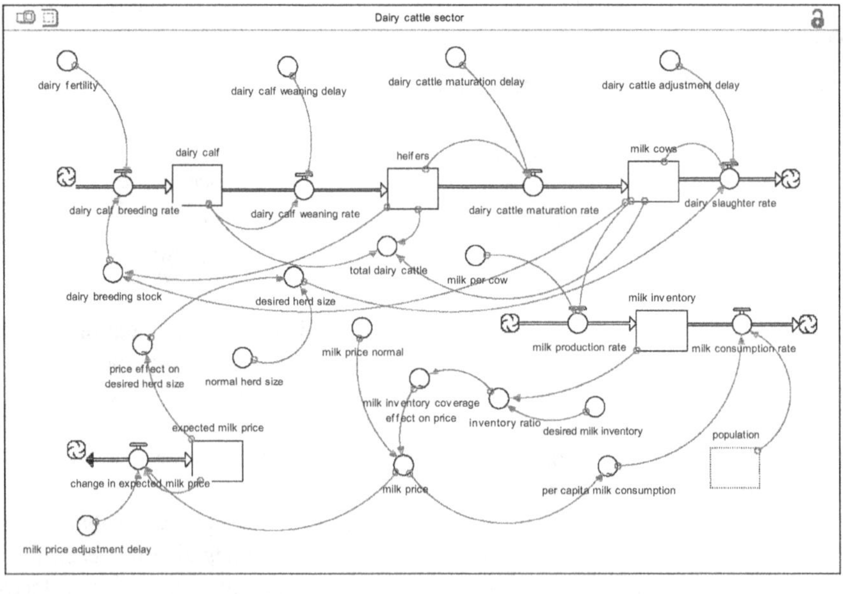

**FIGURE 8.9**  Stock–flow diagram of the dairy cattle sector of the cattle population sub-model.

Dairy cattle are categorized into three age groups: calves, heifers and milk cows. In addition, there is another stock of milk inventory. Figure 8.9 shows the stock–flow diagram of the dairy cattle sector of the cattle population sub-model.

The fundamental equations that correspond to the major variables shown in Figure 8.9 are as follows:

## Dairy calf
Dairy calf is increased by dairy calf breeding rate and decreased by weaning into heifer and is expressed as:

dairy_calf(t) = dairy_calf(t - dt) + (dairy_calf_breeding_rate - dairy_calf_weaning_rate) * dt
INIT dairy_calf = 2000000

## Dairy calf breeding rate
Dairy calf breeding depends on the dairy breeding stock and daily cattle fertility and is expressed as:

## INFLOWS
dairy_calf_breeding_rate = dairy_breeding_stock * dairy_fertility
dairy_fertility = 0.1

## Dairy calf weaning rate
Dairy calf weaning rate depends on the dairy calf and dairy calf weaning delay and is expressed as:

## OUTFLOWS
dairy_calf_weaning_rate = dairy_calf/dairy_calf_weaning_delay
dairy_calf_weaning_delay = 0.25

## Heifers
Heifers are increased by dairy calf weaning rate into heifers and decreased by maturation into dairy cows and is expressed as:

heifers(t) = heifers(t - dt) + (dairy_calf_weaning_rate - dairy_cattle_maturation_rate) * dt
INIT heifers = 3500000

## Dairy cattle maturation rate
Dairy cattle maturation rate depends on the stock of heifers and dairy cattle maturation delay and is expressed as:

## OUTFLOWS
dairy_cattle_maturation_rate = SMTH3(heifers,dairy_cattle_maturation_delay)
dairy_cattle_maturation_delay = 2.25

## Milk cows

Milk cows are increased by dairy cow maturation rate and decreased by dairy cow slaughtering rate and this is expressed as:

milk_cows(t) = milk_cows(t - dt) + (dairy_cattle_maturation_rate - dairy_slaughter_rate) * dt
INIT milk_cows = 4500000

## Dairy cow slaughtering rate

When the milk cows are more than the desired herd, the excess number is slaughtered otherwise no milk cow is slaughtered. It is expressed as:

## OUTFLOWS

dairy_slaughter_rate = MAX(0,(milk_cows - desired_herd_size)/dairy_cattle_adjustment_delay)
dairy_cattle_adjustment_delay = 2

## Desired herd size

Desired herd size depends on the normal herd size and price effect on desired herd size and is expressed as:

desired_herd_size = normal_herd_size * price_effect_on_desired_herd_size
normal_herd_size = 5500000

## Price effect on desired herd size

Price effect on desired herd size depends on the expected milk price. The relationship between desired herd size and expected price of milk is expressed graphically as:

price_effect_on_desired_herd_size = GRAPH(expected_milk_price)
(0.00, 0.87), (10.0, 0.997), (20.0, 1.10), (30.0, 1.14), (40.0, 1.21), (50.0, 1.26), (60.0, 1.28), (70.0, 1.34), (80.0, 1.35), (90.0, 1.40), (100, 1.40)

## Milk inventory

Milk inventory is increased by milk production rate and decreased by milk consumption rate and is expressed as:

milk_inventory(t) = milk_inventory(t - dt) + (milk_production_rate - milk_consumption_rate) * dt
INIT milk_inventory = 120000000

## Milk production rate

Milk production rate depends on the number of milk cows and milk production per cow and is expressed as:

**INFLOWS**

milk_production_rate = milk_cows * milk_per_cow
milk_per_cow = $30 \times 365$

**Milk consumption rate**

Milk consumption rate depends on the population and per capita milk consumption and is expressed as:

**OUTFLOWS**

milk_consumption_rate = population * per_capita_milk_consumption
Per capita milk consumption depends on milk price, and the relationship between the per capita milk consumption and the milk price is expressed graphically as:
per_capita_milk_consumption = GRAPH(milk_price)
(0.00, 143), (10.0, 123), (20.0, 111), (30.0, 105), (40.0, 93.8), (50.0, 90.0), (60.0, 84.8), (70.0, 84.0), (80.0, 72.8), (90.0, 72.8), (100, 65.3)

**Expected price of milk**

Expected price is a stock variable and is represented by an integral equation. Expected price of milk is obtained by averaging the difference between the current market price and the previous expected price over a period of two years and is expressed as:

expected_milk_price(t) = expected_milk_price(t - dt) + (change_in_expected_milk_price) * dt
INIT expected_milk_price = 45

**Change in expected milk price**

Change in expected price of milk depends on the difference between the market price and expected price, and its adjustment delay and is expressed as:

**INFLOWS**

change_in_expected_milk_price = (expected_milk_price - milk_price)/milk_price_adjustment_delay
milk_price adjustment_delay = 2

**Price of milk**

Price of milk is computed from milk price normal and the effect of milk inventory coverage on price and is expressed as:

milk_price = milk_price_normal * milk inventory_coverage_effect_on_price
milk_price_normal = 45

**Milk inventory coverage effect on price**

The price of milk depends on the milk inventory coverage. The relationship between milk price and inventory coverage of milk is expressed graphically as:

milk_inventory_coverage_effect_on_price = GRAPH(milk_inventory_ratio)
(0.00, 1.46), (10.0, 1.13), (20.0, 0.967), (30.0, 0.833), (40.0, 0.705), (50.0, 0.622), (60.0, 0.563), (70.0, 0.502), (80.0, 0.427), (90.0, 0.375), (100, 0.352)

**Milk inventory ratio**

Milk inventory ratio is the ratio of milk inventory to desired milk inventory and is expressed as:

milk_inventory_ratio = milk_inventory/desired_milk_inventory
desired_milk_inventory = 45000000

**Dairy breeding stock**

Dairy breeding stock is the sum of heifers and milk cows.

dairy_breeding_stock = heifers + milk_cows

**Total dairy breeding cattle**

Total dairy breeding cattle is the sum of dairy calf, heifers and milk cows.

total_dairy_cattle = dairy_calf + heifers + milk_cows

### 8.4.4    Fuel Consumption Sub-Model

A fuel consumption sub-model consists of three main components: agricultural wastes, animal wastes and fuelwood. Agricultural wastes and animal wastes are derived from crop production and animal production, respectively, and fuelwood comes from rural forests.

Cooking fuel demand is evaluated on the basis of per capita fuel consumption and rural population. From crop production, the available amount of residue is evaluated by a crop-specific coefficient. Similarly, from cattle production, the available amount of dung is evaluated by a cattle-specific coefficient. Fuelwood comes from rural forests. The felling rate of rural forests should not exceed the allowable cut, which is equal to the regrowth of the rural forest. When the felling rate exceeds the allowable cut, deforestation happens (Munasinghe and Meier, 2008). To avoid deforestation and maintain the required forest cover, an afforestation program is launched.

Improved stoves are used for the efficient utilization of biomass fuels, and biogas digesters using cowdung can provide clean fuel with treated slurry as a fertilizer as a bonus. The best option for the utilization of biomass for cooking fuel is to use agricultural and animal wastes with the introduction of improved stoves and biogas digesters. If agricultural and animal wastes with the introduction of

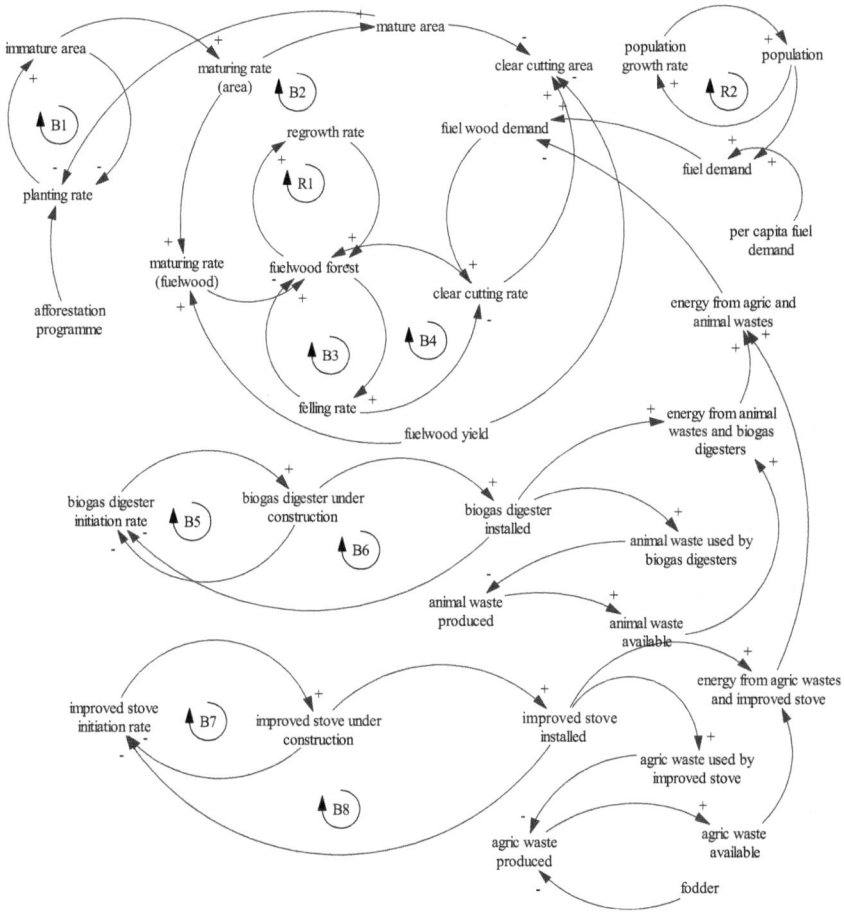

**FIGURE 8.10**   Causal loop diagram of the fuel consumption sub-model.

improved stoves and biogas digesters are not sufficient to meet fuel requirements, fuelwood within allowable cut should be the second option. If the requirements of cooking fuel are still not met, then clear cut of rural forest may be considered to meet cooking fuel requirements.

Figure 8.10 shows the causal loop diagram of the fuel consumption sub-model. The causal loop diagram identifies the major feedback loops of the system, which describe the basic causal mechanisms hypothesized to generate the reference mode of behavior over time. There are four negative feedback loops and two positive feedback loops in the rural forest component. Two negative feedback loops in each of the agricultural waste with improved stoves component and the animal waste with digester biogas component.

In the rural forest component, immature area and planting rate form the negative feedback loop B1, and mature area, planting rate, immature area and maturing rate form the negative feedback loop B2. The three more feedback loops in the rural forest component are two negative feedback loops and one re-enforcing positive feedback loop. The two negative feedback loops are B3 and B4, and the re-enforcing feedback loop is R1. The population and the population growth form re-enforcing feedback loop R2. In the animal waste with biogas digester component, biogas digester initiation rate and biogas digester under construction form the negative feedback loop B5, and biogas digester initiation rate, biogas digester under construction and biogas digester installed form the negative feedback loop B6. In the agricultural waste with improved stove component, improved stove initiation rate and improved stove under construction form the negative feedback loop B7, and improved stove initiation rate, improved stove under construction and improved stove installed form the negative feedback loop B8. All of these loops simulate the dynamic behavior of the fuel consumption system.

The model consists of rural forest section, biogas digester section, improved cooking stove section, agricultural waste section, animal waste section, energy supply section and energy demand section. Figure 8.11 shows the stock–flow diagram of the fuel consumption sub-model.

The fundamental equations that correspond to the major variables shown in Figure 8.11 are as follows:

**Rural forest section**

**Immature rural forest area**
Immature rural forest area is increased by plantation rate and decreased by area maturing and is expressed as:

immature_area(t) = immature_area(t - dt) + (plantation_rate - area_maturing) * dt
INIT immature_area = 100000

**Plantation rate**
Plantation rate depends on the gap between desired plantation and immature area and planning time and is expressed as:

**INFLOWS**
plantation_rate = (desired_plantation - mature_area - immature_area)/planning_time
planning_time = 15
desired_plantation = 3000000

**Area maturing**
Area maturing depends on the immature area and maturation delay and is expressed as:

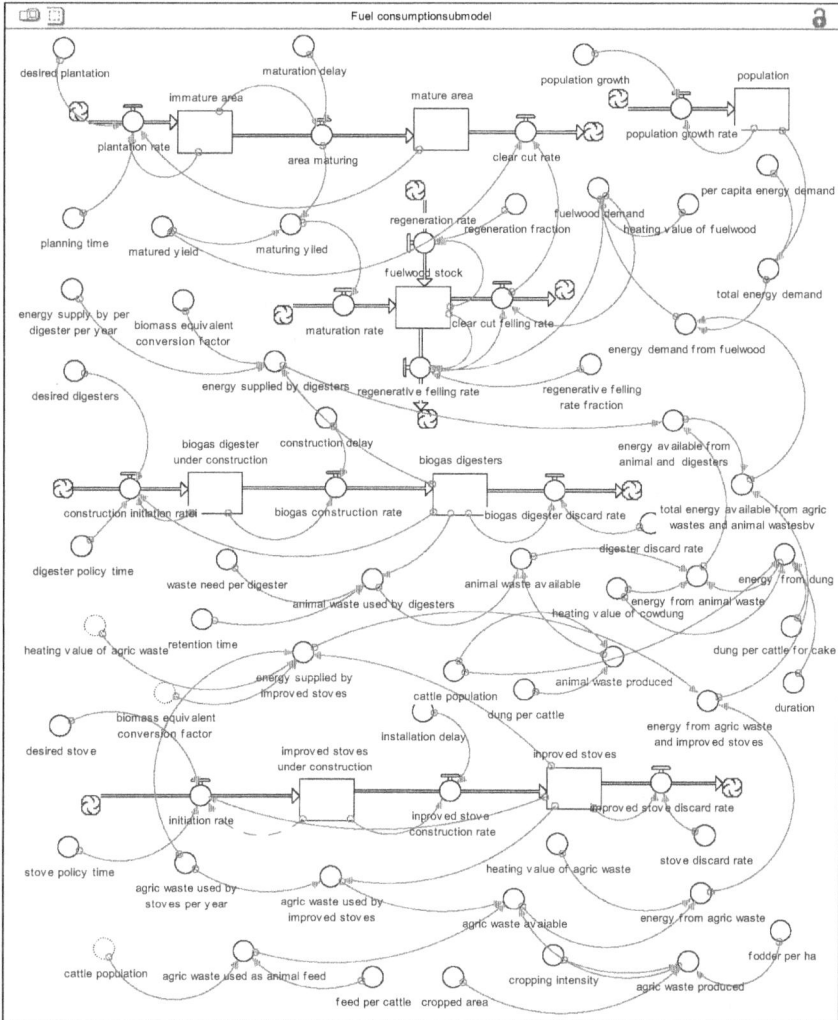

FIGURE 8.11 Stock–flow diagram of the fuel consumption sub-model.

## OUTFLOWS
area_maturing = immature_area/maturation_delay
maturation_delay = 15

### Mature rural forest area
Mature rural forest area is increased by area maturing and decreased by clear cut rate and is expressed as:

mature_area(t) = mature_area(t - dt) + (area_maturing - clear_cut_rate) * dt
INIT mature_area = 200000

**Clear cut rate**
Clear cut rate depends on the clear cut felling rate and matured yield and is expressed as:

**OUTFLOWS**
clear_cut_rate = clear_cut_felling_rate/matured_yield
matured_yield = 116

**Fuelwood stock**
Fuelwood stock is increased by regeneration rate and maturation rate, and decreased by regenerative felling rate and clear cut felling rate and is expressed as:

fuelwood_stock(t) = fuelwood_stock(t - dt) + (regeneration_rate + maturation_rate - regenerative_felling_rate - clear_cut_felling_rate) * dt
INIT fuelwood_stock = 23200000

**Regeneration rate**
Regeneration rate depends on the fuelwood stock and regeneration fraction and is expressed as:

**INFLOWS**
regeneration_rate = fuelwood_stock * regeneration_fraction
regeneration_fraction = 0.02

**Maturation rate**
Maturation rate is the maturing yield which is the product of area maturing and matures yield and is expressed as:

maturation_rate = maturing_yield
maturing_yield = area_maturing * matured_yield

**Regenerative felling rate**
If the fuel demand is greater than the allowable cut, then regenerative felling rate is the allowable cut, otherwise it is the fuelwood demand and is expressed as:

**OUTFLOWS**
regenerative_felling_rate = IF(fuelwood_demand > (fuelwood_stock * regenerative_felling_rate_fraction))THEN(fuelwood_stock * regenerative_felling_rate_fraction)ELSE(fuelwood_demand)
regenerative_felling_rate_fraction = 0.02

**Fuelwood demand**
Fuelwood demand is computed from energy demand from fuelwood and heating value of fuelwood and is computed as:

fuelwood_demand = energy_demand_from_fuelwood/(heating_value_of_fuelwood * 1000)
heating_value_of_fuelwood = 15.1

## Clear cut felling rate
Clear cut felling rate cannot be negative and would be the maximum value of either zero or (fuelwood demand - fuelwood stock × regenerative felling rate) and is expressed as:

clear_cut_felling_rate = MAX(0,(fuelwood_demand - regenerative_felling_rate))

## Biogas digester section

## Biogas digester
Biogas digester is increased by biogas digester construction rate and decreased by biogas digester discard rate and is expressed as:

biogas_digesters(t) = biogas_digesters(t - dt) + (biogas_digester_construction_rate - biogas_digester_discard_rate) * dt
INIT biogas_digesters = 72000

## Biogas digester construction rate
Biogas digester construction rate depends on biogas digester under construction and biogas digester construction delay and is expressed as:

## INFLOWS
biogas_digester_construction_rate = biogas_digester_under_construction/construction_delay
construction_delay = 1

## Biogas digester discard rate
Biogas digester discard rate depends on biogas digester in operation and biogas digester discard rate and is expressed as:

## OUTFLOWS
biogas_digester_discard_rate = biogas_digesters * digester_discard_rate
digester_discard_rate = 0.05

## Biogas digester under construction
Biogas digester under construction is increased by biogas digester construction initiation rate and decreased by biogas digester construction rate and is expressed as:

biogas_digester_under_construction(t) = biogas_digester_under_construction(t - dt) + (construction_initiation_rate - biogas) * dt
INIT biogas_digester_under_construction = 1500

### Biogas digester under construction initiation rate

Biogas digester under construction initiation rate depends on desired number of biogas digester in operation, biogas digester under construction and biogas digester construction policy time and is expressed as:

### INFLOWS

construction_initiation_rate = (desired_digesters-biogas_digester_under_construction-biogas_digesters)/digester_policy_time
digester_policy_time = 10
desired_digesters = 1000000

### Improved stove section

### Improved stoves

Improved stoves are increased by improved stove construction rate and decreased by improved stove discard rate and is expressed as:

improved_stoves(t) = improved_stoves(t - dt) + (improved_stove_construction_rate - improved_stove_discard_rate) * dt
INIT improved_stoves = 1850000

### Improved stove construction rate

Improved stove construction rate depends on improved stove under construction and improved stove installation delay and is expressed as:

### INFLOWS

improved_stove_construction_rate = improved_stoves_under_construction/installation_delay
installation_delay = 0.5

### Improved stove discard rate

Improved stove discard rate depends on improved stove in operation and improved stove discard rate and is expressed as:

### OUTFLOWS

improved_stove_discard_rate = improved_stoves * stove_discard_rate
stove_discard_rate = 0.05

### Improved stove under construction

Improved stove under construction is increased by improved stove construction initiation rate and decreased by improved stove construction rate and is expressed as:

improved_stoves_under_construction(t) = improved_stoves_under_construction(t - dt) + (initiation_rate - improved_stove_construction_rate) * dt
INIT improved_stoves_under_construction = 50000

### Improved stove under construction initiation rate
Improved stove under construction initiation rate depends on desired number of improved stove, improved stove in operation, improved stove under construction and improved stove construction policy time and is expressed as:

### INFLOWS
initiation_rate = (desired_stove - improved_stoves_under_construction - improved_stoves)/stove_policy_time
desired_stove = 3000000
stove_policy_time = 10

### Agricultural waste section

### Agricultural waste available
Agricultural waste available is the agricultural waste produced minus agricultural waste used as animal feed and feedstock in biogas digesters and is expressed as:

agric_waste_available = agric_waste_produced - agric_waste_used_as_animal_feed - agric_waste_used_by_improved_stoves

### Agricultural waste produced
Agricultural waste produced is the product of cropped area, fodder per ha and cropping intensity and is expressed as:

agric_waste_produced = cropped_area * fodder_per_ha * cropping_intensity
cropped_area = 7400000
fodder_per_ha = 3200
cropping_intensity = 2

### Agricultural waste used as animal feed
Agricultural waste used as animal feed is computed from cattle population and feed per cattle and is expressed as:

agric_waste_used_as_animal_feed = cattle_population * feed_per_cattle
cattle_population = 21800000
feed_per_cattle = 3.2 * 365

### Agricultural waste used by improved stoves
Agricultural waste used by improved stoves is computed from the number of improved stoves in operation and agricultural waste used by improved stoves per year and is expressed as:

agric_waste_used_by_improved_stoves = improved_stoves * agric_waste_used_by_stoves_per_year/1000
agric_waste_used_by_stoves_per_year = 5 * 365

**Animal waste section**

**Animal waste available**
Animal waste available is animal waste produced minus animal waste used by biogas digesters and is expressed as:
animal_waste_available = animal_waste_produced - animal_waste_used_by_digesters

**Animal waste produced**
Animal waste produced is the product of cattle population and dung per cattle and is expressed as:

animal_waste_produced = cattle_population * dung_per_cattle
dung_per_cattle = 40 * 365

**Animal waste used by biogas digesters**
Animal waste used by biogas digesters is computed from the number of biogas digesters in operation, animal waste need per biogas digester and retention time and is expressed as:

animal_waste_used_by_digesters = biogas_digesters * waste_need_per_digester/retention_time
retention_time = 20
waste_need_per_digester = 50 * 365

**Energy supply section**

**Energy available animal waste and biogas digester**
Energy available animal waste and biogas digester is the sum of energy available from animal waste and energy supplied by biogas digesters and is expressed as:

energy_available_from_animal_and_digesters = energy_supplied_by_digesters + energy_from_animal_waste

**Energy demand from fuelwood**
Energy demand from fuelwood cannot be negative, and it is the maximum value of either zero or (total energy demand - total energy available) and is expressed as:

energy_demand_from_fuelwood = MAX(0,(total_energy_demand - total_energy_available))

**Energy from agricultural waste**
Energy from agricultural waste is computed from the agricultural waste available and the heating value of agricultural waste and is expressed as:

energy_from_agric_waste = agric_waste_available * heating_value_of_agric_waste * 0.62
heating_value_of_agric_waste = 15.5

### Energy available agricultural waste and improved stoves

Energy available agricultural waste and improved stoves is the sum of energy available from agricultural and energy supplied by improved stoves and is expressed as:
energy_from_agric_waste_and_improved_stoves = energy_from_agric_waste + energy_supplied_by_improves_stoves

### Energy from animal waste

Energy from animal waste is the minimum value of either energy from animal waste or energy from dung and is expressed as:

energy_from_animal_waste = MIN((animal_waste_available * heating_value_of_ cowdung * 0.4),energy_from_dung)
heating_value_of_cowdung = 8.5

### Energy from dung

Energy from dung is computed from cattle population, dung per cattle for cake, duration, and heating value of cowdung and is expressed as:

energy_from_dung = (cattle_population * dung_per_cattle_for_cake * duration * 0.4 * heating_value_of_cowdung)
dung_per_cattle_for_cake = 2.56
duration = 365

### Energy supplied by digesters

Energy supplied by biogas digesters is computed from the number of biogas digesters and energy by per digester per day and is expressed as:

energy_supplied_by_digesters = (biogas_digesters * energy_supply_by_per_ digester_per_year) * biomass_equivalent_conversion_factor
energy_supply_by_per_digester_per_year = 70 * 365

### Energy supplied by improved stoves

Energy supplied by improved stoves is computed from the number of improved stoves, agricultural waste used by stoves per day, and heating value of agricultural waste and is expressed as:

energy_supplied_by_improves_stoves = (improved_stoves * agric_waste_used_ by_stoves_per_day * heating_value_of_agric_waste) * biomass_equivalent_con-version_factor
biomass_equivalent_conversion_factor = 1.25

### Total energy available from agricultural wastes and animal wastes

Total energy available is the sum of energy available from animal waste and biogas digesters as well as the energy from agricultural waste and energy supplied by improved stoves and is expressed as:

total_energy_available_agric_and_animal_wastes =

energy_available_from_animal_and_digesters + energy_from_agric_waste_and_
improved_stoves

**Energy demand section**

**Total energy demand**
Total energy demand is computed from population and per capita energy demand
and is expressed as:

total_energy_demand = population * per_capita_energy_demand
per_capita_energy_demand = 2.8 * 1000 * 0.75

**Population**
Population is increased by population growth rate and is expressed as:

population(t) = population(t - dt) + (population_growth_rate) * dt
INIT population = 165000000

**Population growth rate**
Population growth rate depends on population and population growth factor and is
expressed as:

**INFLOWS**
population_growth_rate = population * population_growth
population_growth = 0.01

## 8.4.5   RURAL ELECTRIFICATION SUB-MODEL

Electricity plays an important role in the socio-economic development of a
country, and per capita consumption of electricity is a measure of physical quality
of life (Bala, 1998). Rural electrification is intended to serve both the economic
and social affairs of rural people. The availability of electricity in rural areas is not
only essential for direct utilization of irrigation pumps, rural cottage industries and
other commercial purposes, it also has a great social benefit that cannot be quan-
tified directly. It invariably helps to increase the general activity of rural people
by increasing effective working time, offers scope for evening and late evening
activity and entertainment, which serves as an indirect effect on population control, makes villages attractive for educated people to live in, and thereby assists in
a reduction of the flight of educated people from villages.

The power grid cannot reach everywhere, yet there are alternatives in the form
of mini-grids for the supply of energy, information and communication using
renewable energies for rural development and economic growth (Schmid, 2003).

Solar homes are designed for rural household electrification far from the grid
connection where extension of the grid connection is neither economic nor available. It is attractive where even a mini or microgrid does not exist. The most
attractive use of a solar home system (SHS) in Bangladesh is the lighting system,

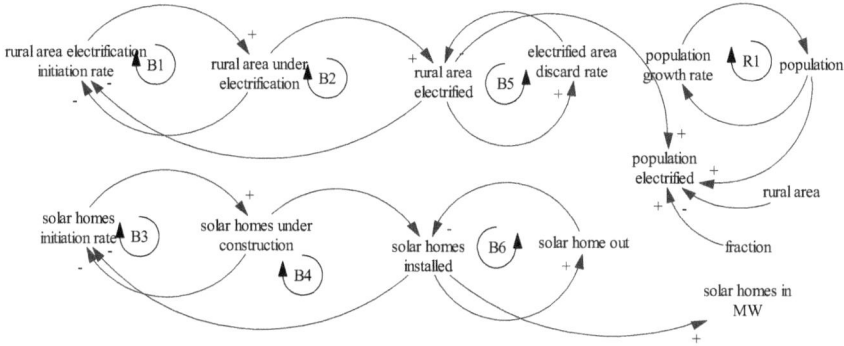

**FIGURE 8.12** Causal loop diagram of the rural electrification sub-model.

and the solar home system is financially attractive for household lighting and entertainment (Mondal, 2010).

The rural electrification system consists of two components: grid-connected rural electrification and solar homes. A causal loop diagram of the rural electrification sub-model is shown in Figure 8.12. The grid-connected component consists of two negative feedback loops and one positive feedback loop. In the grid-connected component, rural electrification initiation rate and rural electrification under construction form the negative feedback loop B1, and rural electrification initiation rate, rural electrification under construction and rural electrification installed form the negative feedback loop B2. Population and the population growth form re-enforcing feedback loop R1. In the solar homes component, solar homes initiation rate and solar homes under construction form the negative feedback loop B3 and solar homes initiation rate, solar homes under construction and solar homes installed form the negative feedback loop B4. Rural area electrified forms the negative feedback loop B5, and electrified area discard rate and solar homes form the negative feedback loop B6 with solar home out. All these loops simulate the dynamic behavior of the rural electrification system.

Rural electrification is categorized as grid-connected rural electrification and off-grid rural electrification using solar homes. Solar homes are used in remote areas far from the grid where grid-connected is not economically feasible or available.

Figure 8.13 shows the stock–flow diagram of the rural electrification sub-model.

The fundamental equations that correspond to the major variables shown in Figure 8.13 are as follows:

## Grid-connected rural electrification

### Grid-connected electrified area
Grid connected electrified area is increased by electrification rate and decreased by area discard rate and is expressed as:

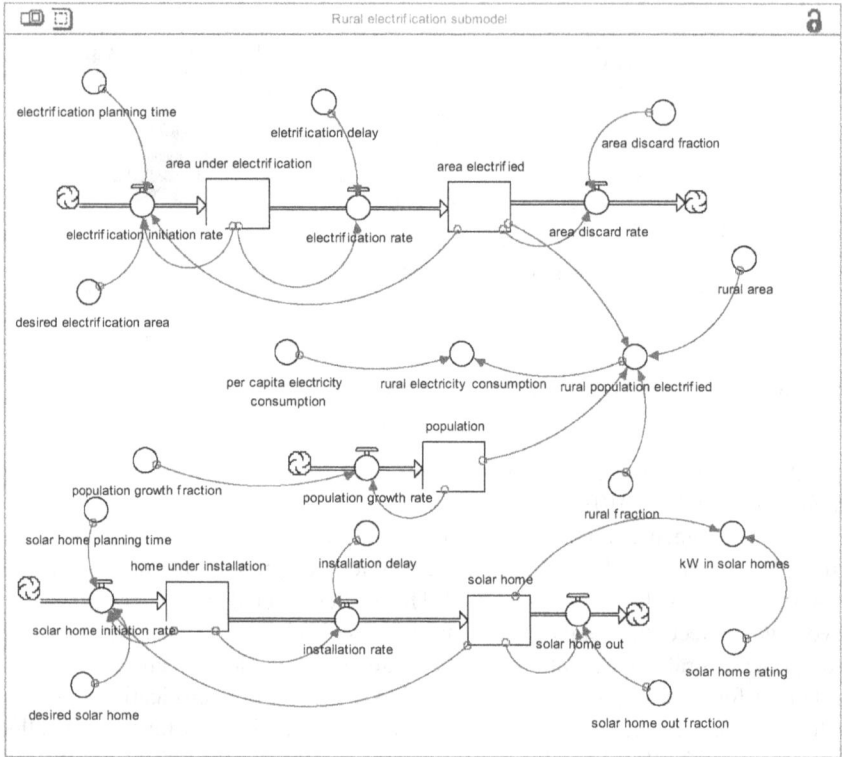

**FIGURE 8.13**   Stock–flow diagram of the rural electrification sub-model.

area_electrified(t) = area_electrified(t - dt) + (electrification_rate - area_discard_rate) * dt
INIT area_electrified = 15000

### Grid-connected electrification rate
Grid connected electrification rate depends on grid connected area under electrification and grid connected electrification delay and is expressed as:

### INFLOWS
electrification_rate = area_under_electrification/eletrification_delay
eletrification_delay = 2

### Grid-connected electrified area discard rate
Grid connected electrified area discard rate depends on grid connected electrified area and area discard fraction and is expressed as:

**OUTFLOWS**
area_discard_rate = area_electrified * area_discard_fraction
area_discard_fraction = 0.002

### Grid connected area under electrification
Grid connected area under electrification is increased by electrification initiation rate and decreased by electrification rate and is expressed as:

area_under_electrification(t) = area_under_electrification(t - dt) + (electrification_initiation_rate - electrification_rate) * dt
INIT area_under_electrification = 5000

### Grid connected area electrification initiation rate
Grid connected area electrification initiation rate depends on desired electrification area, area under electrification, area electrified and electrification planning time and is expressed as:

**INFLOWS**
electrification_initiation_rate = (desired_electrification_area - area_under_electrification - area_electrified)/electrification_planning_time

desired_electrification_area = 45000
electrification_planning_time = 15

### Rural electricity consumption
Rural electricity consumption is computed from rural population electrified and per capita electricity consumption and is expressed as:

rural_electricity_consumption = rural_population_electrified * per_capita_electricity_consumption
per_capita_electricity_consumption = 120

### Rural population electrified
Rural population electrified is computed from population, rural fraction, area electrified and rural area and is expressed as:

rural_population_electrified = (population * rural_fraction) * (area_electrified/rural_area)
rural_area = 45000
rural_fraction = 0.8

### Population
Population is increased by population growth rate and is expressed as:

population(t) = population(t - dt) + (population_growth_rate) * dt
INIT population = 118000000

## Population growth rate
Population growth rate depends on population and its growth fraction and is expressed as:

## INFLOWS
population_growth_rate = population * population_growth_fraction
population_growth_fraction = 0-02

## Solar home connected rural electrification

### Solar home
Number of solar homes is increased by installation rate and decreased by solar home out and is expressed as:

solar_home(t) = solar_home(t - dt) + (installation_rate - solar_home_out) * dt
INIT solar_home = 50000

### Solar home installation rate
Solar home installation rate depends on solar home under installation and installation delay and is expressed as:

## INFLOWS
installation_rate = home_under_installation/installation_delay
installation_delay = 1

### Solar home out
Solar home discard rate depends on solar home installed and solar home out fraction and is expressed as:

## OUTFLOWS
solar_home_out = solar_home * solar_home_out_fraction
solar_home_out_fraction = 0.002

### Solar home under construction
Solar home under construction is increased by solar home initiation rate and decreased by solar home installation and is expressed as:

home_under_installation(t) = home_under_installation(t - dt) + (solar_home_initiation_rate - installation_rate) * dt
INIT home_under_installation = 15000

### Solar home initiation rate
Solar home initiation rate depends on desired solar home, solar home under installation and solar home installed, solar home planning time and is expressed as:

**INFLOWS**
solar_home_initiation_rate = (desired_solar_home - home_under_installation - solar_home)/solar_home_planning_time
desired_solar_home = 450000
solar_home_planning_time = 15

**kW in solar homes**
kW in solar homes is computed from solar homes and solar home rating and is expressed as:

kW_in_solar_homes = solar_home * solar_home_rating
solar_home_rating = 0.032

## 8.4.6 QUALITY OF LIFE SUB-MODEL

Quality of life is used as a measure of the performance of the integrated rural energy system. It is computed from quality of life from food, quality of life from electricity, quality of life from crowding and quality of life standard. Figure 8.14 shows the causal loop diagram of the quality of life sub-model.

Figure 8.15 shows the stock–flow diagram of the quality of life sub-model.

The fundamental equations that correspond to the major variables shown in Figure 8.15 are as follows:

**Quality of life from crowding**
Quality of life from crowding depends on crowding ratio. The relationship between quality of life from crowding and crowding ratio is expressed graphically as:

quality_of_life_from_crowding = GRAPH(crowding_ratio)
(1.00, 0.97), (1.40, 0.845), (1.80, 0.695), (2.20, 0.57), (2.60, 0.51), (3.00, 0.41), (3.40, 0.345), (3.80, 0.295), (4.20, 0.24), (4.60, 0.215), (5.00, 0.2)

**Crowding ratio**
Crowding ratio is a function of time. The relationship between crowding ratio and time is expressed graphically as:

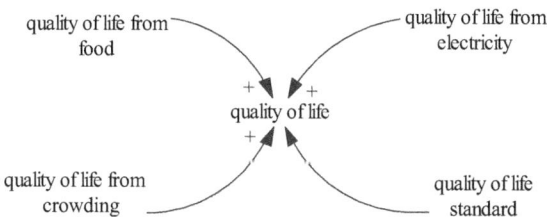

FIGURE 8.14 Causal loop diagram of the quality of life sub-model.

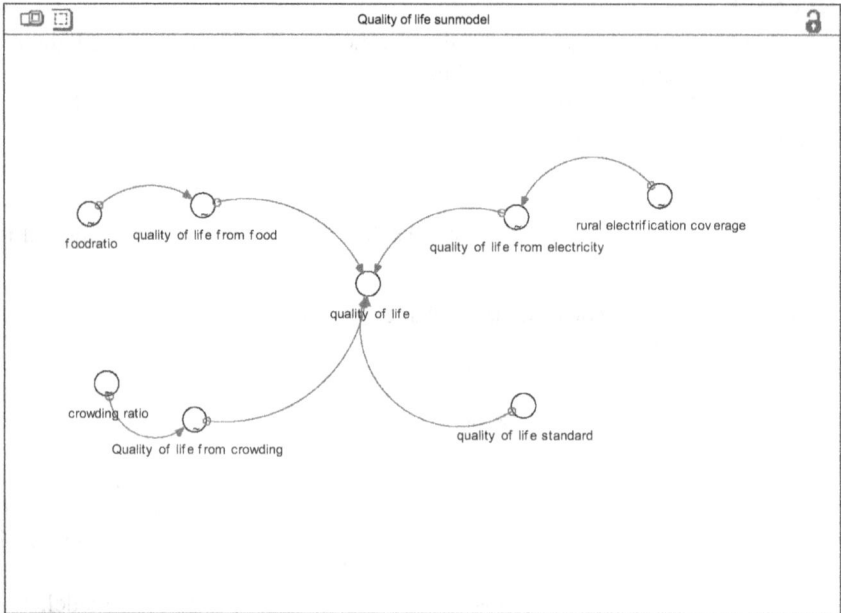

**FIGURE 8.15**    Stock–flow diagram of the quality of life sub-model.

crowding_ratio = GRAPH(TIME)
(0.00, 1.00), (1.25, 1.03), (2.50, 1.08), (3.75, 1.10), (5.00, 1.12), (6.25, 1.16), (7.50, 1.19), (8.75, 1.21), (10.0, 1.24), (11.3, 1.26), (12.5, 1.28), (13.8, 1.30), (15.0, 0.00)

### Quality of life from electricity

Quality of life from electricity depends on rural electrification coverage. The relationship between quality of life from electricity and rural electrification coverage is expressed graphically as:

quality_of_life_from_electricity = GRAPH(rural_electrification_coverage)
(0.00, 1.01), (10.0, 1.14), (20.0, 1.23), (30.0, 1.31), (40.0, 1.36), (50.0, 1.39), (60.0, 1.42), (70.0, 1.44), (80.0, 1.47), (90.0, 1.48), (100, 1.49)

### Rural electrification coverage

Rural electrification coverage increases with time. The relationship between rural electrification coverage and time is expressed graphically as:

rural_electrification_coverage = GRAPH(TIME)
(0.00, 50.3), (1.25, 65.5), (2.50, 81.3), (3.75, 90.5), (5.00, 94.8), (6.25, 95.5), (7.50, 97.3), (8.75, 97.8), (10.0, 99.5), (11.3, 100), (12.5, 100), (13.8, 99.0), (15.0, 99.8)

### Quality of life from food

Quality of life from food depends on food ratio. The relationship between quality of life from food and food ratio is expressed graphically as:

quality_of_life_from_food = GRAPH(foodratio)
(0.00, 1.09), (0.5, 1.14), (1.00, 1.21), (1.50, 1.28), (2.00, 1.35), (2.50, 1.38), (3.00, 1.43), (3.50, 1.46), (4.00, 1.46), (4.50, 1.48), (5.00, 1.50)

**Food ratio**
Food ratio changes with time. The relationship between food ratio and time is expressed graphically as:

foodratio = GRAPH(TIME)
(0.00, 1.17), (1.25, 1.19), (2.50, 1.21), (3.75, 1.22), (5.00, 1.23), (6.25, 1.24), (7.50, 1.25), (8.75, 1.26), (10.0, 1.27), (11.3, 1.27), (12.5, 1.29), (13.8, 1.29), (15.0, 1.30)

## 8.5   MODEL VALIDATION

The initial values and parameters of the model were estimated from primary and secondary data (BBS, 2017). To build up confidence in the predictions of a model, various ways of validating the model, such as checking the structure of the model, comparing the model predictions with historic data, checking whether the model generates plausible behavior and checking the quality of the parameter values, were considered. The time series data of rice production in Bangladesh in 2011–2020 are compared with the simulated rice production to build up confidence in the model, and this is shown in Figure 8.16. The model simulated prediction agrees reasonably well with historical behavior. The model is able to provide a qualitative and quantitative understanding of rice production systems. Hence, the model is reliable, and the validated model was used for policy analysis.

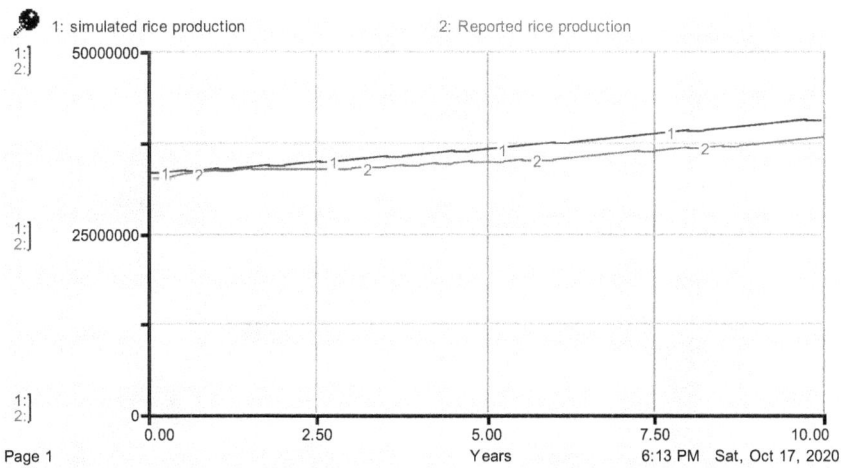

**FIGURE 8.16**   Comparison of simulated and reported rice production.

## 8.6   SIMULATED RESULTS

Figure 8.17 shows model simulated population, irrigated land, rice production and food self-sufficiency ratio for a simulation period of 15 years. Population increases with time while the irrigated area increases until the upper limit (10 years) of irrigated land is reached and then decreases slowly. Rice production follows the irrigated area. Food self-sufficiency ratio follows rice production but is affected by increasing population, resulting in a decreasing trend of food self-sufficiency ratio once the upper limit of irrigated area is reached.

Figure 8.18 shows simulated total cattle for beef, total dairy cattle and milk production rate. Total cattle for beef is almost constant with a little oscillation, but the production capacity is not enough and requires importation. Dairy cattle production oscillates less, and milk production oscillates more rapidly. These oscillations are due to system characteristics (beef cattle and dairy cattle, and gestation maturation delay).

Figure 8.19 shows simulated fuelwood stock, mature area, immature area, planting rate and generative felling rate. Mature area increases with time as long as the maturing rate is greater than the clear cut rate and the fuelwood stock follows the mature area. The planting rate decreases with time since the gap to be adjusted to the desired plantation area decreases. The immature area increases sharply initially because of a higher plantation rate but decreases in a later period because the planting rate becomes less dominant in comparison with the area transfer into mature.

Figure 8.20 shows simulated total energy available from agricultural wastes and animal wastes, total energy demand, biogas digesters and improved cooking stoves. Total energy demand increases with time because of the population increase with

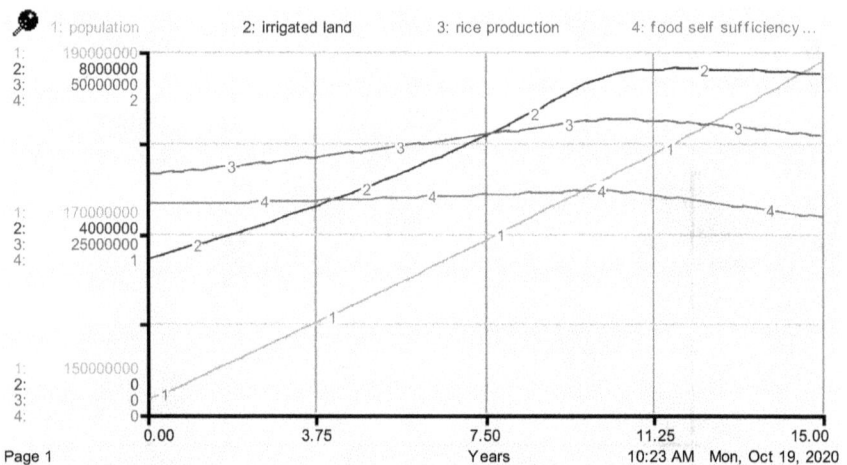

**FIGURE 8.17**   Simulated population, irrigated land, rice production and food self-sufficiency ratio.

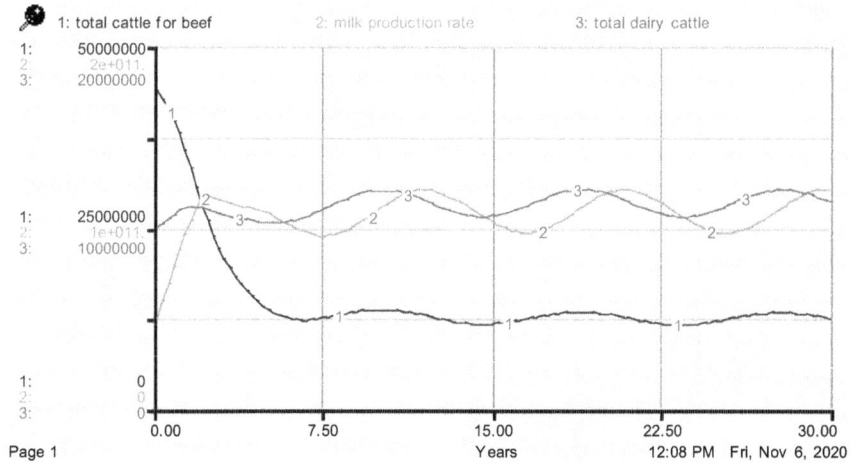

**FIGURE 8.18**   Simulated total cattle for beef, milk production rate and total dairy cattle.

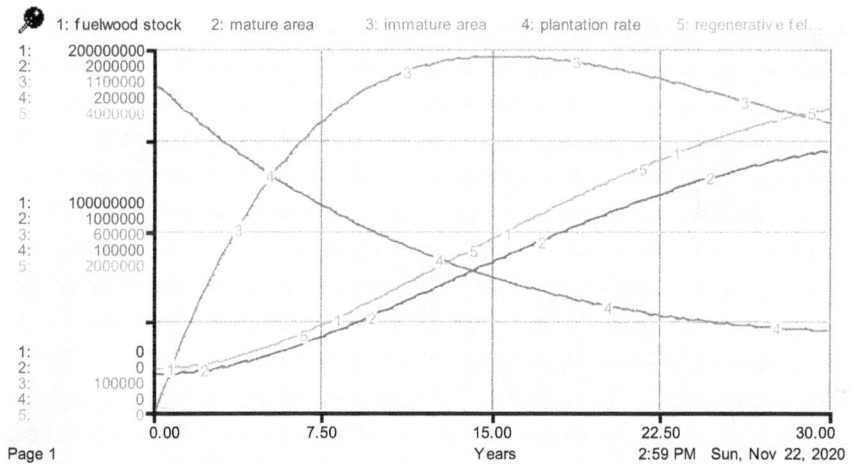

**FIGURE 8.19**   Simulated fuelwood stock, mature area, immature area, planting rate and regenerative felling rate.

time, but total energy available increases initially, which approaches the upper limit set by the upper limits of the availability of animal wastes and agricultural wastes since these are derived from cattle production and crop production, which reached the saturation level. Improved stoves and biogas digesters are also approaching the planned capacity.

Figure 8.21 shows simulated energy available from animal wastes and biogas digesters, energy demand from fuelwood, energy available from agricultural wastes and improved cooking stoves, and total energy demand. Total energy

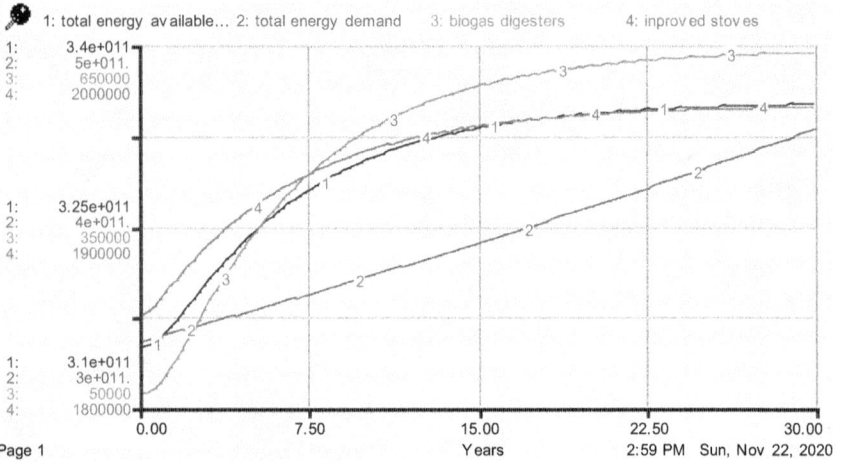

**FIGURE 8.20** Simulated total energy available, total energy demand, biogas digesters and improved cooking stoves.

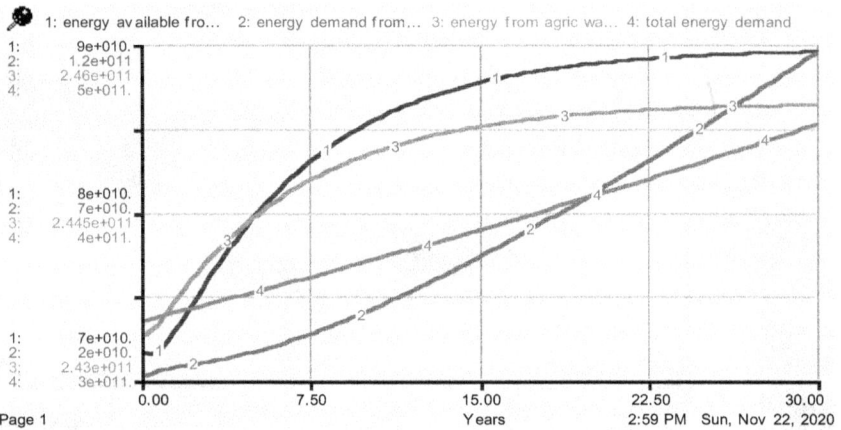

**FIGURE 8.21** Simulated energy available from animal wastes and biogas digesters, energy demand from fuelwood, energy available from agricultural wastes and improved cooking stoves and total energy demand.

demand increases with time because of the population growth, but total energy available from animal wastes and biogas digesters and total energy available from agricultural wastes and improved cooking stoves increase initially, which approach the upper limits set by the upper limits of the availability of animal wastes and agricultural wastes since these are derived from cattle production and crop production. Even with the addition of improved stoves and biogas digesters, the energy available is not sufficient to meet the energy demand; hence fuelwood is required to meet the energy demand.

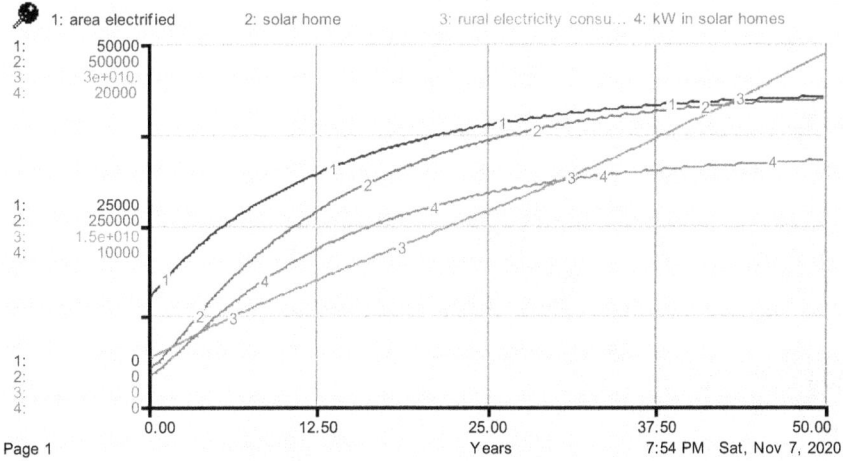

**FIGURE 8.22**  Simulated electrified area, solar homes, rural electricity consumption and kW in solar homes.

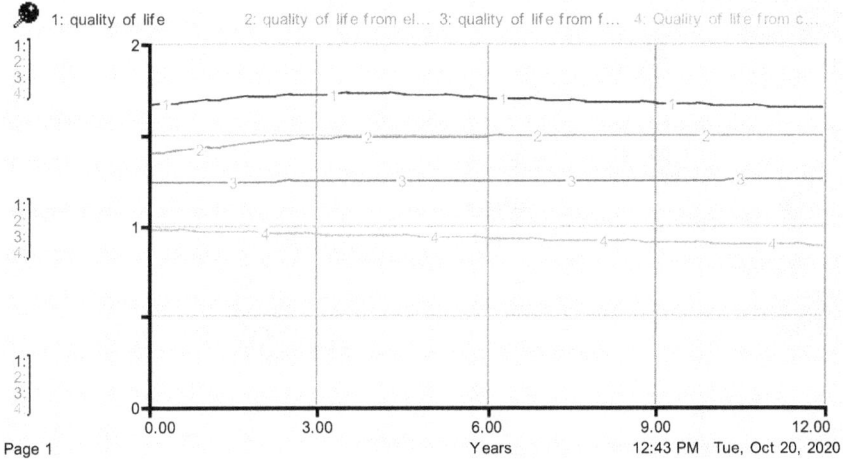

**FIGURE 8.23**  Simulated quality of life, quality of life from electricity, quality of life from food and quality of life from crowding.

Figure 8.22 shows simulated electrified area, solar homes, rural electricity consumption and KW in solar homes. Grid electrification area and solar homes demand increases with time initially, which approaches the upper limit set by planning towards the end of the simulation period.

Figure 8.23 shows simulated quality of life, quality of life from electricity, quality of life from food and quality of life from crowding. Quality of life increases with increase in food self-sufficiency and also with increase in electrified area, but decreases with increase in crowding ratio. The overall effect of the multiplication of these effects shows that the quality of life is decreasing.

## 8.7    CONCLUDING REMARKS

The model should be further refined and updated with the data for the country or region involving all the stakeholders for the practical application of this model for policy planning and analysis. The system dynamics model presented here can be used as a computer laboratory for rural energy planning in developing countries where biomass is still the major share of cooking energy.

## REFERENCES

Alam, M. S., Huq, A. M. Z., & Bala, B. K. (1990). An integrated energy model for a village in Bangladesh. *Energy*. 15: 131–140.

Bala, B. K., Fatimah, M. A., & Kushairi, M. N. (2017). *System Dynamics: Modelling and Simulation*. Springer.

Bala, B. K. (1999). *Principles of System Dynamics* (1st ed.). Agrotech Publishing Academy, Udaipur, India.

Bala, B. K. (1998). *Energy and Environment: Modelling and Simulation*. Nova Science Publishers, USA

Bala, B. K. (1997). Computer modeling of the rural energy system and of $CO_2$ emissions for Bangladesh. *Energy*. 22: 999–1003.

Bala, B. K., & Satter, M. A. (1986a). *Modelling of rural energy systems*. Presented at Second National Symposium on Agricultural Research, BARC, Dhaka 12 February 1986.

Bala, B. K., & Satter, M. A. (1986b). Modelling of rural energy systems for food production in developing countries. Energia and Agricoltura 2 Conferenza Internationale. Sirmione/Brescia (Italia). 3, p. 306, 1986.

BBS. (2017). *Statistical Pocket Book Bangladesh 2017*. Ministry of Planning, Bangladesh.

Conrad, S. H. (2004). *The dynamics of agricultural commodities and their responses to disruptions of considerable magnitude*. Paper presented at the 22nd International Conference of the System Dynamics Society. Oxford, England, July 25–29, 2004.

Forrester, J. W. (1968). *Principles of Systems*. Allen Wright Press, Cambridge, Massachusetts, USA.

Forrester, J. W. (1971). *World Dynamics*. Allen Wright Press, Cambridge, Massachusetts, USA.

Huq, A. Z. M. (1975). *Energy modelling for agriculture units in Bangladesh*. Paper presented at the National Seminar on Integrated Rural Development, Dhaka, 1975.

Maani, K.E., & Cavana, R.Y. (2000). *Systems Thinking and Modelling: Understanding Change and Complexity*. Prentice Hall, New Zealand.

Meadows, D. L. (1970). *Dynamics of Commodity Production Cycles*. Wright-Allen Press, Cambridge, Massachusetts, USA.

Mohapatra, P. K. J., Mandal, P., & Bora, M.C. (1994). *Introduction to System Dynamics Modeling*. Universities Press, India.

Mondal, M. A. (2010). Economic viability of solar home systems: Case study of Bangladesh. *Renewable Energy*. 35: 1125–1129.

Munasinghe, M., & Meier, P. (2008). *Energy Policy Analysis and Modeling*. Cambridge University Press, Cambridge.

Parikh, J., & Kromer, G. (1985). Modeling energy and agriculture interactions – II: Food–fodder–fuel–fertilizer relationships for biomass in Bangladesh. *Energy*. 10(7): 804–817.

Schmid, J. (2003). Minigrids for rural development and economic growth. ISET, University of Kassel, Germany.

Sterman, J. D. (2000). *Business Dynamics: Systems Thinking and Modeling for a Complex World*. McGraw-Hill Higher Education, Boston.

# 9 Simulated Planning of Electric Power Systems

## 9.1 INTRODUCTION

Electrical energy is needed to meet the requirements of development and economic growth. Per capita consumption of electrical energy is also a measure of physical quality of life (Bala, 1998). Conventional energy resources are integrated with renewable energy resources to ensure energy security with a reduction in pollution and contribution to global warming. The optimal policy development and optimal operation planning and management of an integrated electric power system is a complex, dynamic and multifaceted problem depending not only on available technology but also economic and social factors. Computer models are clearly of great value in understanding the dynamics of this complex system. System dynamics, a methodology of computer modeling developed by Forrester, can be used to handle such complex systems (Bala, 2017; Bala, 1999; Forester, 1968; Sterman, 2000; Maani and Cavana, 2000; Mohapatra et al., 1994).

Huq (1958) initiated energy modeling in Bangladesh and the model he proposed was further developed for integrated rural energy systems (Bala et al., 2014; Bala, 2006; Bala, 1997; Bala and Satter, 2006a and 2006b). Nail (1992) reported an integrated model of US energy supply and demand that is used to prepare projections for energy policy analysis in the US Department of Energy's Office of Policy, Planning and Analysis. This model represents one of the real success stories of system dynamics modeling for energy planning. It was implemented at the Department of Energy in 1978 as an in-house analytical tool and has been used regularly for national policy analysis since then. Nail et al. (1992) employed the model to explore a wide range of policy options intended to address the effects of energy use on global warming. Vogstad (2004) reported a system dynamics analysis of the Nordic electricity market: the transition from fossil fuels towards a renewable energy supply within the liberalized electricity market. System dynamics has been used extensively to aid in resource planning in the electric power industry (Ford, 1997), and many applications constitute a major body of work that has proven useful to large and small power companies as well as to government agencies at the local, state and federal level (Gu et al., 2020; Wang et al., 2019; Bastan and Shakouri, 2018; Sani et al., 2018; He et al., 2018. Qurat-ullah, 2013; Vogstad, 2004; Dimitrovsky et al., 2004; Ford and Bull, 1989).

DOI: 10.1201/9781003218401-9

## 9.2    VERBAL DESCRIPTION

The electric power system described here is a typical national power system consisting of electric energy demand, supply of electric power, electricity price and emissions from energy supply contributing to global warming. Electrical energy is consumed by all sectors of the economy, such as residential, industrial and agricultural sectors. Household demand for electricity is increasing with the growth of population and better quality of life, while industrial and agricultural demands for electricity are increasing for economic growth.

Electricity is mainly generated from fossil fuels such as natural gas, oil and coal, and to a limited degree from renewable energy of hydroelectric. The accelerating demand for electrical energy, the rapid depletion of fossil fuels and the emission of gases contributing to global warming have stimulated the gradual transition to renewable energy resources such as nuclear, solar and wind. Thus, the supply system consists of fossil fuels such as natural gas and coal and renewable energy resources such as nuclear and solar. The electric energy supply is also increasing to meet the rapidly growing demand for electricity for the different sectors of the national economy. The price of electricity is settled by the production cost and demand–supply balance of electricity. Figure 9.1 shows the interactions of the major components of an integrated electric power system.

## 9.3    DYNAMIC HYPOTHESIS

Dynamic hypothesis is a conceptual model typically consisting of a causal loop diagram, a stock–flow diagram or a combination thereof, which drives the system's behavior. When the model based on the feedback concept is simulated, the endogenous structure of the model generates the reference mode behavior of the system, which results from the endogenous structure of the system. In system dynamics modeling, causal loop diagrams and stock–flow diagrams are used to describe the basic cause–effect relationships hypothesized to generate reference mode behavior over time (Sterman, 2000). Electric power systems can be represented by causal loop diagrams and stock–flow diagrams, and the simulation models based on the causal loop diagrams and stock–flow diagrams can generate the dynamic behavior of an electric power system. Electricity demand and supply, its price and contribution to global warming in the form of causal loop diagrams and stock–flow diagrams are hypothesized to generate the observed electricity demand, supply, price and emissions to the environment in the reference mode. In essence, the demand for electrical energy increases with the increase in population, and industrial and agricultural development for economic growth and better quality of life and the supply of electricity follows the demand of electricity with a gradual transition to renewable energy to ensure energy security and reduce contributions to global warming.

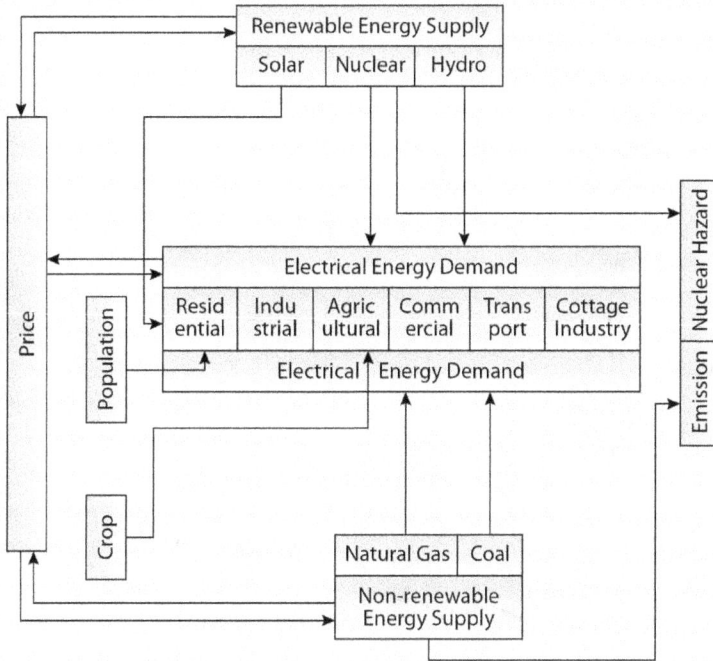

**FIGURE 9.1**    Interactions of the major components of an integrated electric power system.

## 9.4    SYSTEM DYNAMICS MODELING OF INTEGRATED ELECTRIC POWER SYSTEMS

The model is formulated first in the form of causal loop diagrams and then in the form of stock–flow diagrams. The causal loop diagrams show the important cause–effect relationships, and the resulting causal loop diagrams simulate the dynamics of the integrated electric power system. The stock–flow diagrams and corresponding integral finite difference equations show the mathematical structure of the model both in terms of diagrams and equations for computer simulation.

### 9.4.1    CAUSAL LOOP DIAGRAM

Energy planners and researchers are interested in the modeling of integrated electric power systems. The model links energy price with energy demand and energy supply-related factors (Liuguo et al., 2012; Mutingi et al., 2017). The causal loop

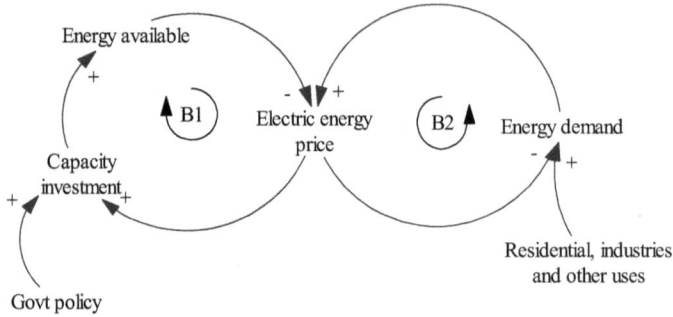

**FIGURE 9.2** Simple causal loop diagram for an integrated power system model.

diagram of a simple electric power system is shown in Figure 9.2. In feedback loop B1, capacity investment depends on price and energy policy, while in loop B2, energy demand depends on industries and population, and energy price.

In a liberalized electricity market, the elasticity of price of demand is high (0.20–0.80), and demand for electricity mainly depends on the price of electricity and the elasticity of price of demand, while the supply of electricity depends on profitability and capacity factor (Vogstad, 2004). However, in developing countries, the demand component mainly accounts for the cause–effect relationships of electric energy demands for residential, industrial, agricultural and other uses that are dependent upon GDP since the price elasticity of electric energy demand is low (-.10). Electric energy supply is an energy mix of fossil fuel and renewable energy-based electricity supply, and it depends on forecasted capacity based on the energy policy to meet the projected energy demand with the gradual transition to renewable energy sources set by the government since the electric power system is owned by the government in developing countries (Khanna and Rao, 2009). The supply component shows the cause–effect relationships of the initiation based on forecasted power, construction and operation of electric power plants to supply electricity to meet electric demands. Each of the power plants passes through a similar process of initiation, construction, operation and retirement since these power plants are designed to meet the increasing demands of electricity with the gradual transition to renewable energy resources to ensure energy security and reduce the contributions to global warming. The price component shows the causal loop diagram of price-setting influenced by demand–supply balance and production cost.

The causal loop diagram of a national integrated power system for developing countries where the price elasticity of electricity demand is low (-0.10) and the power supply owned by the government is shown in Figure 9.3. The total demand for electricity is the sum of household electricity consumption and the electricity demand of industrial, agricultural, etc. Household consumption of electricity increases with population. Population and birth rate form the re-enforcing positive feedback loop R1. The electricity demand of industrial, agricultural, etc., forms

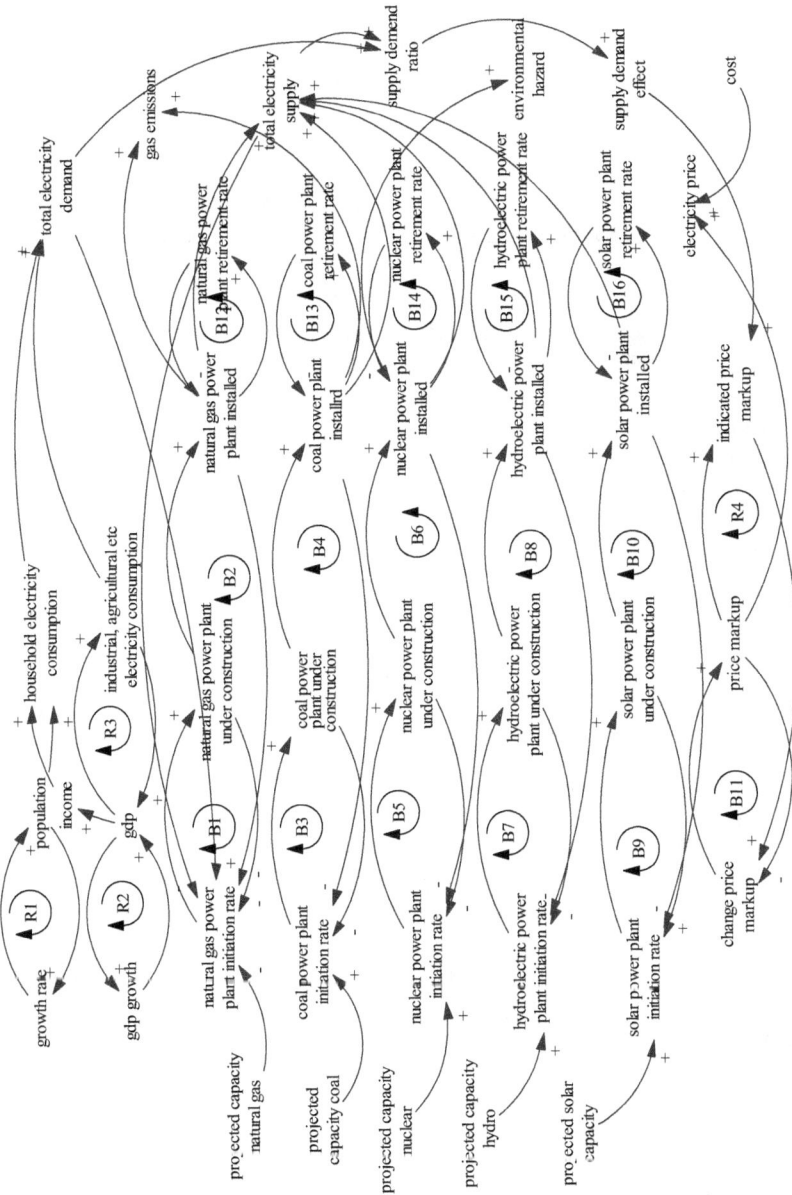

**FIGURE 9.3** Causal loop diagram for national integrated power system model.

the re-enforcing feedback loop R3 with GDP, which itself forms the re-enforcing feedback loop R2 with GDP growth.

Electricity is supplied by five power plants: natural gas power plant, coal-based power plant, nuclear power plant, hydroelectric power plant and solar PV power plant. Each of these power plants consists of three balancing negative feedback loops. Power plant initiation rate and power plant under construction form one negative feedback loop, while power plant under construction and power plant installed capacity form another negative feedback loop. Also, power plant installed and retirement rate form a negative feedback loop. Thus, there are 15 negative feedback loops (B1–B10 and B12–B16) in the supply sector of the national integrated power system tending to the forecast demand or electricity demand.

The price of electricity is determined by the cost of electricity and price markup, which is influenced by the demand–supply balance (Bala et al., 2017). Price markup and change in price markup form the negative feedback loop B11, while indicated price markup, change in price markup and price markup form the positive feedback loop R4. The indicated price markup is computed from markup price and effect of demand–supply ratio on price. Finally, the $CO_2$ emissions from each of the power plants are computed by simply multiplying the installed capacity of each power plant by the corresponding emission factor. Radiation pollution, radiation pollution index and thermal are also computed (WNA, 2020).

## 9.4.2 STOCK–FLOW DIAGRAM

The stock–flow diagram of the national integrated power system consists of five sectors: demand sector, supply sector, price sector, $CO_2$ emission sector and radiation hazard sector. The demand sector shows the stock–flow structure and integral finite difference equations for computation electric energy demands for residential, industrial, agricultural and other uses. Electric energy supply is an energy mix of fossil fuel such as natural gas and coal, and renewable energy resources such as nuclear, hydroelectric and solar power plant-based electricity. The stock–flow structure and integral finite difference equations of the electric energy supply sectors show how the power plants for this energy mix are evolved to meet the supply of the electric demands starting from the initiation of the power plant to its construction and finally its operation to supply electricity. Each of the power plants passes through similar processes of initiation, construction, operation and retirement since these power plants are designed to meet the increasing demands of electricity with a gradual transition to renewable energy resources to ensure energy security and reduce contributions to global warming. The price sector shows the stock–flow and integral finite difference equations of price-setting influenced by demand–supply balance and production cost. The $CO_2$ emission sector shows the stock–flow structure and integral finite difference equations for the computation of $CO_2$ emissions, while the radiation hazard sector shows the stock–flow structure

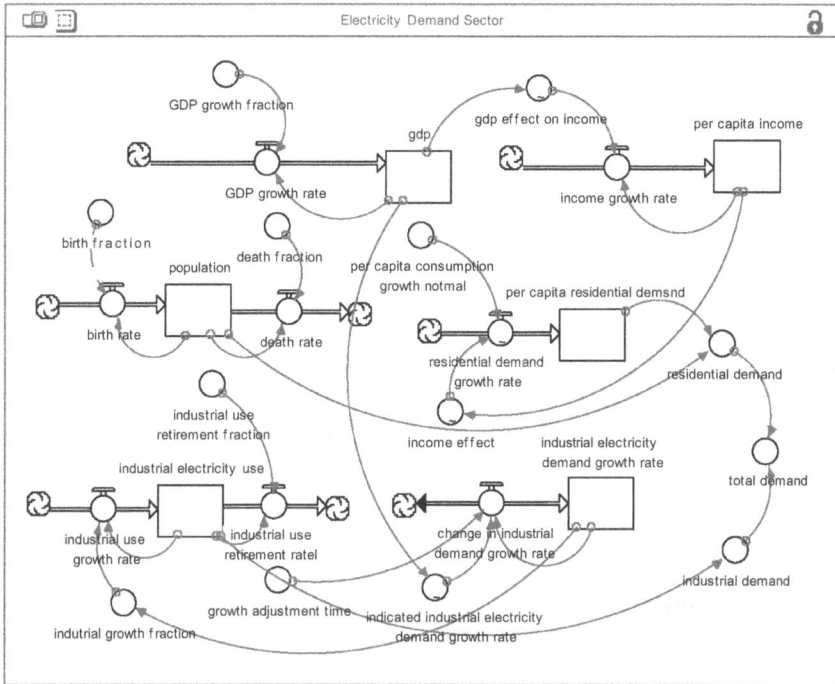

**FIGURE 9.4(a)** Electricity demand sector of the national integrated power system model.

and integral finite difference equations for the computation of thermal pollution and radiation pollution.

A stock–flow diagram of the system dynamics model of a national integrated power system is shown in Figure 9.4. For simplicity of presentation, the model is divided into several sectors: electricity demand sector, electricity supply sector, electricity price sector, $CO_2$ emission sector and radiation hazard sector.

### 9.4.2.1 Electricity Demand Sector

Electricity demand is categorized as residential electricity demand and industrial electricity demand. Figure 9.4(a) shows the stock–flow diagram of the electricity demand sector of the model.

The fundamental equations that correspond to the major variables shown in Figure 9.4(a) are as follows:

### *Industrial Electricity Demand*

GDP increases with the increase in GDP growth rate and is expressed as:

gdp(t) = gdp(t - dt) + (gdp_growth_rate) * dt
INIT gdp = 1855.7

GDP growth rate is computed from GDP and GDP growth fraction:

gdp_growth_rate = gdp * gdp_growth_fraction
gdp_growth_fraction = 0.06

Industrial electricity demand growth rate increases with change in industrial electricity demand growth rate.

industrial_electricity_demand_growth_rate(t) = industrial_electricity_demand_growth_rate(t - dt) + (change_in_industrial_demand_growth_rate) * dt
INIT industrial_electricity_demand_growth_rate = .050
Change in industrial electricity demand growth rate is computed from indicated electricity demand growth rate and industrial electricity demand growth and growth adjustment delay and is expressed as:.

change_in_industrial_demand_growth_rate = (indicated_industrial_electricity_demand_growth_rate-industrial_electricity_demand_growth_rate)/growth_adjustment_time
growth_adjustment_time = 5

Indicated industrial electricity demand growth rate is computed from GDP and the relationship between indicated industrial electricity demand growth rate and GDP is expressed graphically as:

indicated_industrial_electricity_demand_growth_rate = GRAPH(gdp)
(1500, 0.0288), (2850, 0.033), (4200, 0.0384), (5550, 0.0432), (6900, 0.048), (8250, 0.0534), (9600, 0.0588), (10950, 0.063), (12300, 0.0672), (13650, 0.0726), (15000, 0.0756)

The industrial use of electricity is increased by industrial use growth rate and decreased by industrial use retirement rate.

industrial_electricity_use(t) = industrial_electricity_use(t - dt) + (industrial_use_growth_rate - industrial_use_retirement_rate) * dt
INIT industrial_electricity_use = 31430880

Growth of industrial use of electricity is computed from the industrial use of electricity and its growth fraction.

industrial_use_growth_rate = industrial_electricity_use * industrial_use_growth_fraction

Industrial use growth fraction is simply industrial electricity demand growth rate.

industrial_use_growth_fraction = industrial_electricity_demand_growth_rate

Retirement of industrial use of electricity is computed from the industrial use of electricity and its retirement fraction.

industrial_use_retirement_rate = industrial_electricity_use * industrial_use_retirement_fraction

industrial_use_retirement_fraction = 0.0001

### Residential Electricity Demand

Per capita income increases with income growth rate.

per_capita_income(t) = per_capita_income(t - dt) + (income_growth_rate) * dt
INIT per_capita_income = 109.7

Income growth is computed from per capita income and GDP effect on income and is expressed as:

income_growth_rate = per_capita_income * gdp_effect_on_income

GDP effect on income is expressed as a function of GDP graphically as:

gdp_effect_on_income = GRAPH(gdp)
(0.00, 0.0525), (5000, 0.0588), (10000, 0.0681), (15000, 0.0756), (20000, 0.085), (25000, 0.0938), (30000, 0.101), (35000, 0.11), (40000, 0.117), (45000, 0.123), (50000, 1.23)

Per capita residential demand increases with per capita demand growth rate.

per_capita_residential_demand(t) = per_capita_residential_demand(t - dt) + (residential_demand_growth_rate) * dt
INIT per_capita_residential_demand = 109.7

Per capita demand growth rate depends on per capita residential demand growth normal and income and is expressed as:

residential_demand_growth_rate = per_capita_consumption_growth_normal * income_effect
per_capita_consumption_growth_normal = 0.05
Income effect is expressed as a function of per capita income graphically as:
income_effect = GRAPH(per_capita_income)
(0.00, 1.12), (500, 1.19), (1000, 1.24), (1500, 1.29), (2000, 1.34), (2500, 1.35), (3000, 1.38), (3500, 1.41), (4000, 1.43), (4500, 1.45), (5000, 1.45)

Population is increased by birth rate and decreased by death rate.

population(t) = population(t - dt) + (birth_rate - death_rate) * dt
INIT population = 165000000

Birth rate is computed from population and birth fraction.

birth_rate = population * birth_fraction
birth_fraction = 0.015

Death rate is computed as:

death_rate = population * death_fraction
death_fraction = 0.005
industrial electricity demand in MW is:
industrial_demand = (industrial_use)/(0.6 * 8760)

Residential electricity demand in MW is computed from population and per capita residential electricity demand.

residential_demand = (population * per_capita_residential_demand)/(1000 * 0.6 * 8760)

### Total Electricity Demand

Total electricity demand in MW is the sum of industrial electricity demand and residential electricity demand.

total_demand = industrial_demand + residential_demand

### 9.4.2.2   Electricity Supply Sector

Electricity supplies are considered from five power plants: coal-based power plant, hydroelectric power plant, natural gas-based power plant, nuclear power plant and solar PV power plant. Figure 9.4(b) shows the stock–flow diagram of the electricity supply sector of the model.

The fundamental equations that correspond to the major variables shown in Figure 9.4(b) are as follows:

### Coal-Based Power Plant

Coal-based power plant installed capacity increases with the increase in coal-based power plant construction rate and decreases with the retirement rate of coal-based power plant.

coal_installed_capacity(t) = coal_installed_capacity(t - dt) + (coal_construction_rate - retirement_rate_of_coal_based_power_plant) * dt
INIT coal_installed_capacity = 754

Coal-based power plant construction rate is computed from coal-based power plant under construction and coal-based power plant construction delay.

coal_construction_rate = coal_under_construction/coal_construction_delay
coal_construction_delay = 3.5

Coal-based power plant retirement rate is computed from coal-based power plant installed capacity and coal-based power plant retirement fraction.

retirement_rate_of_coal_based_power_plant = coal_installed_capacity * coal_fraction
coal_fraction = 0.0001

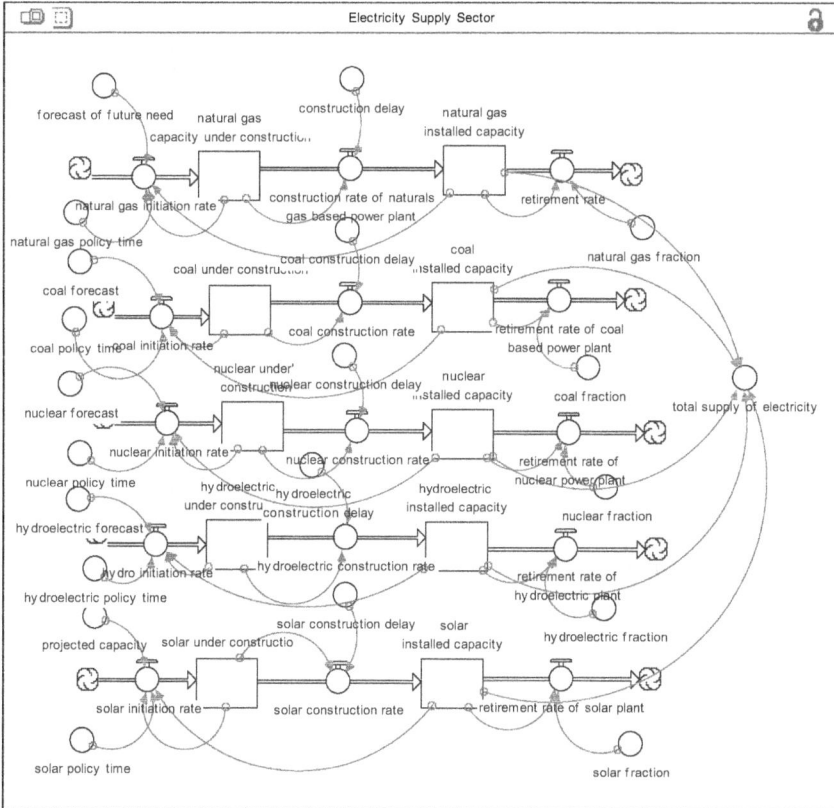

**FIGURE 9.4(b)** Electricity supply sector of the national integrated power system model.

Coal-based power plant under construction increases with the increase in coal-based power plant initiation rate, and it decreases with the construction rate of coal-based power plant.

coal_under_construction(t) = coal_under_construction(t - dt) + (coal_initiation_rate - coal_construction_rate) * dt
INIT coal_under_construction = 3840

Coal-based power plant initiation rate is computed from coal-based power plant forecast, coal-based power plant installed, coal-based power plant under construction and coal-based power plant policy time.

coal_initiation_rate = (coal_forecast - coal_installed_capacity - coal_under_construction)/coal_policy_time
coal_forecast = 4594
coal_policy_time = 30

## Hydroelectric Power Plant

Hydroelectric power plant installed capacity increases with the increase in hydroelectric power plant construction rate and decreases with the retirement rate of hydroelectric power plant.

hydroelectric_installed_capacity(t) = hydroelectric_installed_capacity(t - dt) + (hydroelectric_construction_rate - retirement_rate_of_hydroelectric_plant) * dt
INIT hydroelectric_installed_capacity = 230

Hydroelectric power plant construction rate is computed from hydroelectric power plant under construction and hydroelectric power plant construction delay.

hydroelectric_construction_rate = hydroelectric_under_construction/hydroelectric_construction_delay
hydroelectric_construction_delay = 2.5

Hydroelectric power plant retirement rate is computed from hydroelectric power plant installed capacity and hydroelectric power plant retirement fraction.

retirement_rate_of_hydroelectric_plant = hydroelectric_installed_capacity * hydroelectric_fraction
hydroelectric_fraction = 0.0001

Hydroelectric power plant under construction increases with the increase in hydroelectric power plant initiation rate and decreases with the construction rate of hydroelectric power plant.

hydroelectric_under_construction(t) = hydroelectric_under_construction(t - dt) + (hydro_initiation_rate - hydroelectric_construction_rate) * dt
INIT hydroelectric_under_construction = 0

Hydroelectric power plant initiation rate is computed from hydroelectric forecast, hydroelectric power plant installed, hydroelectric power plant under construction and hydroelectric power plant policy time.

hydro_initiation_rate = (hydroelectric_forecast - hydroelectric_installed_capacity - hydroelectric_under_construction)/hydroelectric_policy_time
hydroelectric_forecast = 400
hydroelectric_policy_time = 30

## Natural Gas Power Plant

Natural gas power plant installed capacity increases with the increase in natural gas power plant construction rate and decreases with the retirement rate of natural gas power plant.

natural_gas_installed_capacity(t) = natural_gas_installed_capacity(t - dt) + (construction_rate_of_natural_gas_based_power_plant - retirement_rate) * dt
INIT Natural_gas_installed_capacity = 10000

Natural gas power plant construction rate is computed from natural gas power plant under construction and natural gas power plant construction delay.
construction_rate_of_natural_gas_based_power_plant = natural_gas_capacity_under_construction/construction_delay_of_ng_power_plant
construction_delay_of_ng_power_plant = 3.5

Natural gas power plant retirement rate is computed from natural gas power plant installed capacity and natural gas power plant retirement fraction.

retirement_rate = natural_gas_installed_capacity * natural_gas_fraction
natural_gas_fraction = 0.0001

Natural gas power plant under construction increases with the increase in natural gas power plant initiation rate and decreases with the construction rate of natural gas power plant.

natural_gas_capacity_under_construction(t) = natural_gas_capacity_under_construction(t - dt) + (natural_gas_initiation_rate - construction_rate_of_natural_gas_based_power_plant) * dt
INIT natural_gas_capacity_under_construction = 3914

Natural gas power plant initiation rate is computed from forecast of future need, natural gas power plant installed, natural gas power plant under construction and natural gas power plant policy time.

natural_gas_initiation_rate = (forecast_of_future_need - natural_gas_capacity_under_construction - natural_gas_installed_capacity)/natural_gas_policy_time
forecast_of_future_need = 22000
natural_gas_policy_time = 30

### Nuclear Power Plant
Nuclear power plant installed capacity increases with the increase in nuclear power plant construction rate and decreases with the retirement rate of nuclear power plant.

nuclear_installed_capacity(t) = nuclear_installed_capacity(t - dt) + (nuclear construction_rate - retirement_rate_of_nuclear_power_plant) * dt
INIT nuclear_installed_capacity = 0

Nuclear power plant construction rate is computed from nuclear power plant under construction and nuclear power plant construction delay.

nuclear_construction_rate = nuclear_under_construction/nuclear_construction_delay
nuclear_construction_delay = 5

Nuclear power plant retirement rate is computed from nuclear power plant installed capacity and nuclear power plant retirement fraction.

retirement_rate_of_nuclear_power_plant = nuclear_installed_capacity * nuclear_fraction

nuclear_fraction = 0.0001

Nuclear power plant under construction increases with the increase in nuclear power plant initiation rate and decreases with the construction rate of nuclear power plant.

nuclear_under_construction(t) = nuclear_under_construction(t - dt) + (nuclear_initiation_rate - nuclear_construction_rate) * dt

INIT nuclear_under_construction = 2400

Nuclear power plant initiation rate is computed from nuclear power plant forecast, nuclear power plant installed, nuclear power plant under construction and nuclear power plant policy time.

nuclear_initiation_rate = (nuclear_forecast - nuclear_installed_capacity - nuclear_under_construction)/nuclear_policy_time

nuclear_forecast = 2400

nuclear_policy_time = 30

### Solar PV Power Plant

Solar PV power plant installed capacity increases with the increase in solar PV power plant construction rate and decreases with the retirement rate of solar PV power plant.

solar_installed_capacity(t) = solar_installed_capacity(t - dt) + (solar_construction_rate - retirement_rate_of_solar_plant) * dt

INIT solar_installed_capacity = 3

Solar PV power plant construction rate is computed from solar PV power plant under construction and solar PV power plant construction delay.

solar_construction_rate = solar_under_construction/solar_construction_delay

solar_construction_delay = 1

Solar PV power plant retirement rate is computed from solar PV power plant installed capacity and solar PV power plant retirement fraction.

retirement_rate_of_solar_plant = solar_installed_capacity * solar_fraction

solar_fraction = 0.0001

Solar PV power plant under construction increases with the increase in solar PV power plant initiation rate and decreases with the construction rate of solar PV power plant.

solar_under_construction(t) = solar_under_construction(t - dt) + (solar_initiation_rate - solar_construction_rate) * dt
INIT solar_under_construction = 200

Solar PV power plant initiation rate is computed from projected capacity, solar PV power plant installed, solar PV power plant under construction and solar PV power plant policy time.

solar_initiation_rate = (projected_capacity - solar_installed_capacity - solar_under_construction)/solar_policy_time
projected_capacity = 500
solar_policy_time = 30

### Total Supply of Electricity
Total supply of electricity is the sum of coal-based power plant installed capacity, hydroelectric power plant installed capacity, natural gas-based power plant installed capacity, nuclear power plant installed capacity and solar PV power plant installed capacity.

total_supply_of_electricity = coal_installed_capacity + hydroelectric_installed_capacity + natural_gas_installed_capacity + nuclear_installed_capacity + solar_installed_capacity + solar_installed_capacity

### 9.4.2.3 Price Sector
Price is computed from cost, price markup and demand–supply balance. Figure 9.4(c) shows the stock–flow diagram of the electricity price sector of the model.

The fundamental equations that correspond to the major variables shown in Figure 9.4(c) are as follows:

Cost of electricity increases or decreases with the change in cost of electricity.

cost_of_electricity(t) = cost_of_electricity(t - dt) + (change_in_cost_of_electricity) * dt
INIT cost_of_electricity = 35

Change in cost of electricity is computed from cost of electricity and cost increase factor.

change_in_cost_of_electricity = cost_of_electricity * cost_increase_factor
cost_increase_factor = 0.002

Price markup of electricity increases with the change in price markup of electricity.

price_markup(t) = price_markup(t - dt) + (change_in_price_markup) * dt
INIT price_markup = 0.5

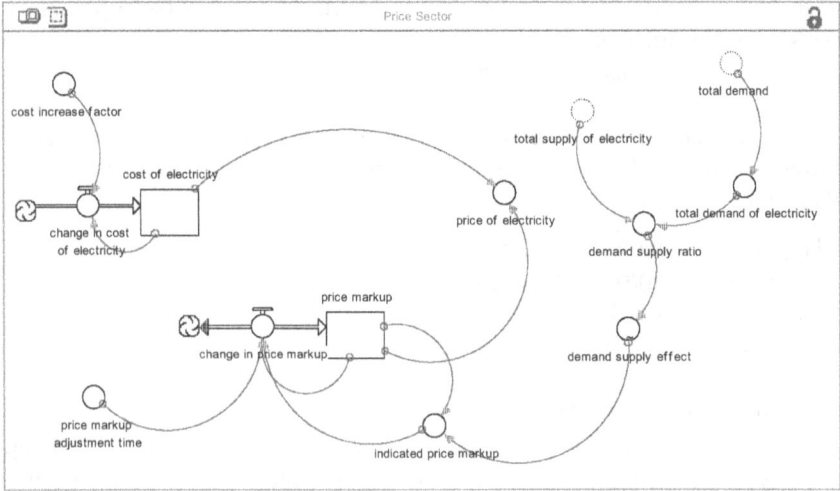

**FIGURE 9.4(c)**   Electricity price sector of the national integrated power system model.

Change in price markup of electricity is computed from indicated price markup, price markup of electricity and price markup adjustment time.

change_in_price_markup = (indicated_price_markup - price_markup)/price_markup_adjustment_time
price_markup_adjustment_time = 3

Indicated price markup of electricity is computed from price markup of electricity and demand–supply effect on price markup.

indicated_price_markup = price_markup * demand_supply_effect

Demand–supply effect on price markup is expressed as a function of demand–supply ratio graphically as:

demand_supply_effect = GRAPH(demand_supply_ratio)
(0.00, 0.525), (0.5, 0.75), (1.00, 1.00), (1.50, 1.15), (2.00, 1.33), (2.50, 1.50), (3.00, 1.60), (3.50, 1.80), (4.00, 2.00), (4.50, 2.20), (5.00, 2.43)

Demand–supply ratio is ratio of total demand of electricity to total supply of electricity.
demand_supply_ratio = total_demand_of_electricity/total_supply_of_electricity

Price of electricity should not be less than the cost of electricity and hence it should be maximum value of the cost of electric and (1 + price mark) × cost.

price_of_electricity = MAX(cost_of_electricity,(1 + price_markup) * cost_of_electricity)

### 9.4.2.4    CO$_2$ Emission Sector

CO$_2$ emission from coal-based power plant and natural gas-based power plant is considered here. Figure 9.4(d) shows the CO$_2$ emission sector of the model.

The fundamental equations that correspond to the major variables shown in Figure 9.4(d) are as follows:

Cumulative CO$_2$ emission is increased by CO$_2$ emission rate

cumulative_CO2_emission(t) = cumulative_CO2_emission(t - dt) + (CO2_emission_rate) * dt
INIT cumulative_CO2_emission = 0

CO$_2$ emission rate is simply CO$_2$ emission.

CO2_emission_rate = CO2_emission

CO$_2$ emission in tonnes is the sum of CO$_2$ from natural gas and CO$_2$ from coal.

CO2_emission = CO2_from_natural_gas + CO2_from_coal

CO$_2$ from coal in tonnes is computed by multiplying coal-based coal power plant installed capacity by emission factor for coal-based power plant.

CO2_from_coal = coal_installed_capacity * 8760 * carbon_footprint_of_coal/1000
carbon_footprint_of_coal = 500

CO$_2$ from natural gas in tonnes is computed by multiplying natural gas-based power plant installed capacity by emission factor for natural gas power plant.

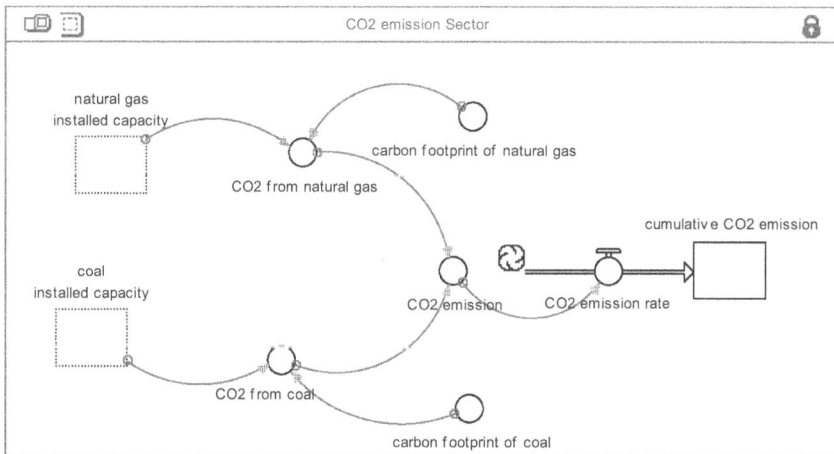

**FIGURE 9.4(d)**   CO$_2$ emission sector of the national integrated power system model.

CO2_from_natural_gas = natural_gas_installed_capacity * 8760 * carbon_foot-print_of_natural_gas/1000
carbon_footprint_of_natural_gas = 880

### 9.4.2.5   Radiation Hazard Sector

Radiation hazards are evaluated in terms of radiation harmful index and thermal pollution. Figure 9.4(e) shows the stock–flow diagram of the radiation hazard sector of the model.

The fundamental equations that correspond to the major variables shown in Figure 9.4(e) are as follows:

Radiation harmful index is computed from radiation pollution as:

radiation_harmful_index = (radiation_pollution * 0.001 * 24 * 365)/100

Radiation pollution is computed from nuclear installed capacity as:

radiation_pollution = IF(nuclear_installed_capacity > 0)THEN(0.035)ELSE(0)

Thermal pollution is computed from nuclear installed capacity as:

thermal_pollution = IF(nuclear_installed_capacity > 0)THEN(1)ELSE(0)

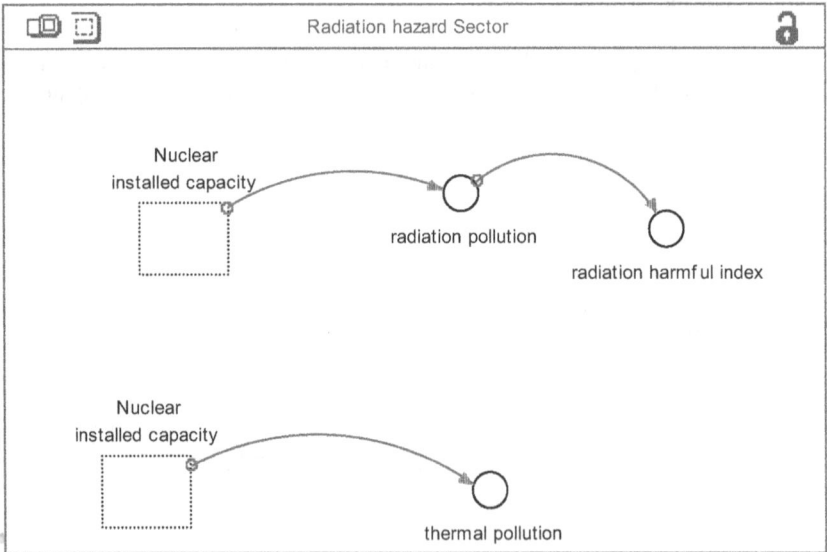

**FIGURE 9.4(e)**   Radiation hazard sector of the national integrated power system model.

## 9.5 MODEL VALIDATION

The initial values and parameter values are estimated from the primary and secondary data collected from the reports and statistical yearbook of Bangladesh for model validation. Tests for building confidence in the model were conducted to demonstrate its potentiality. Tests for building confidence in system dynamics models essentially consist of validation, sensitivity analysis and policy analysis (Bala et al., 2017). Two important notions of building confidence in the system dynamics models are testing and validation of the system dynamics models. Testing means the comparison of a model to empirical reality for accepting or rejecting the model, and validation means the process of establishing confidence in the soundness and usefulness of the model. In a behavioral validity test, emphasis should be on behavioral patterns rather than point prediction (Barlas, 1996).

To build up confidence in the prediction of the model, various ways of validating a model such as model structure, comparing model predictions with historical data, checking whether the model generates plausible behavior and checking the quality of parameter values should be conducted.

Figure 9.5 shows the comparison of simulated electricity demand for Bangladesh from 2005 to 2017 with observed data. The basic patterns and data of the simulated behavior adequately agree with the observed pattern and data. Also, the model generates plausible behavior. An extreme condition test was conducted to check whether the model is capable of coping with extreme conditions and can provide the anticipated behavior. One such condition is the shutdown of the natural gas power plant, and it is anticipated that there will be a large gap between electricity demand and electricity supply and a sudden drop in the demand–supply ratio. Figure 9.6 shows the simulated results of such conditions. The model predictions are exactly as anticipated.

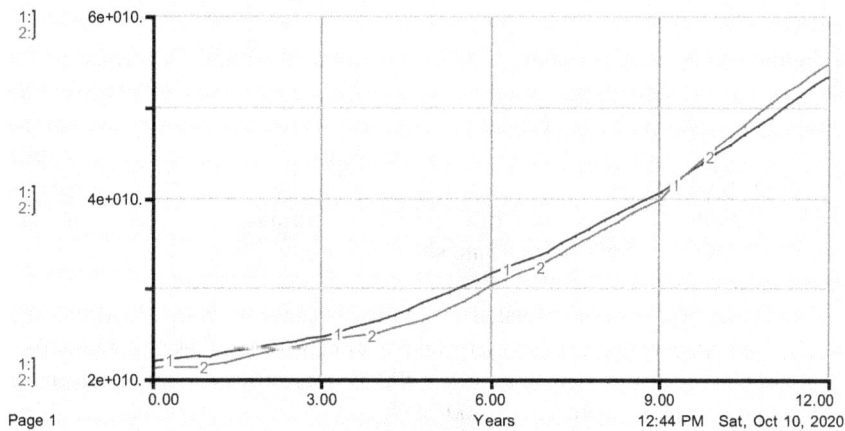

**FIGURE 9.5** Comparison of simulated electricity demand with observed data for Bangladesh from 2005 to 2017.

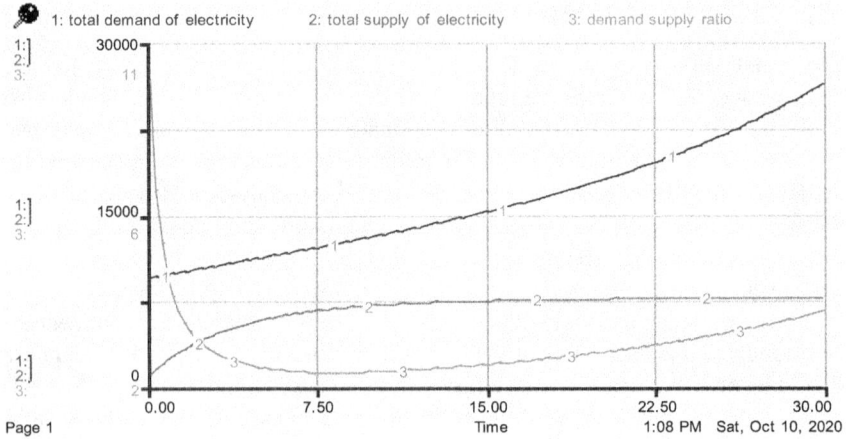

**FIGURE 9.6**   Simulated results for sudden total shutdown of natural gas-based power plant.

## 9.6   SIMULATED RESULTS

Figure 9.7 shows the simulated electric power demand and supply and price of electricity. Electricity demand increases with time to meet the requirements of economic development and better quality of life. The supply of electricity consists of fossil fuels of natural gas and coal, and renewable resources of nuclear and solar, and it follows the electricity demand. The electricity demand is nearer to the supply at the beginning of the simulation due to the fact that the nuclear power plant was under construction at the beginning of the simulation. The price decreases with time mainly due to an increase in supply from nuclear power plants installed, causing a positive supply–demand balance of electricity. However, the supply–demand gap is small at the beginning and at the end of the simulation period.

Figure 9.8 shows the simulated electricity generation and $CO_2$ emissions of electricity generation from natural gas and coal. The emissions of $CO_2$ from the burning of natural gas and coal for the production of electricity are proportional to the electricity generation, and the emission of $CO_2$ from the burning of natural gas for the production of electricity reaches saturation as the electricity generation reaches saturation. The emission of $CO_2$ from the burning of coal for the production of electricity increases with time since electricity generation from coal also increases with time.

Figure 9.9 shows the simulated electricity generation from a hydroelectric power plant, nuclear power plant and solar power plant. Electricity generation from the hydroelectric power plant approaches 400 MW, electricity generation from the nuclear power plant approaches 2400 MW, and electricity generation from the solar power plant approaches 500 MW as planned for a gradual transition to renewable energy resources.

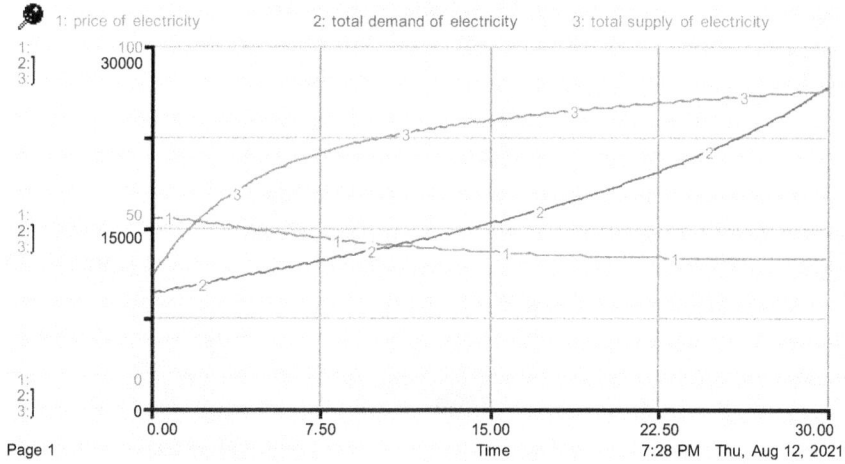

**FIGURE 9.7**  Simulated electric power demand and supply, and price of electricity.

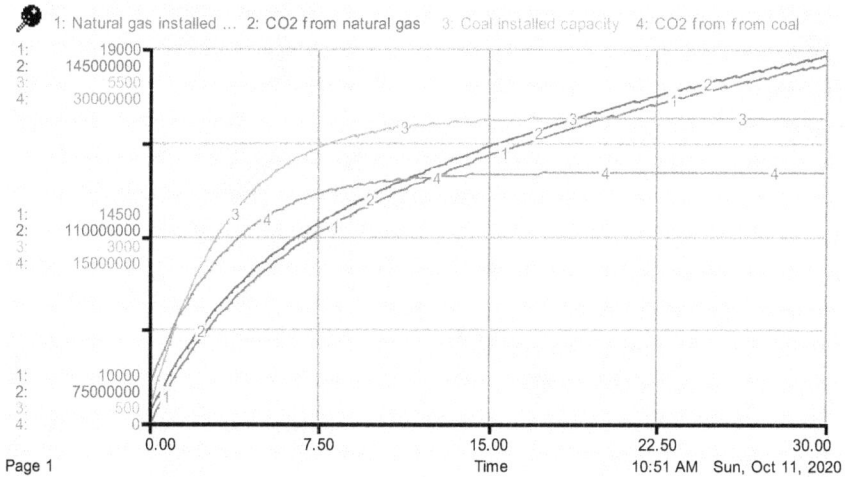

**FIGURE 9.8**  Simulated electricity generation and $CO_2$ emissions from electricity generation from natural gas and coal.

Figure 9.10 shows the simulated radiation harmful index, thermal pollution radiation and radiation pollution from the generation of electricity from a nuclear power plant. The radiation harmful index for the nuclear power plant is 0 and there is no harmful radiation from the electricity generation of a nuclear power plant. Thermal pollution is 1 and radiation pollution is 0.04.

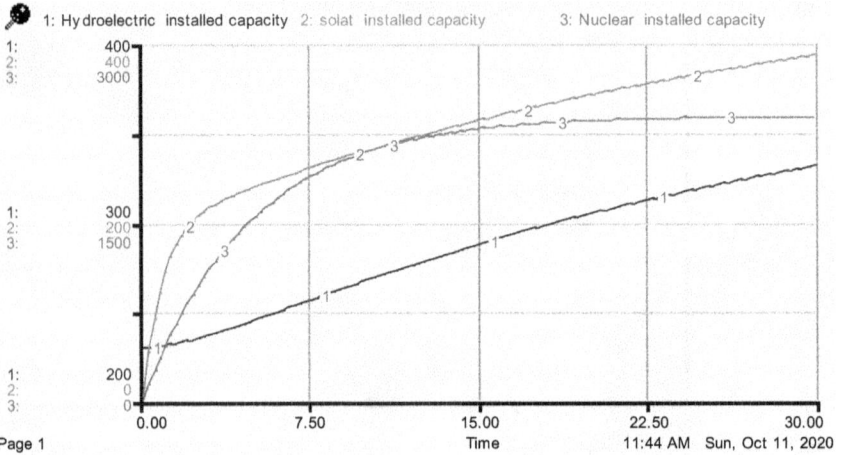

**FIGURE 9.9** Simulated electricity generation from hydroelectric power plant, nuclear power plant and solar power plant.

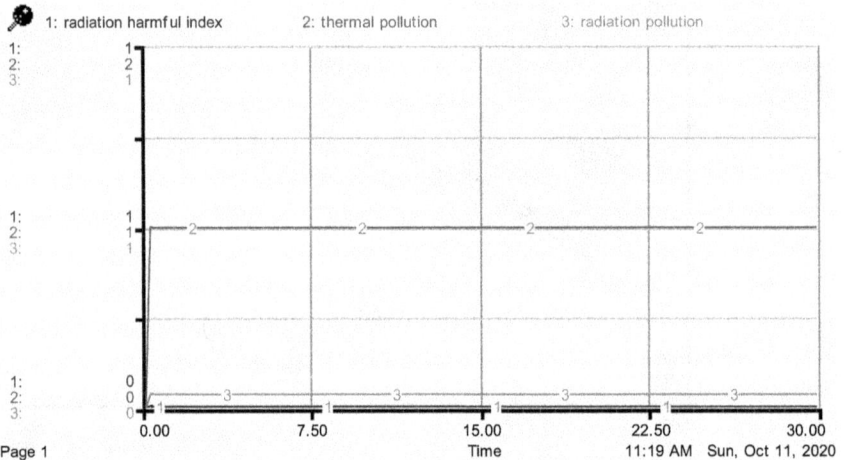

**FIGURE 9.10** Simulated radiation harmful index, thermal pollution and radiation pollution from generation of electricity from nuclear power plant.

## 9.7  POLICY ANALYSIS

Here, the model has been simulated for the gradual transition to renewable energy resources such as nuclear and solar. It can also be simulated for other important policy options, such as:

(1) Gradual transition to renewable energy resources with end-use energy conservation.
(2) Gradual transition to renewable energy resources to ensure energy security with targeted reduction of greenhouse gases.
(3) Targeted production of electricity, such as the development goal of electricity supply with a gradual transition to renewable energy resources with pollution limits.

## 9.8   CONCLUDING REMARKS

A system dynamics model of the national integrated power system can be simulated to develop scenarios of different policy options for a clearer understanding of and greater insight into the national integrated power system. It can be simulated to design a policy to be adopted for the optimal operational policy planning of the national integrated power system. This model can be used as a participatory modeling tool in all stages of the model building and simulation process, involving all the stakeholders of the model. Finally, it can be used as a computer laboratory to study the dynamics of the national integrated power system.

### 9.8.1   LISTING OF STELLA MODEL EQUATIONS

**Electricity demand sector**
gdp(t) = gdp(t - dt) + (gdp_growth_rate) * dt
INIT gdp = 1855.7

**INFLOWS**
gdp_growth_rate = gdp * gdp_growth_fraction
industrial_electricity_demand_growth_rate(t) = industrial_electricity_demand_growth_rate(t - dt) + (change_in_industrial_demand_growth_rate) * dt
INIT industrial_electricity_demand_growth_rate = .050

**INFLOWS**
change_in_industrial_demand_growth_rate = (indicated_industrial_electricity_demand_growth_rate - industrial_electricity_demand_growth_rate)/growth_adjustment_time
industrial_electricity_use(t) = industrial_electricity_use(t - dt) + (industrial_use_growth_rate - industrial_use_retardment_rate) * dt
INIT industrial_electricity_use = 31430880

**INFLOWS**
industrial_use_growth_rate = industrial_electricity_use * industrial_use_growth_fraction

**OUTFLOWS**
industrial_use_retardment_rate = industrial_electricity_use * industrial_use_
retirement_fraction
per_capita_income(t) = per_capita_income(t - dt) + (income_growth_rate) * dt
INIT per_capita_income = 3000

**INFLOWS**
income_growth_rate = per_capita_income * gdp_effect_on_income
per_capita_residential_demand(t) = per_capita_residential_demand(t - dt) + (resi-
dential_demand_growth_rate) * dt
INIT per_capita_residential_demand = 109.7

**INFLOWS**
residential_demand_growth_rate = per_capita_consumption_growth_normal *
income_effect
population(t) = population(t - dt) + (birth_rate - death_rate) * dt
INIT population = 165000000

**INFLOWS**
birth_rate = population * birth_fraction

**OUTFLOWS**
death_rate = population * death_fraction
birth_fraction = 0.015
death_fraction = 0.005
gdp_growth_fraction = 0.06
growth_adjustment_time = 5
industrial_demand = (industrial_electricity_use)/(0.6 * 8760)
industrial_use_retirement_fraction = 0.0001
industrial_use_growth_fraction = industrial_electricity_demand_growth_rate
per_capita_consumption_growth_normal = 0.05
residential_demand = (population * per_capita_residential_demand)/(1000 * 0.6
* 8760)
total_demand = industrial_demand + residential_demand
gdp_effect_on_income = GRAPH(gdp)
(0.00, 0.0525), (5000, 0.0588), (10000, 0.0681), (15000, 0.0756), (20000, 0.085),
(25000, 0.0938), (30000, 0.101), (35000, 0.11), (40000, 0.117), (45000, 0.123),
(50000, 1.23)
income_effect = GRAPH(per_capita_income)
(0.00, 0.855), (3000, 1.03), (6000, 1.10), (9000, 1.20), (12000, 1.23), (15000,
1.27), (18000, 1.34), (21000, 1.41), (24000, 1.43), (27000, 1.47), (30000, 1.48)
indicated_industrial_electricity_demand_growth_rate = GRAPH(gdp)
(1500, 0.0288), (2850, 0.033), (4200, 0.0384), (5550, 0.0432), (6900, 0.048),
(8250, 0.0534), (9600, 0.0588), (10950, 0.063), (12300, 0.0672), (13650, 0.0726),
(15000, 0.0756)

**Electricity supply sector**
coal_installed_capacity(t) = coal_installed_capacity(t - dt) + (coal_construction_
rate - retirement_rate_of_coal_based_power_plant) * dt
INIT coal_installed_capacity = 754

**INFLOWS**
coal_construction_rate = coal_under_construction/coal_construction_delay

**OUTFLOWS**
retirement_rate_of_coal_based_power_plant = coal_installed_capacity * coal_
fraction
coal_under_construction(t) = coal_under_construction(t - dt) + (coal_initiation_
rate - coal_construction_rate) * dt
INIT coal_under_construction = 3840

**INFLOWS**
coal_initiation_rate = (coal_forecast-Coal_installed_capacity-coal_under_con-
struction)/coal_policy_time

**OUTFLOWS**
coal_construction_rate = coal_under_construction/coal_construction_delay
hydroelectric_installed_capacity(t) = hydroelectric_installed_capacity(t - dt) +
(hydroelectric_construction_rate - retirement_rate_of_hydroelectric_plant) * dt
INIT hydroelectric_installed_capacity = 230

**INFLOWS**
hydroelectric_construction_rate = hydroelectric_under_construction/hydroelec-
tric_construction_delay

**OUTFLOWS**
retirement_rate_of_hydroelectric_plant = hydroelectric_installed_capacity *
hydroelectric_fraction
hydroelectric_under_construction(t) = hydroelectric_under_construction(t - dt) +
(hydro_initiation_rate - hydroelectric_construction rate) * dt
INIT hydroelectric_under_construction = 0

**INFLOWS**
hydro_initiation_rate = (hydroelectric_forecast - hydroelectric_installed_capacity
- hydroelectric_under_construction)/hydroelectric_policy_time

**OUTFLOWS**
hydroelectric_construction_rate = hydroelectric_under_construction/hydroelec-
tric_construction_delay

natural_gas_capacity_under_construction(t)     =     natural_gas_capacity_under_
construction(t - dt) + (natural_gas_initiation_rate - construction_rate_of_natural_
gas_based_power_plant) * dt
INIT natural_gas_capacity_under_construction = 3914

**INFLOWS**
natural_gas_initiation_rate = (forecast_of_future_need - natural_gas_capacity_
under_construction - natural_gas_installed_capacity)/natural_gas_policy_time

**OUTFLOWS**
construction_rate_of_natural_gas_based_power_plant  =  natural_gas_capacity_
under_construction/construction_delay_of_ng_power_plant
natural_gas_installed_capacity(t) = natural_gas_installed_capacity(t - dt) + (con-
struction_rate_of_natural_gas_based_power_plant - retirement_rate) * dt
INIT natural_gas_installed_capacity = 10000

**INFLOWS**
construction_rate_of_natural_gas_based_power_plant  =  natural_gas_capacity_
under_construction/construction_delay_of_ng_power_plant

**OUTFLOWS**
retirement_rate = natural_gas_installed_capacity * natural_gas_fraction
nuclear_under_construction(t) = nuclear_under_construction(t - dt) + (nuclear_
initiation_rate - nuclear_construction_rate) * dt
INIT nuclear_under_construction = 2400

**INFLOWS**
nuclear_initiation_rate = (nuclear_forecast - nuclear_installed_capacity - nuclear_
under_construction)/nuclear_policy_time

**OUTFLOWS**
nuclear_construction_rate  =  nuclear_under_construction/nuclear_construction_
delay
nuclear_installed_capacity(t) = nuclear_installed_capacity(t - dt) + (nuclear_con-
struction_rate - retirement_rate_of_nuclear_power_plant) * dt
INIT nuclear_installed_capacity = 0

**INFLOWS**
nuclear_construction_rate  =  nuclear_under_construction/nuclear_construction_
delay

**OUTFLOWS**
retirement_rate_of_nuclear_power_plant = nuclear_installed_capacity * nuclear_
fraction

solar_under_construction(t) = solar_under_construction(t - dt) + (solar_initiation_
rate - solar_construction_rate) * dt
INIT solar_under_construction = 200

**INFLOWS**
solar_initiation_rate = (projected_capacity - solar_installed_capacity - solar_
under_construction)/solar_policy_time

**OUTFLOWS**
solar_construction_rate = solar_under_construction/solar_construction_delay
solar_installed_capacity(t) = solar_installed_capacity(t - dt) + (solar_construc-
tion_rate - retirement_rate_of_solar_plant) * dt
INIT solar_installed_capacity = 3

**INFLOWS**
solar_construction_rate = solar_under_construction/solar_construction_delay

**OUTFLOWS**
retirement_rate_of_solar_plant = solar_installed_capacity * solar_fraction
coal_construction_delay = 3.5
coal_forecast = 4594
coal_fraction = 0.0001
coal_policy_time = 30
construction_delay_of_ng_power_plant = 3.5
forecast_of_future_need = 22000
hydroelectric_forecast = 400
hydroelectric_fraction = 0.0001
hydroelectric_policy_time = 30
hydroelectric_construction_delay = 2.5
natural_gas_fraction = 0.0001
natural_gas_policy_time = 30
nuclear_construction_delay = 5
nuclear_forecast = 2400
nuclear_fraction = 0.0001
nuclear_policy_time = 30
projected_capacity = 500
solar_construction_delay = 1
solar_fraction = 0-0001
solar_policy_time = 30
total_supply_of_electricity = coal_installed_capacity + hydroelectric_installed_
capacity + natural_gas_installed_capacity + nuclear_installed_capacity + solar_
installed_capacity

**CO2 emission sector**

cumulative_CO2_emission(t) = cumulative_CO2_emission(t - dt) + (CO2_emission_rate) * dt

INIT cumulative_CO2_emission = 0

**INFLOWS**

CO2_emission_rate = CO2_emission

carbon_footprint_of_coal = 500

carbon_footprint_of_natural_gas = 880

CO2_emission = CO2_from_natural_gas + CO2_from_coal

CO2_from_coal = coal_installed_capacity * 8760 * carbon_footprint_of_coal/1000

CO2_from_natural_gas = natural_gas_installed_capacity * 8760 * carbon_footprint_of_natural_gas/1000

**Price sector**

cost_of_electricity(t) = cost_of_electricity(t - dt) + (change_in_cost_of_electricity) * dt

INIT cost_of_electricity = 35

**INFLOWS**

change_in_cost_of_electricity = cost_of_electricity * cost_increase_factor

price_markup(t) = price_markup(t - dt) + (change_in_price_markup) * dt

INIT price_markup = 0.5

**INFLOWS**

change_in_price_markup = (indicated_price_markup - price_markup)/price_markup_adjustment_time

cost_increase_factor = 0.002

demand_supply_ratio = total_demand_of_electricity/total_supply_of_electricity

indicated_price_markup = price_markup * demand_supply_effect

price_markup_adjustment_time = 3

price_of_electricity = MAX(cost_of_electricity,(1 + price_markup) * cost_of_electricity)

total_demand_of_electricity = total_demand

demand_supply_effect = GRAPH(demand_supply_ratio)

(0.00, 0.525), (0.5, 0.75), (1.00, 1.00), (1.50, 1.15), (2.00, 1.33), (2.50, 1.50), (3.00, 1.60), (3.50, 1.80), (4.00, 2.00), (4.50, 2.20), (5.00, 2.43)

**Radiation hazard sector**

radiation_harmful_index = (radiation_pollution * 0.001 * 24 * 365)/100

radiation_pollution = IF(nuclear_installed_capacity > 0)THEN(0.035)ELSE(0)

thermal_pollution = IF(nuclear_installed_capacity > 0)THEN(1)ELSE(0)

# REFERENCES

Barlas, Y. (1996). Formal aspects of model validity and validation in system dynamics. *System Dynamics Review*. 12(3): 183–210.

Bala, B. K. (1997). *Computer modelling of energy and environment: The case of Bangladesh*. Proceedings of 15th International System Dynamics Conference, Istanbul, Turkey, August 19–22, 1997.

Bala, B. K. (1999). *Principles of System Dynamics* (1st ed.). Agrotech Publishing Academy, Udaipur, India.

Bala, B. K. (1998). *Energy and Environment: Modelling and Simulation*. Nova Science Publishers, USA

Bala, B. K. (2006). Computer modelling of energy and of environment for Bangladesh. *International Agricultural Engineering Journal*. 15: 151–160.

Bala, B. K., Alam, M, S., & Denath, N. (2014). Energy perspective of climate change: The case of Bangladesh. *Strategic Planning for Energy and the Environment, Development*, 33(3): 6–22.

Bala, B. K., Fatimah, M. A., & Kushairi, M. N. (2017). *System Dynamics: Modelling and Simulation*. Springer.

Bala, B. K., & Satter. M. A. (1986a). *Modelling of rural energy systems*. Presented at Second National Symposium on Agricultural Research, BARC, Dhaka 12 February 1986.

Bala, B. K., & Satter, M. A. (1986b). *Modelling of rural energy systems for food production in developing countries*. Energia and Agricoltura 2 Conferenza Internationale. Sirmione/Brescia (Italia). 3, p. 306, 1986.

Bastan, M., & Shakouri, H. (2018). *A system dynamics model for policy evaluation of energy dependency*. Proceedings of the International Conference on Industrial Engineering and Operations Management Paris, France, July 26–27, 2018

Dimitrovsky, A., Gebremicael, M., Tomsovic, K., Ford, A., & Vogstad, K. (2004). *Comprehensive long term modeling of the dynamics of investment and growth in electric power system*. Paper presented at EPNES Workshop. Puerto Rico. July 12–14, 2004.

Ford, A. (1997). System dynamics and the electric power industry. *System Dynamics Review*. 13(1}: 57–85

Ford, A., & Bull, M. (1989). Using system dynamics for conservation policy analysis in the Pacific Northwest. *System Dynamics Review*. 5: 1–16.

Forrester, J. W. (1968). *Principles of Systems*. Allen Wright Press, Cambridge, Massachusetts, USA.

Gu, C., Ye, X., Cao, Q., Guan, W., Peng, C., Wu, Y., & Zhai, W. (2020). System dynamics modelling of urbanization under energy constraints in China. *Scientific Reports: Nature Research*. 103:9956 (1–16).

He, Y. X., Jiao, J., Chen, R. J., & Shu, H. (2018). The optimization of Chinese power grid investment based on transmission and distribution tariff policy: A system dynamics approach. *Energy Policy*. 113: 112–122.

Huq, A. Z. M. (1975). *Energy modelling for agriculture units in Bangladesh*. Paper presented at the National Seminar on Integrated Rural Development, Dhaka, 1975.

Khanna, M., & Rao, N. D. (2009). Supply and demand of electricity in the developing world. *Annual Review of Resource Economics*. 1: 567–597.

Liuguo, S., Shijing, Z., & Jianbai, H. (2012). Pricing simulation platform based on system dynamics. *Systems Engineering Procedia*. 5:445–453.

Maani, K. E., & Cavana, R.Y. (2000). *Systems Thinking and Modelling: Understanding Change and Complexity*. Prentice Hall, New Zealand.

Mohapatra, P. K. J., Mandal, P., & Bora, M. C. (1994). *Introduction to System Dynamics Modelling*. Universities Press, India.

Mutingi, M., Mbohwa, C., & Kommula, V. P. (2017). System dynamics approaches to energy policy modelling and simulation. *Energy Procedia*. 121:530–539.

Nail, R. F. (1992). A system dynamics model for national energy policy planning. *System Dynamics Review*. 8: 1–19.

Nail, R. F., Belanger, S., Klinger, A., & Petersen, E. (1992) An analysis of the cost effectiveness of U.S. energy policies to mitigate global warming. *System Dynamics Review* 1992: 8, 111–128.

Qurat-ullah, H. (2013). Understanding the dynamics of electricity generation capacity in Canada: A system dynamics approach. *Energy*. 59: 285–294.

Sani, K., Siallagan, M., Putro, U.S., & Mangkusubroto, K. (2018). Indonesia energy mix modelling using system dynamics. *International Journal of Sustainable Energy Planning and Management*. 18: 29–52.

Sterman, J. D. (2000). *Business Dynamics: Systems Thinking and Modeling for a Complex World*. McGraw-Hill Higher Education, Boston.

Vogstad, K. (2004). *A system dynamics analysis of the Nordic electricity market: The transition from fossil fuelled toward a renewable supply with a liberalized electricity market*. PhD thesis. Norwegian University of Science and Technology.

Wang, J., Wu, J., & Che, Y. (2019). Agent and system dynamics-based hybrid modeling and simulation for multilateral bidding in electricity market. *Energy*. 180: 444–456.

World Nuclear Association. (2020). *Comparison of Lifecycle Greenhouse Gas Emissions of Various Electricity Generation Sources*. WNA Report. Page 6.

# 10 Operational Planning of Electrical Power Systems and Smart Grids

## 10.1 INTRODUCTION

Energy is needed for economic and social development, and per capita consumption of energy is a measure of physical quality of life. Per capita consumption of electrical energy is also a measure of physical quality of life and an indicator of economic development (Bala, 1998). However, the production of electrical energy using fossil fuels causes environmental pollution and global warming. Furthermore, electrical energy is essential for economic and social development, but is capital intensive. Hence, electrical power systems should be planned to operate optimally and efficiently within the constraints of the availability of energy sources with a minimum contribution to global warming.

The energy crisis in the 1970s created an upsurge of interest in the optimal planning of electrical power systems among researchers. The US made a huge investment into developing a rational basis of optimal energy planning in the mid-70s and the Brookhaven Energy Systems Optimization Model (BESOM) was developed. The first version of BESOM was a linear programming optimization of US energy systems. Indeed, many of today's energy models are derivatives of BESOM. For example, the MARKAL model was developed under the sponsorship of the International Energy Agency (IEA) as a collaborative project between a German research institute and Brookhaven National Laboratory. In Bangladesh, energy modeling and planning was initiated by Huq (1975) and further developed for integrated rural energy systems (Bala and Satter, 1986). Bala (2006) simulated the integrated energy systems of Bangladesh. In India, the optimal power system planning model was developed by Parikh (1997) for the Government of India. Bala et al. (2014) simulated the integrated energy systems of Bangladesh and identified the policy options for energy security. Mohamed and Eltamaly (2018) reported the smart grid application for sizing optimal components of hybrid renewable energy systems and optimization of the system to supply load at five sites of Saudi Arabia. Several pieces of research have been reported on the optimal operational planning and modeling of smart grids (Babu, 2019; Melhern, 2018; Ranganathan and Nygard, 2017; Ranganathan, 2013; Xu, 2016). Wang et al., (2018) reported

DOI: 10.1201/9781003218401-10

the operational optimization of smart microgrids in the presence of distributed generation and demand response, and the operational optimization problem was solved using a genetic algorithm. The genetic algorithm was used to implement the objective function and the demand response strategy. The optimization results demonstrate that it can significantly reduce the power consumption of the grid, and the model has certain practicability.

## 10.2　LINEAR PROGRAMMING MODEL

Power system operations planning is of crucial importance in order to minimize loss due to inefficient utilization of power and shortages of power. To ensure optimal power system operation for better utilization of available resources to reduce shortages of power and pollution from the use and production of electrical energy and cost reductions, we must explore alternative ways to operate and manage existing power systems for better efficiency and cost reductions. To address the problem of power system operations planning, the problem may be stated as a linear programming model with the system cost as the objective function to be minimized subject to a set of constraints. Hence, the problem is stated as:

Minimize

$$Z = \sum_s \sum_k (\text{generation})_{sk} (\text{generation cost})_s \\ + \sum_i \sum_k (\text{unmet energy})_{ik} (\text{unmet energy cost})_k \quad (10.1)$$

where

s = power plants such as natural gas, coal, nuclear, and hydroelectric.

k = time blocks of load duration curves.

i = states/regions

subject to

**Constraint 1:** Generation capacity availability

$$(\text{generation})_{sk} \leq (\text{availability})_s (\text{capacity})_s (\text{duration})_k \quad (10.2)$$

**Constraint 2:** The demand minus unmet energy must be equal to the sum of generation capacity available

$$(\text{demand})_{ik} \leq \sum_s (\text{generation})_{sk} (1 - (\text{auxiliary})_s) + (\text{unmet energy})_{ik} \quad (10.3)$$

**Constraint 3:** Non-negativity constraint

The decision variables of generation and unmet energy must be greater than or equal to zero.

There are currently a number of solvers available for optimal solutions to general linear programming problems. However, we need a solver that will scale extremely well and produce solutions in near real-time. Parikh (1997) used this linear programming model for the operational planning of the Indian power system. The optimized cost objective and average generation cost for operational planning of the Indian power system was reduced from Rs 37 million thousands of actual cost objective to Rs 20 million thousands of optimal cost objective, and Rs 650/MWh of average generation cost of electricity generation to Rs 350/MWh of optimized cost, respectively. This clearly demonstrates the potentiality of optimal operational planning power systems using the linear programming technique. The integrated optimal operation of the electric power system of India is thus shown to be a promising option to reduce system operating costs and prevailing energy shortages.

## 10.3   MARKAL MODELING

MARKAL (MARKet ALlocation) is a linear programming model developed by the Energy Technology Systems Analysis Program (ETSAP) of the IEA and is a widely used dynamic technique. It can depict both the energy supply and demand sides of energy systems, and it can aid policy planners and managers in policy planning by providing optimization scenarios of different policy options.

MARKAL modeling is based on primary and secondary data, and the MARKAL simulator developed by ETSAP under the auspices of the IEA is a bottom-up, partial equilibrium, linear programming, and least-cost optimization system. Thus, it is an ideal scenario simulator and a perfect tool for policy analysis. The modeling framework requires the full spectrum of processes from the supply of primary fuels through to the conversion technologies to meet the end-user demand sectors. The structure of MARKAL is shown in Figure 10.1. Energy carriers in MARKAL interconnect the conversion and consumption of energy, and demand for energy services may be disaggregated by sectors. The optimization routine in MARKAL selects from each of the sources, energy carriers and transformation technologies to produce the least-cost solution subject to a variety of constraints. As a result of this integrated approach, supply-side technologies are matched to energy service demands.

In MARKAL, all power plants connected to the grid are considered to match the centralized national grid, and it is assumed that consumption of electricity will never exceed generation levels and electricity demand will increase throughout the study horizon. The national average for the transmission and distribution loss of electricity is taken into account. Currency can be specified in US dollars ($) or local currency, and primary and secondary fuel costs can be obtained from the EIA. Capital, operating and variable cost, as well as technical efficiencies and availability factor for various technologies, are adopted as the Energy Technology Reference Indicator for the projection period. Transmission and distribution cost is not accounted for in the model, seasonal and daily load fluctuations are not considered, and the power sector has no financial constraints due to active investments by the private sector.

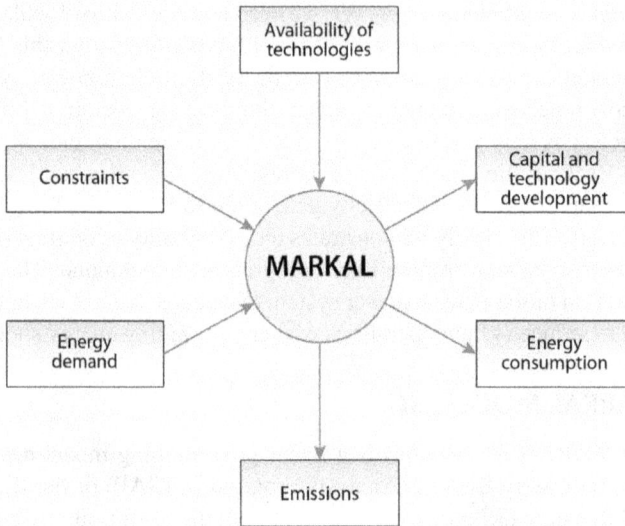

**FIGURE 10.1** Structure of the MARKAL model.

The steps in modeling using MARKAL are as follows:

(1) Description of the integrated energy systems structure based on research reports and studies and primary and secondary data.
(2) Formulation of optimization of integrated energy systems using MARKAL.
(3) Analysis of scenarios of optimal energy systems and energy planning.

### 10.3.1 MARKAL OPTIMIZATION

To ensure optimal power system operation for the better utilization of available resources, the generation cost of power and pollution from the use and production of electrical energy needs to be minimized. The problem of power system operations may be stated as:

Minimize

$$Z = \sum_{s} \sum_{k} (\text{generation})_{sk} (\text{generation cost})_{s} \qquad (10.4)$$

subject to
**Constraint 1:** Generation capacity availability

$$(\text{generation})_{sk} \leq (\text{availability})_{s} (\text{capacity})_{s} (\text{duration})_{k} \qquad (10.5)$$

The electricity generation is from an integrated power system consisting of non-renewable energy resources such as oil, gas and coal, and from renewable energy resources such as solar and wind.

**Constraint 2:** The demand must be less than or equal to the sum of generation capacity available

$$(\text{demand})_{ik} \leq \sum (\text{generation})_{sk} (1 - (\text{auxiliary})_s) \qquad (10.6)$$

**Constraint 3:** Pollution must be within limits permitted

$$\sum_s (\text{generation})_s (CO_2 \text{ production multiplier})_s \leq (CO_2 \text{ limitation})_s \quad (10.7)$$

**Constraint 4:** Non-negativity constraint
The decision variables of generation, unmet energy and $CO_2$ production must be greater than or equal to zero.
Some of the energy policies that can be optimized using MARKAL are:

(1) Policy for the optimal energy mix for integrated energy systems including solar, wind and nuclear.
(2) Policy for private and public investments and energy conservation.
(3) National energy policy and policy for energy security for Vision 2041.

Here the optimized results of the integrated electric power systems of Bangladesh using MARKAL are discussed to illustrate optimal operation planning. Optimization by MARKAL of the power system in Bangladesh from 2005 to 2035 focusing on renewable energy integration for the target reduction of 10%, 20% and 30% of $CO_2$ emissions reduces the conventional energy use of 10%, 21% and 32% from the base year, respectively (Mondal et al., 2010). These $CO_2$ emission targets enhance the use of renewable energy resources for electricity generation by about 431 TWh, 709 TWh and 995 TWh, respectively. The share of electricity generation from renewable energy resources increases by about 19%, 27% and 35% in 2035 for 10%, 20% and 30% reduction of $CO_2$ emissions, respectively.

## 10.4   OPTIMAL POWER SYSTEM OPERATION WITH SCADA

In modern electric power systems, large generators are used to supply electric power through step-up transformers into high-voltage interconnected transmission lines. The power is transmitted over long distances and then stepped down through a series of distribution transformers to final circuits for delivery to consumers. The present-day electricity grid has evolved as a result of rapid urbanization and infrastructure development. However, the growth of electric power systems is influenced by economic, political and geographic factors.

Since energy resources are limited in supply and demands for energy increase with time, the optimization of available energy resources is essential. Conventional power generation resources such as coal, oil and gas are depleting rapidly and creating environmental problems. These suggest a gradual transition to renewable resources such as solar and wind. Hence, there is a need to optimize energy use and reduce contributions to global warming. The automation of power systems is the solution for achieving this goal, and every sector of the power system, from generation to customer, is being automated today to achieve optimal use of energy and resources.

An overriding question is how to maintain system security during the operation of a power system. Essentially, system security practices are designed to keep the system operating when components fail to work properly (Wood et al., 2014). For example, a generating unit is taken out when auxiliary equipment fails. By proper arrangement of the spinning reserve, we can use the remaining units to make up the deficit without too low a frequency drop, or we can do load-shedding. Similarly, when a transmission line is damaged by a storm, it is taken out by automatic re-laying. If in committing and dispatching generation, proper regard for transmission flows is maintained, the remaining transmission lines can take the increased loading and still remain within limits. Also, all equipment in an electric power system is designed in such a way that it can be disconnected from the network.

A SCADA system is a technology that enables a user to collect data from one or more distant facilities or send limited control instructions to those facilities, and SCADA systems are designed to prevent faults and maintain expensive upstream plants and equipment without any damage. Figure 10.2 shows the SCADA system structure for generation, transmission, distribution and consumers of electricity in a power industry where it supervises several operations, including protection, controlling and monitoring (Sayed and Gabbar, 2016). Most SCADA systems monitor and make slight changes to function optimally. SCADA systems usually are used for large geographically extended electricity distribution operations. However, the application of a SCADA system is not totally effective and has covered about 15%–20% of the distribution system. Primarily, SCADA systems have been implemented into transmission systems.

SCADA systems are used for large-scale geographical distribution or generation systems and are applicable for large-scale renewable systems such as solar and wind. A SCADA system is the heart of large-scale renewable power generation, as well as the Distribution Management System (DMS) architecture.

SCADA systems usually monitor and make slight changes to function optimally. They are considered to be closed-loop control systems and operate with comparatively little human interface. Modern SCADA systems essentially replace manual labor to operate electrical distribution tasks and manual processes in distribution systems with automated equipment. SCADA systems maximize the efficiency of power distribution systems by providing a real-time view into operations, data-trending and logging, maintaining desired voltages, currents and power factors. One of the key processes of a SCADA system is the ability to supervise the whole

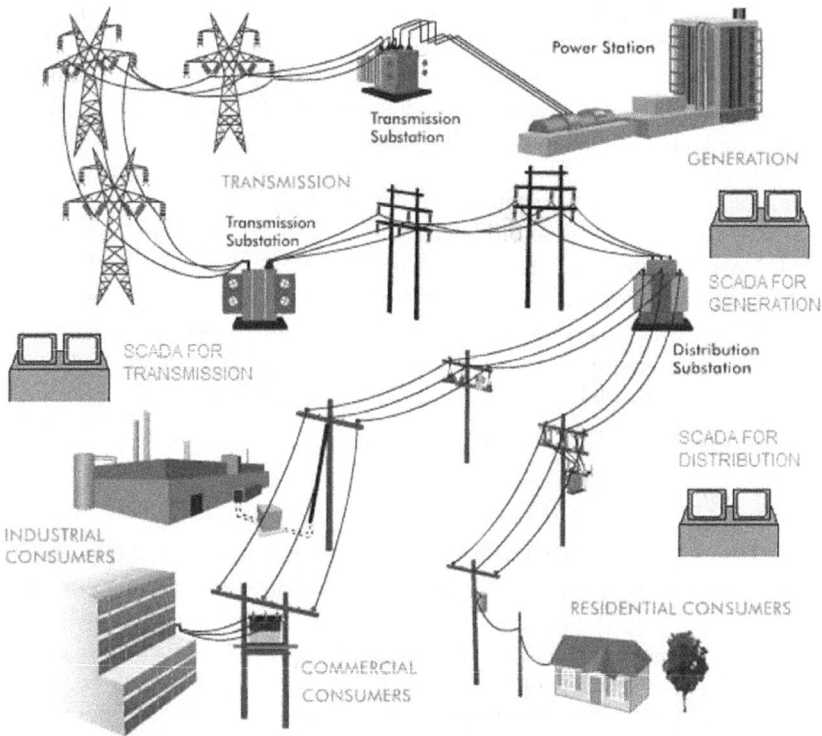

**FIGURE 10.2**  SCADA for electrical power industry (Sayed and Gabbar, 2016).

system in a real-time environment. This is accomplished by data acquisitions communicated at regular intervals. As an automation system, a SCADA system is used to acquire data from the instruments and sensors located at remote sites and to receive and transmit the data at master station central sites for controlling or monitoring purposes. The collected data can be displayed and viewed at host computers at the central site. Based on these data, automated or operator-driven supervisory control commands are sent to remote substation control devices, often referred to as field devices.

Essentially, a SCADA system is mainly responsible for conveying measurement information and control messages. The operation of the generation and transmission systems is monitored and controlled by a SCADA system. These link the various elements through communication networks and connect the transmission substations and generators to a manned control center that maintains system security and facilitates integrated operation. In larger power systems, regional control centers serve an area, with communication links to adjacent area control centers. In addition to this central control, all the generators use automatic local governor and excitation control. Local controllers are also used in some transmission circuits for voltage control and power flow control, for example, using phase shifters (sometimes known as quadrature boosters)

A SCADA system performs automatic monitoring, protecting and controlling of various equipment in distribution systems using intelligent electronic devices (IEDs) or remote terminal units (RTUs). It restores the power service during fault conditions and maintains desired operating conditions. A SCADA system also improves the reliability of the supply by reducing the duration of outages and provides cost-effective operation of the distribution system. Therefore, a distribution SCADA system supervises the entire electrical distribution system. The major functions of a SCADA system can be categorized into the following types:

- substation control;
- feeder control; and
- end-user load control.

## 10.4.1 What is SCADA?

A SCADA system is defined as a collection of equipment that will provide an operator at a remote location with sufficient information to determine the status of particular equipment or processes and initiates actions to monitor and control equipment or processes without being physically present (Thomas and McDonald, 2015). SCADA implementation involves two major activities: data acquisition (monitoring) of a process or equipment; and supervisory control of the process, thus leading to complete automation. Thus, complete automation of a process can be achieved by automating the monitoring and control actions.

The major functions of the SCADA system are supervision, control, optimization and management of the generation and transmission of electrical energy. The main components of this system are RTUs, which are connected directly to sensors, meters, loggers or process equipment to collect data automatically. RTUs are located near the monitored process to transfer data to the controller unit when requested. They have integral software, data-logging capabilities, a real-time clock (RTC) and a battery backup, and most of the RTUs are time redundant. RTU devices are complete remote terminal units that consist of the transceivers, encoders and processors, and ensure proper functioning when a primary RTU stops working. Meter readings and equipment status reports are handled by programmable logic controllers (PLCs).

A SCADA system consists of equipment and procedures to control remote stations from a master control station, and SCADA equipment and procedures consist of digital control equipment, sensing and telemetry equipment, and two-way communications between the master stations and the remotely controlled stations. SCADA digital control equipment consists of control computers and terminals for data display and entry. The sensing and telemetry equipment used include sensors, digital-to-analog and analog-to-digital converters, actuators, and relays used at the remote station to sense operating and alarm conditions and to remotely activate equipment such as circuit breakers. The communications equipment used includes modems (modulator/demodulator) for transmitting digital data, and communications links (radios, phone lines, and microwave links, or powerlines).

In essence, a SCADA system is a technology that enables a user to collect data from one or more distant facilities or send limited control instructions to those facilities. It is used to ensure continuity and safety of operation. The SCADA system plays an indispensable role in a modern power system operation because of its flexibility, reliability, efficiency and cost-effectiveness.

## 10.4.2 Functions of SCADA

The basic functions of a SCADA system are data acquisition, remote control, a human-machine interface (HMI), historical data analysis and report writing, and these functions are common to generation, transmission and distribution systems. Some of the typical functions a SCADA system can perform to achieve these goals are as follows:

1. Control and indication of the position of a two- or three-position device (e.g., a switch or circuit breaker).
2. Indication of position without control (e.g., transformer fans on or off).
3. Control without indication (e.g., capacitors switched in or out).
4. Set point control of remote control station (e.g., nominal voltage for an automatic tap changer).
5. Alarm sensing (e.g., fire or performance of a non-commanded function).
6. Permit operators to initiate operations at remote stations from a central control station.
7. Initiate and recognize sequences of events (e.g., routing power around a bad transformer by opening and closing circuit breakers or sectionalizing a bus with a fault on it).
8. Data acquisition from metering equipment, usually via an analog/digital converter and a digital communication link.

Today, all routine substation functions are remotely controlled. For example, a complete SCADA system can perform the following substation functions:

1. Automatic bus sectionalizing.
2. Automatic reclosing after a fault.
3. Synchronous check.
4. Protection of equipment in a substation.
5. Fault reporting.
6. Transformer load balancing.
7. Voltage and reactive power control.
8. Equipment condition monitoring.

9. Data acquisition.
10. Status monitoring.
11. Data logging.

All SCADA systems have two-way data and voice communication between the master and the remote stations. Modems at the sending and receiving ends modulate, that is, put information with the carrier frequency, and demodulate, that is, remove information from the carrier, respectively.

## 10.4.3 Basics of SCADA

A SCADA system is an automation field. It is basically the transfer of data between a SCADA central host computer and a number of RTUs or PLCs, and the central host and the operator terminals. Basic SCADA functions include data acquisition, remote control, an HMI, historical data analysis, and report writing, which are common to generation, transmission, and distribution systems. A SCADA system collects information, transfers the information back to a central site, then alerts the home station, carrying out necessary analysis and control, and displaying the information in a logical and organized fashion.

A SCADA system essentially consists of five components: an RTU, a communication system, a master station, an HMI and a database (Ancillotti et al., 2013). Figure 10.3 shows the basics of a SCADA system. RTUs or PLCs link the control system to field sensing devices for collecting data from the field devices and

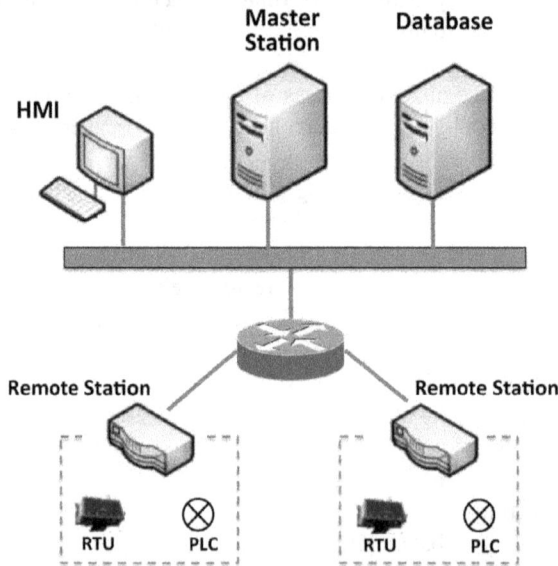

**FIGURE 10.3**    SCADA architecture in modern power grids (Ancillotti et al., 2013).

passing commands from the control station to the field devices. A communication system transfers data between the field data interface devices and control units and the computers in the SCADA central host. The central host computer server or servers are sometimes called a SCADA center, master station, or master terminal unit (MTU), where the operator monitors the system and makes control decisions to be conveyed to the field. The HMI is the interaction between the operator and the machine. Databases are used for storing historical data, measuring trends and deriving forecasts. MTUs are higher-level units including supporting applications, HMIs, data storage and data acquisition. All automation systems essentially have these five components. A SCADA system allows a utility operator to monitor and control field devices that are distributed among various remote sites.

### 10.4.4 COMPONENTS OF A TYPICAL SCADA SYSTEM

A SCADA system is an integral part of modern power systems. It will typically include a central computer, an operator interface, mass data storage, control software, and an integrated communication network. It is also a combination of telemetry and data acquisition and essentially collects information, transfers it back to the central site, carries out necessary analysis and control and then displays it on a number of operator screens or displays. The required control actions are then sent back to the process. Figure 10.3 shows the SCADA system architecture in modern power grids. The major components of the SCADA system are RTUs, master stations/MTUs, communications systems, operator workstations (HMIs) and databases.

**Remote terminal units (RTUs)**
RTUs are the eyes, ears and hands of a SCADA system and are equipped with internal computational and optimization facilities. The RTU is the main component and is directly connected to the various sensors, meters and actuators associated with a control environment. RTUs are real-time PLCs.

An RTU functions as a slave in a master/slave architecture. RTUs collect data from field devices, then process and send the data to the master station through the communication system to the monitoring of the power system. They also receive control commands from the master station and transmit these commands to the field devices.

**Master station/master terminal unit (MTU)**
A central host computer server or servers, sometimes called a SCADA center, master station or MTU, is a collection of computers, peripheral, and appropriate input and output (I/O) systems that enable the operators to monitor the state of the power system (or a process) and control it, and is one of the five components of a SCADA system. In a small system, it refers to a single computer responsible for communicating with field equipment. In a large system, a master station consists of multiple servers, distributed software applications and disaster recovery sites.

Apart from field equipment like RTUs/PLCs, master station servers and software communicate with HMI software running in the workstation of the control room or other place, as well as with RTUs by reading and writing operations during scheduled scanning. In addition, they also perform control, alarming, networking with other nodes, etc. An MTU is equivalent to a master unit in a master/slave architecture. It presents data to the operator through the HMI, collects data from the distant site, and transmits control signals to the remote site.

**Communications system**

Communication means the communication method between the MTU and the remote controllers. The communication system refers to the communication channels employed between the field equipment and the master station. The bandwidth of the channel limits the speed of communication. The communication system plays a vital role in the SCADA system implementation, and the communication network transfers data among the central host computer servers, the field data interface devices and the control units. The medium of transfer can be cable, radio, telephone, satellite, etc., or any combination of these.

**Human machine interface (HMI)**

An HMI is the interface required for the interaction between the master station and the operators or users of the SCADA system. The HMI presents data to the operator and provides for control inputs in a variety of formats, including graphics, schematics, windows, pull-down menus, touchscreens and so on. A human operator monitors the SCADA system and performs supervisory control functions for remote field operations.

At operator workstations, the operator monitors and controls the system, and these are the computer terminals consisting of standard HMI software, which are networked with a central host computer. These workstations are operator terminals that request and send information to the host client computer in order to monitor and control the remote field parameters.

**Database**

The heart of any SCADA system is the central control computer. The central computer is used to control the SCADA system data communications as well as provide a usable operator interface for monitoring and implementing any control decisions. The central computer may also be used to manage the real-time and historical database. Databases are used for storing historical data, measuring trends and deriving forecasts for use in the operation of the SCADA system.

## 10.4.5  SCADA CONTROL CENTER

The SCADA control center is the center for monitoring and controlling field sites over long-distance communications networks, including monitoring alarms and processing status data. Based on information received from remote stations, automated or operator-driven supervisory commands are sent to remote station control devices called field devices. Field devices control local operations such as opening and closing valves and breakers, collecting data from sensor systems, and monitoring the local environment for alarm conditions. Here, digital codes are used for information exchange with various error detection schemes to ensure that all data are received correctly. The RTU codes remote station information into the proper digital form to transmit and to convert the signals received from the master into the proper form for each piece of remote equipment.

When a SCADA system is in operation, it periodically scans all routine alarm and monitoring functions by sending the proper digital code to interrogate or poll each device. The polled device sends its data and status to the master station. The total scan time for a substation can be from 30 seconds to several minutes, subject to the speed of the SCADA system and the substation size. When an alarm takes place, it interrupts a normal scan. When an alarm is received, the computer polls the device at the substation that indicated the alarm.

At the control center, an alarm can trigger a computer-initiated sequence of events, for example, a breaker action to sectionalize a faulted bus. Each of the activated equipment has one code to activate it to make it listen and another code to cause the controlled action to take place. Also, when some alarm conditions sound, it indicates that an action is needed by an operator. In such situations, the operator initiates the action needed using a keyboard or a CRT. The computers used must have sufficient memory to store all the data, codes for the controlled devices, and programs for automatic responses to abnormal events.

## 10.4.6   SCADA COMMUNICATION SYSTEM

There are three generations of SCADA system architectures. The first generation uses a wide area network (WAN) for communication between MTUs, which execute decision-making, and RTUs, which serve end-users. The second generation uses local area networks (LANs) to communicate between MTUs and RTUs. The third generation uses WAN and internet protocol (IP).

The components of SCADA architecture include the following: (i) on-field devices, such as RTUs, PLCs, intelligent electronic devices (IEDs), and process automation controllers (PACs); (ii) monitoring and controlling equipment, such as HMIs, historians, controllers for SCADA, and real-time data processors; and (iii) communications, such as Inter-Control Center Communications Protocol (ICCP), Odyssey Commutation Processor (OCP), Ethernet, wireless networks, serial network connections, and Modbus and DNP3 protocols. The terminal controller unit is responsible for communicating, analyzing data, and displaying occurring events to the users and the service providers. The devices are generally controlling and controlled devices that run on embedded operating systems to communicate data using various controlling protocols, such as Modbus and DNP3.

A SCADA communication system consists of a central host computer and a number of RTUs, operator terminals and/or programmable logic controllers (PLCs)

(Gao et al., 2014; Gaushell and Block, 2002). Some key components include a SCADA meter, RTUs, PLCs and communication infrastructure (Gaushell and Block, 2002). A SCADA meter is used to collect data from (acquiring) and send commands to (control) a plant; RTUs are used for connecting to sensors in the plants, converting sensor signals to digital data, and sending digital data to the supervisory system; PLCs are used as field devices because they are more economical, flexible, and configurable than special-purpose RTUs; and communication infrastructure is used for connecting the supervisory system to the RTUs and/or PLCs. A typical SCADA communication system generally consists of a master station and many other distributed RTUs (Gaushell and Block, 2002). Figure 10.4 shows a typical SCADA communication system. The RTUs are interconnected to the master station through a variety of communication channels, such as radio links, leased lines, fiber optics and others (Gaushell and Block, 2002). A SCADA master performs state estimation based on information received from RTUs and sends it back to the RTUs. Essentially, the communication system transfers data between the field data interface devices, the control units and the computers in the SCADA central host. The system can be radio, telephone, cable, satellite, etc., or any combination of these. However, one of the greatest communication challenges is that the channel limits the speed of data acquisition and control that can be performed. Furthermore, random noise on the channel is another problem that hinders SCADA communication (Gaushell and Block, 2002).

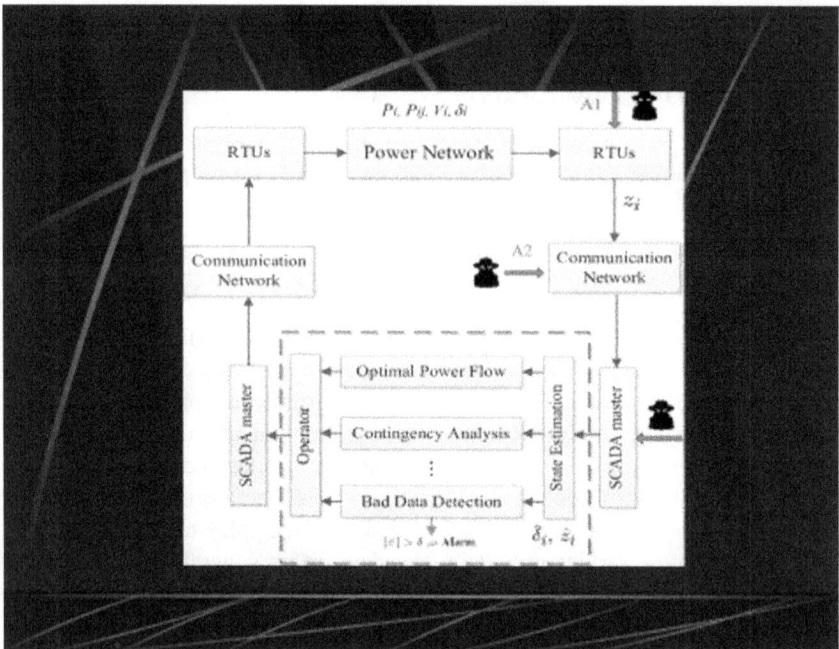

**FIGURE 10.4**   SCADA communication links.

To ensure SCADA systems are well maintained, security measures should be given special importance (Creery and Byres, 2005). Attacks on a SCADA system can cause a threat to people's safety, a loss of productivity, and even environmental damage (Creery and Byres, 2005). Some basic network systems (e.g., ports, hubs, switches, routers, firewalls, and the Simple Network Management Protocol (SNMP)) are also general, electrical power grid components that are at risk of being attacked.

The interconnection of microprocessors used in a SCADA system has seen an increasing trend in recent times, and this interconnection makes the system less secure (Creery and Byres, 2005). PLCs and distributed controlled systems (DCSs) used as process controllers have been replaced by IEDs, which are generally applied to control power meters, control power stations, and trace heat (Creery and Byres, 2005). Power meters, wireless LANs, IEDs, relay networks, and master control centers (MCCs) are interconnected in SCADA systems when setting up power grids (Creery and Byres, 2005). With all these devices interconnected, the network of a SCADA system is less isolated and, thus, more prone to attack (Creery and Byres, 2005). A proper understanding of SCADA communication systems is necessary to analyze the various security threats that need to be addressed and ensure the continuity, reliability and safety of the operation of SCADA communication systems.

### 10.4.7   SCADA in Generating Stations

Power generating stations can provide an optimum solution for each and every operation with flexible and advanced control structures using PLC hardware and advanced communication links along with SCADA software and hardware. The SCADA system provides an integrated set of control, supervision and management functions for power generation stations, including (Sayed and Gabbar, 2016):

- Control of generator voltage using governor and excitation controllers.
- Control of circuit breakers using protection relays.
- Synchronization of functions between generators.
- Control of transformers and tap-changers on the basis of the status of the electrical network.

### 10.4.8   SCADA in Power Distribution Systems

Distribution management systems deal with the electric power system from the distribution substation to the different loads with the use of medium- and low-voltage cables and transmission lines. Most of the power distribution or utility companies rely on manual labor to carry out the distribution tasks, such as interrupting the power to loads, hourly parameter checking, fault diagnosis, etc.

Traditionally, the distribution network has been passive with limited communication between elements. Some local automation functions are used, such as on-load tap changers and shunt capacitors for voltage control and circuit breakers or auto-reclosers for fault management. These controllers operate with only local meas-urements, and wide-area coordinated control is not used. Implementing a SCADA system into the power distribution not only reduces the manual labor operation but also facilitates smooth automatic operations while minimizing disturbance.

PLCs in electrical substations continuously monitor substation components and transfer monitor data to a centralized PC-based SCADA system. In the event of a power outage or failure, the SCADA system detects the fault type and location without waiting for calls from customers. The SCADA system gives an alarm or event to the operators so that they can identify and analyze it. The SCADA system in substations automatically controls the circuit breakers and switches that exceed parameter limits, thereby continuous inspection of network status and parameters can be performed regularly without a line worker. Some of the SCADA functions in the power distribution system include (Sayed and Gabbar, 2016):

- Improvement of power system quality by maintaining power factor and har-monic content.
- Limitation of peak power demand.
- Continuous monitoring and controlling of various electrical components in normal and abnormal conditions.
- Trending and alarming to enable operators to fix outage problems.
- Previous data and viewing from remote locations.
- Quick response to customer service interruptions.
- Motor control with motor control centers.
- Power control, including tie-line control, peak shaving and load sharing.
- Load-shedding based on fast, slow and frequency.

### 10.4.9 SCADA Optimization of Integrated Energy Systems

A SCADA system receives data from local controllers in the field (i.e., power gen-erators and smart homes). These data are sent back as control commands to the local controllers after processing for optimization. Here, we present a SCADA-supported smart grid for optimization of the demand side of a microgrid.

The energy management problem in the microgrid is modeled to minimize the day-ahead energy cost of all consumers by scheduling smart home appliances, power consumption and energy drawn. The objective function and all the con-straints related to the implemented systems are presented as:

**Objective function**
The objective function intends to minimize the global electricity cost of a set of smart homes, including in the microgrid. It is formulated as (Soetedio et al., 2017):

$$\text{Min f(cost)} = \sum_{t}^{T} \left\{ \begin{array}{l} \left[ (P_{grid}(t) \times C_{Grid}(t) \right] \\ + \left[ P_{PV}(t) \times C_{PV}(t) \right] \\ + \left[ P_{W}(t) \times C_{W}(t) \right] \end{array} \right\} \tag{10.8}$$

where
$P_{grid}$ = power from the grid
$P_{PV}$ = power from solar PV
$P_{W}$ = power from wind
$C_{grid}$ = cost of grid power
$C_{PV}$ = cost of solar power
$C_{W}$ = cost of wind power

subject to

**Constraint 1:** Domestic balance

$$P_{Grid}(t) + P_{PV}(t) + P_{W}(t) = \sum_{i} \left( D_{appl}(i,t) \right) + D_{t}(t) \tag{10.9}$$

where
$D_{appl}$ = power consumption by appliances
$D_{t}$ = thermal load demand

The balance of power between the smart microgrid production and the main grid import must be the same as the total electrical power demand.

**Constraint 2:** Global balance

$$\sum_{h} \left( P_{Grid}(t,h) \right) = P^{M}_{Grid}(t) \tag{10.10}$$

where
$P^{M}_{grid}$ = power for microgrid from the grid

The exchange of power between the smart microgrid and the main grid must be the same.

**Constraint 3:** Electric grid – The limit of the amount of electric power imported from the grid

$$0 \leq P_{Grid}(t) \leq P^{max}_{Grid} \tag{10.11}$$

where

$P^{max}_{grid}$ = maximum power for microgrid from the grid

**Constraint 4:** Photovoltaic system – The limit of the power generated by a PV system

$$0 \le P_{PV}(t) \le P^{max}_{PV} \qquad (10.12)$$

where

$P^{max}_{PV}$ = maximum power from PV

The generated output power from the photovoltaic system is

$$P_{PV}(t) \le A_{PV} \times \eta_{PV} \times SI(t) \qquad (10.13)$$

where

$A_{PV}$ = solar PV area
$SI_t$ = solar insolation
$\eta_{PV}$ = efficiency of solar PV

**Constraint 5:** Wind turbine system – The limit of the power generated by a wind system

$$\begin{cases} P_W(t) = 0 & \text{If } v_f < v_{ci} \text{ and } v_f > v_{co} \\ P_W(t) = P_{rated} & \text{If } v_r \le v_f \le v_{co} \\ P_W(t) = P_{rated} \times \dfrac{v_f - v_{ci}}{v_r - v_{ci}} & \text{If } v_{ci} \le v_f \le v_r \end{cases} \qquad (10.14)$$

where

$P_{rated}$ = rated power of wind system
$v_{ci}$ = cut-off speed of wind turbine
$v_{co}$ = cut-in speed of wind turbine
$v_f$ = forecasted wind speed
$v_r$ = rated speed of wind turbine

**Constraint 6:** Non-negativity constraint
   The decision variables such as grid generation, solar PV generation and wind energy generation must be greater than or equal to zero.
   The proposed algorithm ensures that the total energy generated satisfies the load requirements; otherwise, the size of the wind energy system or PV system must be increased by a certain value. Conversely, if the total energy generated

is greater than the load requirements, the size of the wind energy system or PV system must be reduced by a certain value. The cycle is repeated for each year until the generated energy just satisfies the load requirements within the specified constraints.

The optimization provides optimal economic generation with penetration into renewable energy with a reduction in pollution and contribution to global warming. A number of pieces of software have been developed and are available for producing optimal solutions using linear programming. However, the software chosen needs to scale extremely well and produce solutions in near real-time.

## 10.4.10 Advanced SCADA Concepts

The competitive business of electrical utilities, due to deregulation, requires a reexamination of SCADA utility operations, not the process itself. Present-day business dictates the incorporation of modern SCADA system hardware and software into the corporation-wide management information systems strategy to maximize the benefits to the utility. Until now, we have considered the minimization of costs with restrictions on power availability and pollution.

Today, the dedicated islands of automation provide an opportunity to incorporate the information system. It is expected that tomorrow's advanced SCADA systems will be a function of workstation-based applications interconnected through a wide area network (WAN) to create a virtual system. This arrangement will provide SCADA applications to a host of other applications such as substation controllers, automated mapping/facility management systems, trouble call analysis, crew dispatching, and demand-side load management (LM). The WAN will provide the traditional link between the utility's energy management system (EMS) and SCADA processors. The workstation-based applications will also provide an opportunity for flexible expansion and economic system reconfiguration.

Also, unlike the centralized database of the most exciting SCADA systems, the advanced SCADA system database will exist in dynamic pieces that are distributed throughout the network. Modifications to any interconnected elements will be immediately available to all users, including the SCADA system. The SCADA system will have to become a more powerful and involved partner in the process of economic delivery and quality of service to the end-user.

In most applications today, the SCADA system and the energy management system (EMS) exist only in the transmission and generation sides of the power system. In the future, economic dispatch algorithms will include demand-side (load) management and voltage control/reduction solutions. The control and its hardware and software resources will cease to exist. Finally, a SCADA system plays an important role in modern power system operations due to its flexibility, reliability, efficiency and cost-effectiveness. All personnel in the electric power industry should be familiar with the operation of SCADA.

## 10.5 SMART GRID

In a conventional electric power system, centralized generating stations are used to generate bulk electric power, which is then transmitted to consumers through a one-way transmission and distribution system called the grid. Electric power systems generate electrical power at large central generating stations that step up the voltages to supply electric power to a high-voltage interconnected network known as the transmission grid. Each individual generator unit, whether it is hydropower, natural gas or fossil fuel, may be as large as 1000 MW. The transmission grid is used to transmit the electrical power, sometimes over considerable distances, and then the voltages are stepped down through a series of distribution transformers to the final circuits for delivery to the end customers.

High-quality electricity is a necessity in modern society for innumerable applications that require quality power, such as electronic manufacturing, microprocessors, and many sensitive devices. A smart grid can supply electricity from generation to consumer using digital technology to save energy, reduce cost and increase reliability. It connects everyone to abundant, affordable, clean, efficient and reliable electric power anytime, anywhere, addressing energy independence and global warming issues. Asset optimization includes proactive equipment maintenance via equipment condition monitoring and produces many advantages for the power industry. Focused maintenance can be done on equipment, and asset optimization reduces outages and risks of failure as the assets are monitored and assessed continuously.

The present revolution in communication systems, particularly stimulated by the internet, offers the possibility of much greater monitoring and control throughout the power system and hence more effective, flexible and lower cost operation. The smart grid is an opportunity to use new information and communication technologies (ICTs) to revolutionize the electrical power system. However, due to the huge size of the power system and the scale of investment made in it over the years, any significant change will be expensive and requires careful justification. Figure 10.5 shows the smart grid vision of an electric power research institute, and the model set up for the smart grid includes smart generation, smart transmission, smart storage and smart sensors to isolate the fault (EPRI, 2011).

Man-made greenhouse gases are leading to dangerous climate change. Hence, energy should be used more effectively, and electricity should be used with a minimum contribution to global warming. The effective management of loads and reduction of losses and wasted energy requires accurate information, while the use of large amounts of renewable generation requires the integration of the load in the operation of the power system to help balance supply and demand. Smart meters are an important element of the smart grid as they can provide information about the loads and hence the power flows throughout the network. Once all the parts of the power system are monitored, its state becomes observable and many possibilities for control emerge.

The anticipated future decarbonized electrical power system is likely to rely on generation from a combination of renewables, nuclear generators and fossil-fuelled

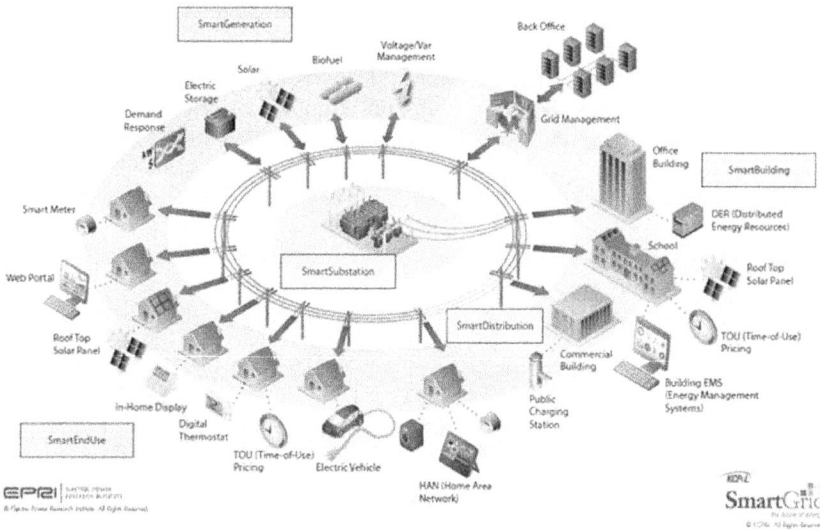

**FIGURE 10.5** Smart grid vision of electric power research institute (EPRI, 2011).

plants with carbon capture and storage. It is hard to design such a power system that can be operated cost-effectively without the monitoring and control provided by a smart grid.

## 10.5.1 WHAT IS A SMART GRID?

A smart grid is an electrical generation and distribution system that is fully networked, instrumented and automated. According to the European Technology Platform, a smart grid is an electricity network that can intelligently integrate the actions of all users connected to it – generators, consumers and those that do both – in order to efficiently deliver sustainable, economical and secure electricity supplies (European Commission, 2006). According to the US Department of Energy (U.S. Department of Energy, 2009): "A smart grid uses digital technology to improve reliability, security, and efficiency (both economic and energy) of the electric system from large generation, through the delivery systems to electricity consumers and a growing number of distributed-generation and storage resources."

All these requirements dictate the modernization of the electric grid by installing intelligent electronic devices (IEDs), electronic switches, smart meters and advanced communication, and data acquisition and interactive software with real-time control to optimize the operation of the whole electrical system and make for more efficient utilization of the grid assets. Such a grid is called a smart grid.

Most of the present electric power grid infrastructures were built more than 50 years ago and are now complex spider webs of power lines with obsolescent technology and outdated communications. In addition, there is a growing demand for low carbon emissions, renewable energy resources, improved reliability, and security. It is expected that the smart grid should integrate renewable energy sources, especially wind and solar, with conventional power plants in a coordinated and intelligent way so that it not only improves system reliability and service continuity, but also effectively reduces energy consumption and significantly reduces carbon emissions.

## 10.5.2  Functions of a Smart Grid

Based on the Smart Grid System Report of July 2009 of the United States Department of Energy, the functions of a smart grid are (U.S. Department of Energy, 2009):

1. **Optimization of asset utilization and operating efficiency**: The smart grid optimizes the utilization of existing and new assets, improves load factors, lowers system losses, maximizes operational efficiency and reduces cost. Advanced sensing and robust communications ensure early problem detection and corrective actions.
2. **Assurance of power quality for a range of needs:** The smart grid ensures load sensitivities with power quality, and consumers will have the opportunity to purchase varying grades of power quality at different prices and at different times.
3. **Accommodation of all generation and storage options**: The smart grid will integrate all types of electrical generation (conventional and renewable energy) and storage systems, including small-scale power plants (such as solar and wind) that serve their loads, known as distributed generation.
4. **Informed participation of customers**: The smart grid will provide an opportunity for consumers to become active participants in the grid. Well-informed customers will modify consumption based on balancing their demands and resources with the electric system's capability to meet those demands
5. **Market participation of buyers and sellers**: The smart grid will provide an opportunity for the market participation of buyers and sellers based on the supply and demand interactions of markets and real-time prices.
6. **Operational resiliency to disturbances, attacks and natural disasters**: The smart grid has the ability to withstand and recover from disturbances in a self-healing manner to prevent or mitigate power outages and maintain reliability, stability and service continuity. It will operate resiliently against attacks and natural disasters. It incorporates new technologies and higher cyber security for the protection of the entire electric system and, thus, reduces physical and cyber vulnerabilities, and enables rapid recovery from disruptions.

Therefore, the next-generation electricity grid, known as the smart grid or "intelligent grid", is designed to address the major shortcomings of the existing power grid. Basically, the smart grid should provide the electric power utility industry with full visibility and penetrative control and monitoring over its assets and services. It is required to be self-healing and resilient to system abnormalities.

Furthermore, the smart grid should provide an improved platform for utility companies to engage with each other and perform energy transactions across the whole system. It is expected to provide tremendous operational benefits to power utilities around the world because it can provide a platform for enterprise-wide solutions that can deliver far-reaching benefits to both utilities and their end customers.

### 10.5.3   Why Use a Smart Grid?

Since about 2005, there has been increasing interest in the smart grid. It is recognized that ICT can offer a significant opportunity to modernize the operation of electrical systems to decarbonize them at a realistic cost if it is monitored and controlled effectively. In addition, a number of more detailed reasons have now emerged to stimulate interest in the smart grid (Ekanake et al., 2012).

#### 10.5.3.1   Ageing of Transmission and Distribution Equipment and Lack of Circuit Capacity

Power systems have expanded rapidly since 1950 globally, and the transmission and distribution equipment they use is now very old and beyond its design life and requires replacement. The capital costs of replacing old equipment are very high, whether there is manufacturing capacity available or skilled staff to run it. Thus, there is an opportunity to innovate power equipment with new designs and operating practices.

In many countries, overhead lines are required to meet increased electricity demand or to connect renewable generation, but these have been delayed up to 10 years due to difficulties obtaining rights-of-way and environmental permits. Therefore, some of the existing power transmission and distribution lines are operating near capacity limits, and so some renewable generation cannot be connected. This calls for more intelligent methods of increasing the power transfer capacity of circuits dynamically and rerouting power flows through optimally loaded circuits.

#### 10.5.3.2   Thermal Constraints of Transmission and Distribution Lines and Equipment

The thermal constraints of transmission and distribution lines and equipment are the upper limits of their power transfer capabilities. When power equipment carries overcurrent, it becomes overheated, and its insulation deteriorates rapidly. This reduces the life of the equipment and increases the incidence of faults. If an overhead line carries too much current, the conductor will lengthen, the sag of the catenary will increase, and the clearance to the ground will be reduced. Any reduction in the clearance of an overhead line to the ground has important implications resulting in an increase in the number of faults and causing a danger to public safety. Thermal constraints depend on environmental conditions that change throughout the year. Hence, the use of dynamic ratings can increase circuit capacity at times when needed.

### 10.5.3.3   Operational Constraints of Voltage and Frequency Limits, Distributed Generation and Renewable Energy Generation

Power systems should operate within rated voltage and frequency limits. If the voltage exceeds its upper limit, the insulation of the components of the power system and consumer equipment may become damaged and may cause short circuits. Too low a voltage may cause customer equipment to malfunction and may cause overcurrent and tripping of some lines and generators. The capacity of many traditional distribution circuits is limited by the variations in voltage that occur between times of maximum and minimum load, and so the circuits are not loaded near to their thermal limits. Although reduced loading of circuits leads to low losses, it requires greater capital investment.

The frequency of the power system is governed by the second-by-second balance of generation and demand. Any imbalance is reflected as a deviation in the frequency from 50 or 60 Hz or excessive flows in the tie lines between the control regions of very large power systems. System operators should maintain the frequency within strict limits, and when it varies, response and reserve services should be called upon to bring the frequency back to within its operating limits (Erinmez et al., 1999). Under emergency conditions, some loads are disconnected to maintain the stability of the system.

Since about 1990, there has been a revival of interest in connecting generation to the distribution network. This distributed generation can cause overvoltages at times of light load, thus requiring the coordinated operation of the local generation, on-load tap changers and other equipment used to control voltage in distribution circuits.

Renewable energy generation, for example, wind power and solar PV power, has a varying output that cannot be predicted with certainty hours ahead. A large central fossil-fuelled generator may require six hours to start up from cold. Some generators on the system, for example, a large nuclear plant, may operate at a constant output for technical or commercial reasons. Thus, maintaining the supply–demand balance and keeping the system frequency within limits becomes very difficult. Part-loaded generation "spinning reserve" or energy storage can address this problem but with a consequent increase in cost. Therefore, power system operators increasingly seek frequency response and reserve services from the load demand. In the future, the electrification of domestic heating loads (to reduce emissions of $CO_2$) and electric vehicle charging is expected to lead to a greater capacity of flexible loads. This would help maintain network stability and reduce the requirement for reserve power from part-loaded generators and the need for network reinforcement.

### 10.5.3.4 Security of Efficient and Reliable Supply of Electricity

Modern society requires an increasingly efficient and reliable electricity supply. The traditional approach to improving reliability was to install additional redundant circuits at considerable capital cost and environmental impact. An alternative to disconnecting faulty circuits is to use intelligent post-fault reconfiguration so that after the (inevitable) faults in the power system, the supplies to customers are maintained, but without the expense of multiple circuits that may be only partly loaded for much of their lives. Fewer redundant circuits result in better utilization of assets but higher electrical losses.

The increasing demand for energy and the need to integrate renewable energy to increase energy security and minimize contributions to global warming, along with its automation to control and manage grid systems, have stimulated researchers into developing the smart grid. The smart grid makes power systems more environmentally friendly and also ensures energy security using renewable energy. It also makes power systems more reliable during instability and faults using information and communication technologies for collecting information and actuating processes of automation. The integration of solar and wind creates instability in the power grid, and using ICT-incorporated automation to control and manage it is a challenging task for power system engineers. A grid system must simulate two-way communication between the utility and consumers, and such simulation is essential for a clear understanding of the grid system to support their planning and operation.

### 10.5.4 Basic Concept of a Smart Grid

The basic concept of a smart grid is to add monitoring, analysis, control and communication capabilities to national electrical delivery infrastructure to maximize the throughput of the system while reducing energy consumption. The smart grid will allow utilities to move electricity around the system as efficiently and economically as possible.

The smart grid can be defined as a system that employs digital information and control technologies to facilitate the deployment and integration of distributed and renewable resources, smart consumer devices, automated systems, electricity storage and peak-shaving technologies.

Recent technological advancements help to connect and integrate isolated technologies to achieve better and more efficient energy management. The advanced communication system empowers the consumer and increases the participation of the end-user in the economic dispatch of electricity along with the main utility. Figure 10.6 shows a typical smart grid setup with complete connectivity to each and every source and load. Connectivity plays a vital role in making the grid smart by facilitative two-way energy flow (Sayed and Gabbar, 2016). In a conventional grid, energy flow is unidirectional, that is, from the utility to consumers. Whereas with the help of advanced technology, end-users are allowed to produce or store power through distributed generation and participate in the energy ecosystem. With

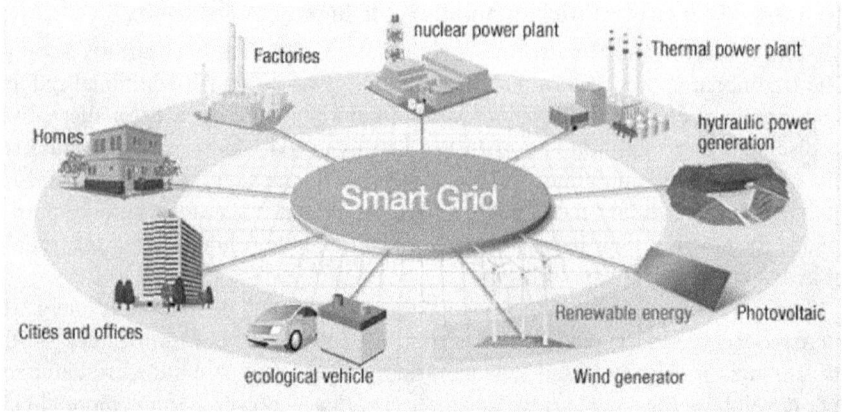

**FIGURE 10.6**   Smart grid architecture (Sayed and Gabbar, 2016).

the help of historical data and accurate forecasting mechanisms, it is possible to implement efficient demand and response programs by committing to reduce load when demand is high and allowing direct control of loads. Variable tariff programs can also be another aspect to encourage consumers to schedule their electricity consumption to avoid surges in demand during peak periods by increasing the tariff. Through the integration of smart building devices and systems, intelligent schemes can be used to perform automated load management to achieve desired energy efficiency. The cloud database system enables the end-user to access their consumption pattern and energy pricing data. This will help consumers predict energy needs, sell excess power, and isolate the sources of problems (Caamano et al., 2009).

### 10.5.5  Components of a Smart Grid

Smart grids are designed to improve the stability and reliability of generation, intelligent controls and energy mix, including renewable energy resources such as solar and wind. The fundamental components of a smart grid are the transmission line subsystem component, monitoring and control technology component, storage component, intelligent grid distribution subsystem component, demand-side management component and smart device interface component, and these are discussed as follows (Momoh, 2012).

### 10.5.5.1  Transmission Subsystem

The transmission system interconnects all major substations and load centers and is the backbone of an integrated power system. The ultimate goal of transmission planners and operators is efficiency and reliability at an affordable cost. Transmission lines must be capable of tolerating dynamic changes in load and contingency without any service disruption. To ensure performance, reliability and quality of supply standards are preferred following a contingency. Strategies for achieving better performance from smart grids at the transmission level are the design of analytical tools and advanced technology, with intelligence for performance analysis such as dynamic optimal power flow, robust state estimation, real-time stability assessment, and reliability and market simulation tools. Real-time monitoring based on phasor measurement units (PMUs), state estimators sensors, and communication technologies are the transmission subsystem's intelligent enabling tools for developing smart transmission functionality.

### 10.5.5.2   Monitoring and Control Technology

Intelligent transmission systems/assets provide a smart, intelligent network, self-monitoring and self-healing, and adaptability and predictability of generation and demand robust enough to handle congestion, instability, and reliability issues. This new resilient grid should be able to withstand shock (durability and reliability) and should be reliable enough to provide real-time changes in its use. Monitoring and control technologies are needed to address these changes.

### 10.5.5.3   Intelligent Grid Distribution Subsystem

The distribution system is the final stage in the transmission of power to end consumers. Primary feeders are used to supply power to small industrial customers, and secondary distribution feeders supply power to residential and commercial consumers. At the distribution level, intelligent support schemes for monitoring capabilities for automation using smart meters, communication links between consumers and utility control, energy management components, and AMI are introduced, and the automation function is equipped with self-learning capabilities, including modules for fault detection, voltage optimization and load transfer, automatic billing, restoration and feeder reconfiguration, and real-time pricing.

### 10.5.5.4   Demand-Side Management

Demand-side management and energy efficiency options are effective means for modifying consumer demand to reduce operating expenses from expensive generators and defer capacity addition. Demand-side management options also provide reduced emissions from fuel production, lower costs, and contribute to the reliability of supply. These options have an overall impact on the utility load curve. A standard protocol for customer delivery with two-way information highway technologies as the enabler is needed. Plug-and-play, smart energy buildings and smart homes, demand-side meters, clean air requirements, and customer interfaces for better energy efficiency should be included for demand-side management.

### 10.5.5.5   Storage of Generated Energy

Renewable energy supply is a variable energy source, and there is a disjoint between peak availability and peak consumption. Hence, it is necessary to store generated energy for later use. Energy storage technologies available are pumped hydro, advance batteries, flow batteries, compressed air, super-conducting magnetic energy storage, super-capacitors, and flywheels. Market mechanisms for handling renewable energy resources, distributed generation, environmental impact and pollution are important considerations at the generation level. Market mechanisms for handling renewable energy resources, distributed generation, environmental impact and pollution need to be considered and introduced in the design of smart grid components at the generation level

### 10.5.5.6   Smart Devices Interface

Smart devices for monitoring and control are used for generation component real-time information processes. These devices are seamlessly integrated into the operation of centrally distributed energy systems and district energy systems.

### 10.5.6   COMMUNICATION IN A SMART GRID

The smart grid is the next generation of power distribution grid, and the governments of many countries and companies are supporting research into smart grid applications. The smart grid uses two-way communication, digital technologies, advanced sensing and computing infrastructure and software abilities to provide improved monitoring, protection and optimization of all grid components, including generation, transmission, distribution and consumption (Baimel et al., 2016). The smart grid ensures energy security and reduces greenhouse gas emissions with the application of advanced and controlled large-scale integration of renewable energy sources. This large-scale integration requires the application of advanced distributed control algorithms to avoid unexpected frequency and voltage fluctuations (Baimel et al., 2016; Carrascoo et al., 2006; Grubb, 2003; Lund, 2007). By using energy storage systems (Baimel et al., 2016; Ibrahim et al., 2008), communication between the grid and customers, and advanced algorithms for forecasting generation and loading of the grid (Baimel et al., 2016; Pai and Hong, 2005; Azadeh et al., 2008; Azadeh and Tarverdian, 2007), the smart grid ensures full coordination between generated and consumed energy. This reduces energy losses, peak demand and energy costs.

Two-way communications allow energy consumers to receive accurate real-time prices and bills. The grid operator can receive consumers' real-time information about the amount of consumed energy. The reliable real-time information flow between all grid components is essential for a smart grid's successful operation. This can be implemented by a reliable and effective communication infrastructure, which can be wired or wireless. The advantages of a wireless infrastructure compared to a wired infrastructure are low costs and simple connection to distant and unreachable areas. The disadvantages are interference with other signals and electromagnetic fields and dependence on batteries.

### 10.5.6.1   Smart Grid Communication Infrastructure

The smart grid is the next generation of power distribution grid, and it essentially consists of power infrastructure and information infrastructure. The smart grid integrates electrical grids and communication infrastructure and forms an intelligent electricity network working with all components to deliver sustainable electricity supplies (Ma et al., 2013). The power infrastructure uses generators, transformers, circuit breakers, etc., to distribute electricity with minimal disturbance, while the information infrastructure is the communication network. The smart grid uses two-way communications, digital technologies, advanced sensing and computing infrastructure and software abilities to provide improved monitoring, protection, and optimization of all grid components, including generation, transmission, distribution and consumers. The two-way communication links of the communication architecture and electrical flow of power infrastructure of a smart grid are shown in Figure 10.7. Electrical flows involve the traditional subsystems of the electrical grid, whereas communication flows almost create a mesh topology between every domain, which illustrates the outstanding importance of communications in the smart grid (Lopez, 2014). These are elaborated as follows:

A smart grid is a platform consisting of generation, transmission, distribution, customer, service provider, operations and market domains to provide various applications. The generation domain is responsible for generating electricity, and the transmission system is responsible for transmitting the generated electricity over long distances to distribution systems. The distribution domain provides electricity to customers. The service provider domains provide services to customers and

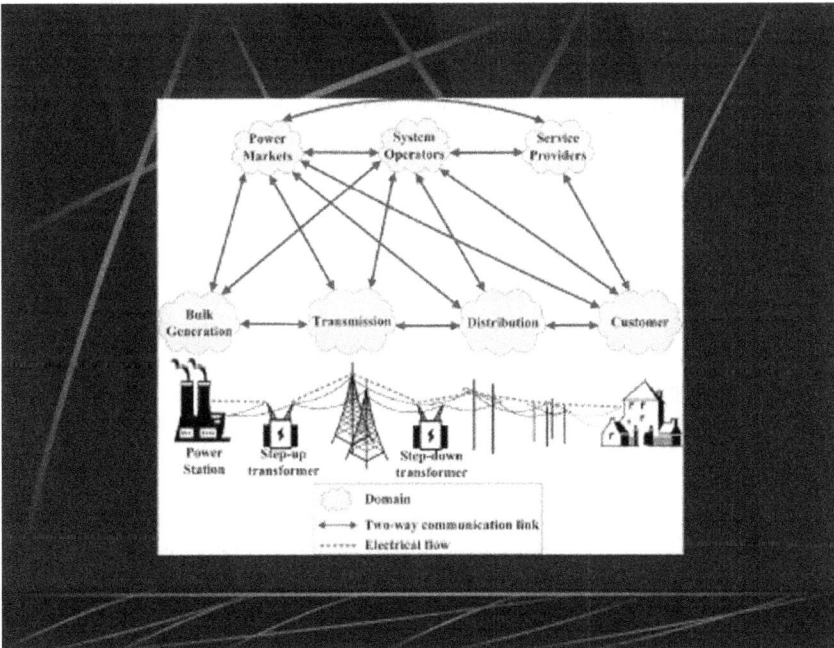

FIGURE 10.7  Smart grid communication links.

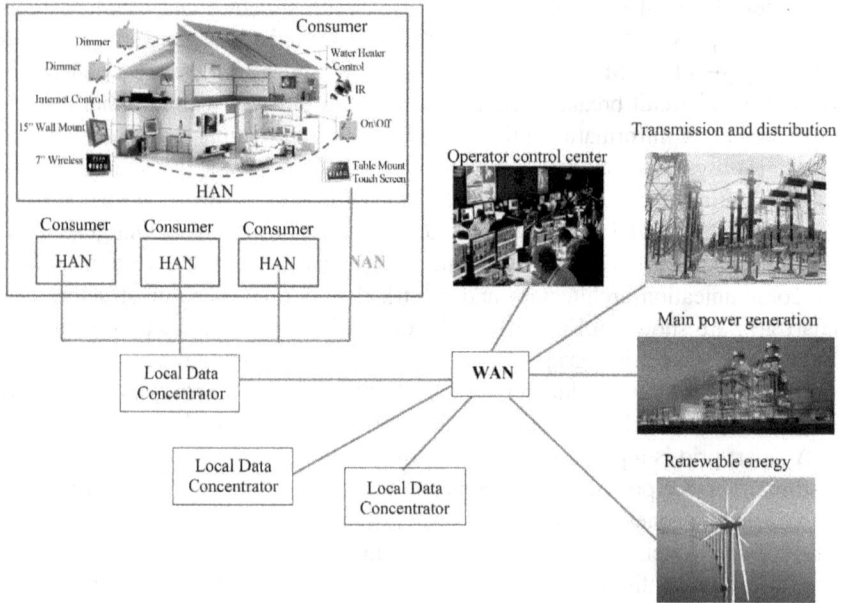

**FIGURE 10.8**   Smart grid communication infrastructure (Baimel et al., 2016).

utilities. The operations domains manage the supply of electricity and are responsible for the operation of the power system. The market domain exchanges price and balances supply and demand within the power system. Each domain can be interactive with other domains through different communications area networks to serve the requirements of the different smart grid applications (Kuzlu et al., 2014).

The communication infrastructure of the smart grid can be based on three types of network: home area network (HAN), neighborhood area network (NAN) and wide area network (WAN) (Baimel et al., 2016). Each network is different in monitoring and measuring power flow and provides control messages to the equipment used. A schematic diagram of the smart grid communication infrastructure based on these networks is shown in Figure 10.8. Power line communication (PLC) is specifically designed for smart grid communication.

HAN applications include home automation and building automation, and are deployed and operated within a small area (1–100 m), usually a house or a small office. The main component of a HAN network is the smart meter that measures the energy used by the customer. These applications do not require data to be transmitted at high frequency, and all applications occur inside residential, commercial or industrial buildings. Hence, a HAN requires a communication network of low bandwidth and low data rate. Communication technologies that provide a data rate of up to 100 kbps with a short coverage distance (up to 100 m) are generally sufficient. Hence, the decentralized communication infrastructure of the smart grid is easily prone to attack. Thus, security is essential for a smart grid network and is required for availability, integrity and confidentiality. In a typical implementation, a HAN consists of a broadband internet connection shared between multiple users through a wired or wireless modem. This enables the communication and sharing of resources between computers, mobiles and other devices over a network connection. In a smart grid implementation, all smart home devices that consume energy and smart meters can be connected to a HAN. The data from these devices is acquired and transmitted through the HAN to the smart meters. The HAN allows for more efficient home energy management and can be implemented by ZigBee or Ethernet technologies. ZigBee, Wi-Fi, Z-Wave, power-line carrier, Bluetooth and Ethernet are widely used to support HAN applications.

In NAN applications such as smart metering, demand response and automation, data are required to transmit from a large number of customers and field devices to a data concentrator/substation or vice versa. Therefore, a NAN is deployed and operated within an area of 100 m to 10 km, and these applications require communication technologies that support a higher data rate (100 kbps to 10 Mbps). Several HANs can be connected to one NAN, and they transmit data of energy consumed by each house to the NAN network. The NAN network delivers this data to local data centers (LDCs) for storage. The NAN aggregates the data from the smart meter and provides command messages to the end network. This infrastructure is called an advanced metering infrastructure (AMI). Data storage is important for analysis for energy generation-demand pattern recognition. A NAN has up to a 2 Kbps transmission data rate. It can be implemented by PLC, Wi-Fi, and cellular technologies.

WAN applications include wide-area control, monitoring and protection, and require a large number of data points at much higher frequencies to allow for the stable control of a power system. Hence, a WAN is deployed and operated within a vast area of 10–100 km and consists of several NANs and LDCs. Moreover, the communication of all smart grid components, including operator control center, main and renewable energy generation, transmission and distribution, is based on WAN. The WAN collects data on power use from an aggregator, power flow data from phasor measurement unit (PMU), market-related data, etc. It communicates with the master terminal unit (MTU), which acts as a main control center. The WAN requires communication of a high data rate and high bandwidth to avoid congestion. It has a very high transmission data rate (10 Mbps to 1 Gbps). Optical communication is commonly used as the communication medium between the transmission/distribution substations and the utility control center due to its high capacity and low latency. Cellular and Wi-Fi are also used due to their wide coverage range and high data throughput. Satellite communications can also be used to provide redundant communications at critical transmission/distribution substation sites as a means of backup communication in a remote location.

Finally, as wireless communication technologies provide lower installation costs, more rapid deployment, higher mobility and flexibility than their wired counterparts, wireless technologies are recommended in most smart grid applications (Kuzlu et al., 2014).

### 10.5.7  Smart Microgrid

A microgrid system can be defined as a low or medium voltage electric power system that includes renewable energy systems, an energy storage system, controllable loads and an energy management system (Erol-Kantarci et al., 2011), and the size of a microgrid system can range from a single household to a large geographic area such as a campus. With the development of microgrid core technology, the smart microgrid has gradually become the focus of microgrid research. A microgrid can be made smarter by integrating advanced sensing technologies, control methods and communication techniques. A smart microgrid can be defined as a low-voltage distribution network with distributed energy resources such as distributed generation units and distributed storage units and loads (Gupta et al., 2013).

The basic concept of a smart grid is the automation of monitoring, analysis, control, and communication technologies to the national grid to optimize the grid system. Figure 10.9 illustrates the concept of a smart microgrid (Ahmed et al., 2015) and shows that a smart microgrid exploits digital information and control technologies for the integration of distributed and renewable resources, smart consumer devices, automated systems, electricity storage and peak-shaving technologies. It is essentially the application of digital information technology to optimize electrical power generation, delivery and end-use, and this integrates the interaction of geographically dispersed equipment and hence enables coordinated operations to be performed through better communications and control. Thus, it is a set of advanced technologies, concepts, topologies and approaches that allow generation, transmission and distribution to be replaced by organically intelligent, fully integrated services with the efficient exchange of data, services and transactions. Hence, smart microgrids can benefit customers by providing uninterruptible power, enhancing reliability, reducing transmission loss, and supporting local voltage and frequency.

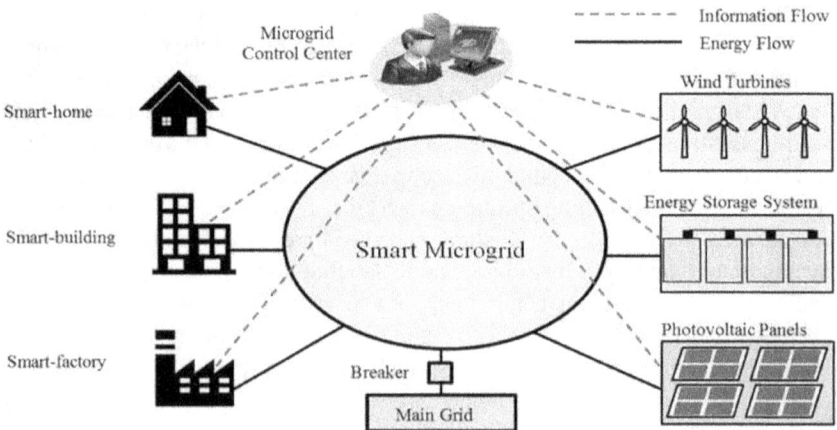

**FIGURE 10.9**   Microgrid architecture overview (Ahmed et al., 2015).

## 10.5.8 Microgrid and Smart Grid

One of the many new processes of the smart grid for power delivery is the microgrid, which is a low-voltage distribution network. Microgrids are autonomous energy management systems under the control of a single administrative authority and are capable of operating in parallel to or in intentional or accidental islanded mode from the existing power grid. They usually include distributed and renewable energy sources as well as some level of energy storage subsystems.

In comparison to microgrids, smart grids have digital information and control, dynamic optimization of grid operation and resources, distributed resources (similar to microgrids or, more specifically, smart microgrids), demand responses, demand-side resources, energy efficiency (EE) resources, smart metering systems, smart integration (real-time response and timely information on consumption) and advanced electricity storage. Moreover, smart grids have other advantages such as faster protection and controlled self-heals, and they are more robust, more renewable, more efficient, of higher power quality, more reconfigurable and have a higher capacity factor. The smart grid roadmap seeks to achieve its targets by evolving a "smart grid", an electrical grid with automated, ICT-based secure systems that can provide the capability for two-way power and information flows, monitor power flows from points of generation to points of consumption, and control power flows or manage loads to match generation in real-time. Anticipating a rapid proliferation of distributed and renewable resources, this roadmap recognizes that it will be imperative to incorporate smarter automation, smart meters and information technology (IT) systems into the grid in order to manage onsite solar photovoltaic (PV) systems, electric vehicles (EV), energy storage, and other distributed energy resources. Such IT systems can also enable utilities to better integrate intermittent renewable resources; provide increased visibility, predictability and event control of generation and demand (bringing flexibility to generation and consumption); and enable utilities to reduce the cost of peak demand, reduce outages and increase system efficiencies.

Research has been conducted into understanding the differences between a microgrid and a smart grid (Momoh, 2012). Basically, a microgrid is a local island grid that can operate as a stand-alone or grid-connected system. It is powered by gas turbines or renewable energy and includes special purpose inverters and a link for plug-and-play to the legacy grid. Special purpose filters overcome harmonics problems while improving power quality and efficiency. In summary, think of the microgrid as a local power provider with limited advanced control tools and the smart grid as a wide area provider with sophisticated automated decision support capabilities.

## 10.5.9 Microgrid and Renewable Energy Sources

A microgrid is defined as a group of interconnected loads and distributed energy resources within clearly defined electrical boundaries that act as a single controllable entity with respect to a grid. Microgrids thus comprise low-voltage distribution systems with renewable energy resources (micro-turbines, fuel cells, photovoltaic generation, etc.) with storage devices (e.g., flywheels, energy capacitors, and batteries) and flexible loads. The operation of microsources in the network can provide distinct benefits to the overall system performance if managed and coordinated efficiently.

The concept of a microgrid thus applies to a local distribution system that encompasses generation, storage, and controllable loads. The generation will be a few kilowatts generally below megawatt (MW) level and can supply part or complete loads as required. Figure 10.9 is a smart microgrid with renewable energy resources, and the microgrid system includes PV panels, small-scale wind turbines and energy storage units. The microgrid can operate in two modes: it operates in grid-connected mode most of the time, however, there are times when the microgrid will operate in isolation mode, supplying partial or full load depending on the capacity of the distributed energy resource (DER). There are also isolated microgrids that operate without grid interfaces at remote locations where there is no access to grid supply.

A microgrid is a self-sufficient energy system that serves a discrete geographic area, such as a remote village, college campus, hospital complex or neighborhood. There are several types of distributed energy sources (e.g., solar panels, wind turbines, combined heat and power, generators) for producing electricity. Many microgrids possess energy storage in the form of batteries, and some also have electric vehicle charging stations.

To ensure energy security and reduce pollution due to production and the use of energy, a gradual transition to renewable energy is advocated. The real challenge to the power industry does not lie in a 100% efficient renewable energy source but in how to integrate several types of energy sources into the grid that are less predictable. Renewable energy sources such as wind and solar could be directly connected to the grid. Communication systems are crucial technologies for grid integration of renewable energy resources. Two-way communication is the fundamental infrastructure that enables the accommodation of distributed energy generation and assists in the reconfiguration of network topology for more efficient power flow. The communication infrastructure is considered the fundamental element that allows the monitoring and control of the operation of renewable energy systems. In addition, it enables the transfer of both measured information and control signals between the renewable energy systems and the control center (Yu et al., 2011). There are several communication options available for the grid integration of renewable energy resources. These options include a hybrid mix of technologies such as fiber optics, copper-wire lines, power-line communications, and a variety of wireless technologies.

For localized renewable sources (such as residential), an optimal method for integrating them into the grid would be by utilizing the microgrid concept. To accommodate this variability, the grid has to be smart, and the full potential of renewable energy sources should be exploited, leading to a completely sustainable electricity supply system (Jacob and Nithiyananthan, 2009).

## 10.5.10 MICROGRID AND SMART CITIES

The smart city idea has been in the spotlight during the past decade and is suggested as an optimal solution for overcoming urbanization problems by researchers. Here the smartness is defined as the desire to improve the quality of life within the cities and the residents living there from several points of view by utilizing information and communication technology. However, employing ICT in city setups does not constitute a smart city (Hollands, 2008). The concept of a smart city is mainly how socio-economic, environmental and technical systems are expected to evolve the current urban infrastructure to a more sustainable, fair and resilient urban structure (Warsi et al., 2019; Batty et al., 2012). Smartness not only refers to technological issues but includes the smart integration of all infrastructure and socio-economic functions, paving the way for desirable development. This desirable development can be attained only through the smart use of human, financial and technical resources considering environmental, demographic, social, infrastructural and economic challenges. The main features of smart cities in terms of electricity are assured supply of electricity using renewable sources of energy and energy-efficient techniques. In fact, two technological dimensions that mainly characterize smart cities are electricity grids and information and communication networks (Warsi et al., 2019). A smart microgrid with a smart electricity network, as shown in Figure 10.9, can lead to the integration of the microgrid in a smart city. It is expected that microgrids will play a crucial role in the development of the smart city. The changing of dump infrastructure into smart infrastructure lies beneath adaptive feedback loops as a result of smart grid digital communication (Farmanbar et al., 2019; Chen, 2010).

One of the key features of a smart city in terms of power scenarios is smart energy and reliable electricity networks. Smart energy and electricity networks lead to the integration of microgrids in smart cities that not only facilitate the integration of renewable energy sources but also enable new energy-related value-added services (Coelho, 2017). In particular, the transformation of urban infrastructures into smart cities will be mainly driven by the deployment of microgrids. Distribution grids provide indispensable basic services to other infrastructures such as water, transportation, etc., and the introduction of microgrids with advanced information and communication technologies makes them an exemplary guide for the implementation of smart cities. The introduction of information and communication technologies in traditional grids has changed the dynamics of the operation of electricity systems, facilitating interactions among all stakeholders that result in new means for the creation of social welfare and economic values.

Microgrids has many positive aspects in the context of smart cities as they provide the capacity to manage vastly distributed generation, enable large quantities of renewable sources in almost every building and city spot, and provide for the increase in demand and peer-to-peer exchanges of electricity among end-users (Jin, 2015). However, incorporating microgrids into smart cities requires proper planning and investment. Deficits in any of them will result in deficiencies in the power system. The system should be flexible enough for transformation into a new smart energy system.

Not least, microgrids and smart cities are about anticipating situations with the efficient forecast of weather conditions and providing real-time solutions. In order to achieve adequacy and reliability of a highly distributed system, detailed planning and a new approach for optimization are required (Liang and Shahidehpour, 2014). In comparison to technical issues whose solutions can be found by existing and foreseeable technologies, the main challenges lie in institutional, market and social arrangements that should accompany the deployment of microgrids in smart cities. The socio-techno-economic implementation of microgrids in smart cities can only be achieved by ensuring retail electricity markets with fair practices, trustworthy management of consumers, commercial information and active participation of consumers in the energy market. Nevertheless, until now, little has been reported on how to face the challenges in energy systems for the smart cities of the future. The benefits of microgrid projects have been analyzed by Morris et al. (2012). Liu (2015) reported microgrid applications and the integration of renewable energy systems in smart cities, and Hernández et al. (2014) provides an insight into the future trends in smart grids, microgrids and smart buildings.

## 10.5.11 SMART GRID OPTIMIZATION OF INTEGRATED ENERGY SYSTEMS

Smart Grid optimization is essentially the optimization of generation, transmission, distribution and demand management. Communication and control in a fully-realized smart electrical grid involve generation, transmission, distribution and consumers, wireless network systems working cooperatively, and supporting information and data among many types of sensors, and the goal is to intelligently automate corrective actions when a disruption in grid operations occurs.

Smart grid technology promises to revolutionize the way in which electricity is produced, delivered and utilized. A fundamental problem in building open-distributed systems is designing mechanisms that compute optimal system-wide solutions effectively despite the self-interest of individual microgrids. To produce efficient and reliable grid operations, an optimization model that generates high-performance electric flows in the grid is desirable. This can essentially be a distributed linear programming technique for an electric utility's resource-allocation problem (Ranganathan, 2013).

The smart grid supplies a part of the load profile of commercial consumers and a part of the load profile of plug-in hybrid electric vehicles (PHEVs) through wind and solar virtual power plants (VPPs) responsive loads, distributed generators (DGs) and energy storage systems. One of the main challenges and essentials of a power system is flexibility in generation scheduling. The flexibility of the system can be enhanced by using a smart grid comprising demand response, hybrid/diesel generation units and energy storage systems. Alirezazadeh et al. (2020) used mixed-integer linear programming and mixed-integer non-linear programming to solve the unit commitment problem and smart grid scheduling, respectively, and showed that the model can optimize the costs of the system and the system be-comes more flexible.

The energy management problem in the microgrid is modeled such as to min-imize the day-ahead energy costs of all consumers by scheduling the power con-sumption and energy drawn by smart home appliances. The objective function and all the constraints for the optimization of smart microgrids can be presented as follows (Ranganathan, 2013; Xu, 2016; Melhern, 2018):

### 10.5.11.1  Objective Function and Constraints

The objective function intends to minimize the global electricity cost of the set of smart homes included in the microgrid. It is formulated as:

$$\text{Min f(cost)} = \sum_{t}^{T} \left\{ \begin{array}{l} \left[ (P_{grid}(t)) \times dt \times C_{Grid}(t) \right] \\ + \left[ (P_{PV}(t)) \times dt \times C_{PV}(t) \right] \\ + \left[ (P_{W}(t)) \times dt \times C_{W}(t) \right] \end{array} \right\} \tag{10.15}$$

where

$P_{grid}$ = power from the grid
$P_{PV}$ = power from solar PV
$P_{W}$ = power from wind
$C_{grid}$ = cost of grid power
$C_{PV}$ = cost of solar power
$C_{W}$ = cost of wind power

subject to
**Constraint 1:** Domestic balance

$$P_{Grid}(t) + P_{PV}(t) + P_{W}(t) = \sum_{i} \left( D_{appl}(i,t) \right) + D_{t}(t) \tag{10.16}$$

where

$D_{appl}$ = power consumption by appliances
$D_{t}$ = thermal load demand

The balance of power between the smart microgrid production and the main grid import must be the same as the total electrical power demand.

**Constraint 2:** Global balance

$$\sum_{h} \left( P_{Grid}(t,h) \right) = P^{M}_{Grid}(t) \tag{10.17}$$

where

$P^{M}_{grid}$ = power for microgrid from the grid

The exchange of power between the smart microgrid and the main grid must be the same.

**Constraint 3:** Electric grid – The limit of the amount of electric power imported from the grid

$$0 \leq P_{Grid}(t) \leq P^{max}_{Grid} \tag{10.18}$$

where

$P^{max}_{grid}$ = maximum power for microgrid from the grid

**Constraint 4:** Photovoltaic system – The limit of the power generated by a PV system

$$0 \leq P_{PV}(t) \leq P^{max}_{PV} \tag{10.19}$$

The generated output power from a photovoltaic system is:

$$P_{PV}(t) \leq A_{PV} \times \eta_{PV} \times SI(t) \tag{10.20}$$

where

$A_{PV}$ = solar PV area
$SI_t$ = solar insolation
$\eta_{PV}$ = efficiency of solar PV

**Constraint 5:** Battery constraint: photovoltaic system – Battery voltage should be within the limits of the allowable battery voltage (Mohamed and Eltamaly, 2018)

$$E_{Bmin} \leq E_B \leq E_{Bmax} \tag{10.21}$$

where

$E_{Bmin}$ = voltage at minimum SOC (state of charge of battery)
$E_B$ = voltage at normal SOC (state of charge of battery)
$E_{Bmax}$ = voltage at maximum SOC (state of charge of battery)

**Constraint 6:** Wind turbine system – The limit of the power generated by wind system is

$$
\begin{cases}
P_W(t) = 0 & \text{If } v_f < v_{ci} \text{ and } v_f > v_{co} \\
P_W(t) = P_{rated} & \text{If } v_r \leq v_f \leq v_{co} \\
P_W(t) = P_{rated} \times \dfrac{v_f - v_{ci}}{v_r - v_{ci}} & \text{If } v_{ci} \leq v_f \leq v_r
\end{cases} \tag{10.22}
$$

where

$P_{rated}$ = rated power of wind system
$v_{ci}$ = cut-off speed of wind turbine
$v_{co}$ = cut-in speed of wind turbine
$v_f$ = forecasted wind speed
$v_r$ = rated speed of wind turbine

**Constraint 7:** Reliability constraint – LOLP (loss of load probability) should be less than or equal to allowable LOLP (Mohamed and Eltamaly, 2018)

$$\sum LOLP \le LOLP_{index} \tag{10.23}$$

where

$$LOLP = \frac{\sum_0^t deficit\,load\,time}{8760}$$

**Constraint 8:** Non-negativity constraint
The decision variables such as grid generation, solar PV generation and wind energy generation must be greater than or equal to zero.
The proposed algorithm ensures that the total energy generated must satisfy the load requirements; otherwise, the number of wind turbines or size of area of solar PV ought to be increased by specific values and vice versa. The cycle begins again until the demand requirements are satisfied, accomplishing the objective function and constraints. Optimization can be executed using MATLAB/Simulink software or writing computer programs to solve this problem. Optimization provides optimal economic generation with penetration into renewable energies with a reduction in pollution and contribution to global warming. Systems with high penetrations of renewable energy sources may experience deviations in voltage or reactive power that cannot be handled by local voltage and frequency controls. It is necessary to make sure that there are no large recirculating reactive currents between the sources for local stability and reliability.
The inverter output voltage and frequency are dependent on the real and reactive power of the microsource. Active and reactive power for an integrated system can be realized as:

$$P = \frac{3EV\sin(\partial)}{\omega L} \tag{10.24}$$

$$Q = \frac{3V(E\cos(\partial)-V)}{\omega L} \tag{10.25}$$

where V and E are the grid and inverter output voltage, respectively. L is the magnitude of the inductive reactance. A voltage source inverter (VSI) controls both the magnitude and phase of its output voltage.

Consequently

$$P = \sin(\partial)\frac{E\,V}{X}$$

(10.26)

$$Q = (E\cos(\partial) - V)\frac{V}{X}$$

(10.27)

Hence, the relationship between the inverter voltage E, system voltage V and inductive reactance X determines the flow of real and reactive power from the system. The real-time values of the active power (P) and reactive power (Q) can be used to implement the droops for frequency and voltage.

In conclusion, optimization can provide the perfect balance between reliability, efficiency and cost, and it ranges from generation to transmission and distribution to end-users.

### 10.5.12 OPTIMIZATION OF INTEGRATED ENERGY SYSTEMS BY GENETIC ALGORITHM

Classical optimization techniques and computational methods are currently being applied in smart grid design, planning and operations, and these are linear programming, non-linear mixed-integer programming (MIP), dynamic programming (DP) and Lagrangian relaxation methods, but they are limited for use in the smart grid due to the static network of the programs they can solve (Chapter 3). Future work that accounts for the predictive and stochastic nature of the smart grid involves modeling components to account for predictivity and stochasticity, and selecting new optimization methods such as adaptive dynamic programming (ADP). Based on natural genetics, evolution computation techniques can solve combinatorial optimization problems. The techniques in this category, including particle-swarm, ant-colony, genetic algorithms and artificial intelligence, learn or adapt to new situations, generalize, abstract, discover, and associate (Momoh, 2012). To facilitate computational methods, we can incorporate a hybrid of intelligent systems such as Pareto algorithm systems. Pareto optimality and genetic algorithms are discussed in Chapter 3 and Chapter 7, respectively.

Microgrids can solve the problem of distributed generation integration into the grid and are of great importance in studying the optimal energy management of microgrids with distributed generation. Bala and Siddique, (2009) reported optimal design and control strategy of stand-alone mini-grid using genetic algorithm. Awais et al., (2015) formulated the load scheduling problem as a cost minimization problem and solved it using a genetic algorithm Bharati et al., (2017) used a genetic algorithm fitness function load redistribution, and it benefited 21.91%. Wang et al., (2018) reported operational optimization in a smart microgrid in the presence of a distributed generation and demand response using genetic algorithms.

In practice, the tools for smart grid operation can be used by engineers and operators who lack knowledge and training. Thus, advanced tools must be readily interpretable, user-friendly, and self-teaching. These advanced techniques have a host of characteristics that make them viable for application in the smart grid environment, such as handling the non-linear modeling of time-varying dynamics of the power system, and handling dynamics and stochasticity (Momoh, 2012). Pareto's genetic algorithm method for the optimization of a smart grid is one such design technique described in this section.

The smart grid operational optimization problem can be solved using genetic algorithms. Genetic algorithms, a computational model first mentioned by Holland, (1975), simulate the natural selection of Darwin's theory of biological evolution and the evolutionary process of the genetic mechanism and is a method of searching the optimal solution by simulating the natural evolution process. Genetic algorithms have been discussed in Chapter 7.

Genetic algorithms can be used to implement the objective function of a smart microgrid and demand response scheduling strategy discussed in subsection 10.5.11. Genetic algorithms are used to transform the objective function because of its randomness, and since the searched Pareto non-dominated solution is randomly distributed around the optimization point, realize the economic operation optimization of the objective interval. The conversion method of the objective function is as follows (Wang et al., 2018):

$$\begin{cases} a \leq F(X) \leq b \\ X \leq S \leq R \end{cases} \tag{10.28}$$

$$Q = -\left( F(X) - \frac{a+b}{2} \right)^2 \tag{10.29}$$

where
  F(X) = the objective function
  $X = (X_1\ X_2\ X_3 \dots X_m)$
  S = decision space of X
  Q = converted objective function.
  a and b are the upper and lower limits of the objective function

Considering the distributed output and load demand, the upper and lower limits of each decision variable are determined. According to the operation strategy of the microgrid, the proposed model is solved by genetic algorithm.

The implementation steps of the genetic algorithm to solve the economic operation optimization problem of the microgrid based on demand response are as follows (Wang et al., 2018):

(1) Determine the optimization variables. System optimization variables include photovoltaic output, the charging and discharging current of the storage battery, the discharge depth of the battery, the gas turbine output, the tie-line power and the electricity price information.

(2) Input the basic data. Optimization objectives, system control strategies, optimization period, optimization time interval, distributed energy re-sources unit in the microgrid system, energy storage unit parameters, initial optimization time t = 0, etc.
(3) Include the system operation simulation (e.g., for 1500 min a day) for the daily output of the photovoltaic system, the gas turbine unit, the discharge depth of the storage battery, the charging and discharging current, and the tie-line power according to the system operation scheme and the annual load data of the case study.
(4) Calculate the objective function and iterate for optimization.
(5) Output the optimal decision variable results (optimization curve and calcu-lation result) and the optimal value of the problem optimized, and thus, the algorithm ends.

A genetic algorithm is a continuous iterative process. Each iteration will update the individual extremum and group extremum, and the adaptability of the solution is compared with the previous solution results. If the difference between the two solutions is less than or equal to the set minimum deviation value K or the number of iterations has reached the set maximum number of iterations $N_{max}$ (i.e., Equation 10.30) is satisfied, the algorithm stops the iteration and outputs the optimization result (Wang et al., 2018):

$$\begin{cases} H_i - H_{i-1} \leq K \\ N \geq N_{max} \end{cases} \tag{10.30}$$

where
$H_i$ and $H_{i-1}$ are the result of iteration i and i-1, respectively

Some typical results of the operational optimization of a smart microgrid are pre-sented to illustrate the application of a genetic algorithm in the operational opti-mization of a microgrid (Wang et al., 2018). The model was simulated using the Pareto genetic algorithm, which optimally schedules the entire microgrid demand response resources under different demand response prices without having a nega-tive impact on customer comfort, which reduces the user's cost of electricity and decreases the peak load and energy consumption, simultaneously. Finally, the simulated result shows a reduction in power consumption. The model was simu-lated for three electricity pricing policies to develop scenarios to understand the power output patterns for these electricity pricing policies, which are (1) fixed price, (2) electricity price based on demand response time of use price (TOU); and (3) electricity price based on demand response real-time price (RTP). The user's priority on distributed generation and the simulated results for these electricity policies are as follows:

In the first case, the user does not participate in the demand as the electri-city price is the regional fixed price, and the priority is for the use of distributed generation and energy storage discharge. The simulated results show that when the electricity is surplus, the electric energy is transported to the power grid, and when the electricity power is insufficient, electric energy can be purchased from the grid.

In the second case, the operation status is a TOU system, and the daily electricity price is higher. The simulated results show that to save on electricity purchase cost, the output of the gas turbine and the degree of the daily energy storage discharge are improved as compared with the first case.

In the third case, the status is an RTP system, and the output of each part of the system is more sensitive to the change of electricity price due to the higher electricity price and the rising state. The simulated results show that the improvement of the output power of each microsource and the connect line power with the distribution network is more obvious than either the first case or the second case. These essentially demonstrate that the TOU pricing policy system is better than the fixed-price pricing policy, and the RTP pricing policy system is better than the TOU pricing policy system.

This example of the optimal operational planning of a smart grid using a Pareto genetic algorithm demonstrates the potentiality of the optimal operational planning of a smart grid using a Pareto genetic algorithm, and fast Pareto genetic algorithms are available for the optimal operational planning of smart grid (Deb et al., 2002; Eskandari and Geiger, 2020).

## 10.5.13 SMART GRID/SCADA INTEGRATION

The integration of a SCADA system into the smart grid is not a difficult task. Integration of a SCADA system into the smart grid by both electrical and data networks allows the utility to remotely monitor, supervise and control the utility control over the entire electrical utility network, as depicted in Figure 10.10 (Sayed and Gabbar, 2016). The SCADA system empowers the electricity consumer to manage their own demand for energy and control costs. It allows the grid to be self-healing by automatically responding to power quality issues, power outages, and power system faults. The SCADA system optimizes the grid assets by monitoring and optimizing them while minimizing operations and maintenance costs. To adequately deliver and administer the products and services made possible by the smart grid, intelligence and control need to exist along the entire power supply chain. This includes electricity generation and transmission from beginning to delivery end-points at the customer's side and includes both fixed and mobile devices in the smart grid architecture.

Digital communication on a smart grid occurs over a variety of devices, technologies and protocols that include wired and wireless telephone, voice and dispatch radio, fiber optics, power-line carriers, and satellite. SCADA software allows for dynamic grid management that involves monitoring the required number of control points. To be fully efficient and operational, monitoring is conducted for every power line and piece of equipment in the distribution system, in addition to allowing consumers to monitor and control their own devices and usage. This

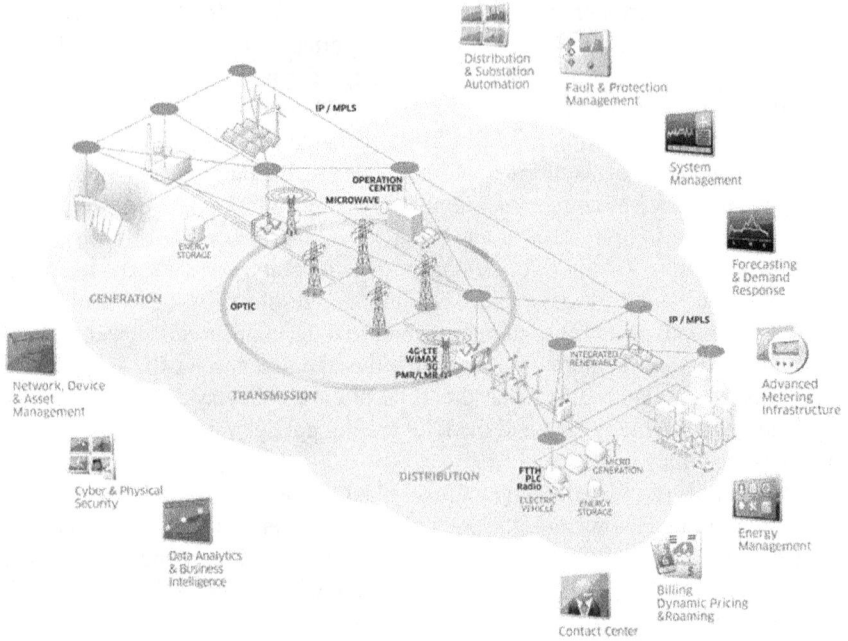

**FIGURE 10.10**   SCADA/Smart Grid Integration (Sayed and Gabbar, 2016).

results in a considerable amount of data to be organized, analyzed and used for both manual and automated decision software that are classified into two basic categories: decentralized and back-office. Decentralized software is necessary to handle the magnitude of devices and data collection and computation, which precludes a centralized data collection solution. Back-office software is part of a utility's line of business software solutions necessary to conduct the business of the organization. The net effect on these solutions by the deployment of IEDs and two-way communications is a more powerful and effective solution set for both the utility and the consumer.

## 10.5.14   SMART GRID OF THE FUTURE

Farhangi (2010) predicts that the smart grid of the future will be interconnected through dedicated highways for power exchange, data and commands. However, it is expected that not all microgrids will have the same capabilities and needs and will be subject to the load diversity, geography, economics, and mix of the primary energy resources.

The necessary advanced metering infrastructure (AMI) systems now being established will facilitate the evolution of the smart grid. However, due to the high costs involved, it is foreseeable that new and old grids may coexist for some time; eventually, though, it is expected that the new system will replace the old grid.

Thus, during the transition period, there will be a hybrid system. The new power grid will appear as a system of organically integrated collections of smart grids with extensive command-and-control functions implemented at all levels (IEEE PES, 2010).

## 10.5.15 CHALLENGES OF A SMART GRID

Today's electric grids were designed to operate as vertical structures, but system operators are now facing new challenges due to the penetration of renewable energy resources, rapid technological change, and different types of market players and end-users. The smart grid is an evolution of existing power systems with close interplays among energy, control and communication infrastructure. A smart grid equipped with communication support schemes and real-time measurement techniques can enhance resiliency and forecasting and protect against internal and external threats.

The design framework of the smart grid is based upon unbundling and restructuring the power sector and optimizing its assets. The new smart grid will be capable of handling uncertainties in schedules and power transfers across regions, accommodating renewable energy resources, optimizing the transfer capability of transmission and distribution networks, meeting the demands for increased quality and reliable supply, managing and resolving unpredictable events and uncertainties in operations, and planning more aggressively (Momoh, 2012).

With the smart grid concept, small distributed energy sources can be integrated into an urbanized network enabling real-time optimization and facilitating interactions with other infrastructures using information and communication technologies for collecting information and actuating processes of automation. Power processing is done using both non-conventional and conventional energy generation for loads, and actuation is done by power electronic devices and systems. The realization of such types of interactive, resilient, and sustainable models is a challenge.

It is expected that in the future, the electric power grid will make a transition to a smart grid within the next few decades. However, there are some fundamental challenges to this transition (Amin and Wollenberg, 2005):

1. Lack of transmission capacity to meet increasing loads.
2. Difficulties of grid operation in a competitive market.
3. Power-system planning and operation in a competitive market.
4. Determination of the optimum energy mix and placement of sensing, communication and control hardware.
5. Coordination of centralized and decentralized controls.

Smart grids are not really different from the modern grids that are in operation nowadays. Of course, there exists a need for sharing communication, infrastructures, filling in product gaps, and leveraging existing technologies to a greater extent while moving ahead for a higher level of integration to realize the synergies across enterprise integration.

A smart grid is not something that can be installed and turned on the next day. Rather, it is an integrated solution of technologies for incremental benefits for capital expenditures, operation and maintenance expenditures, and customer and societal benefits.

A well-designed smart grid built on existing infrastructure can provide a greater level of integration at the enterprise level and has a long-term focus. It is not definitely a one-time solution but a change in how utilities look at a set of technologies that can enable both strategic and operational processes. It is the means to leverage benefits across applications and remove the barriers that are created by past company practices (More and McDonnell, 2007). Smart grid software solutions such as those of Esyasoft and startups are now becoming available for smart grid distribution of electricity.

## REFERENCES

Ahmed, M. A., Kang, Y. C., & Kim, Y. C. (2015). Communication network architectures for smart-house with renewable energy resources. *Energies*. 8: 8716–8735.

Alirezazadeh, A., Rashidinejad, M., Abdollahi, A., Afzali, P., & Bakhshai, A. (2020). A new flexible model for generation scheduling in a smart grid. *Energy*. 191. 1164.38.

Amin, S. M., & Wollenberg, B. F. (2005). Toward a Smart Grid: Power Delivery for the 21st Century. *IEEE Power and Energy Magazine*, 3, 34–41.

Ancillotti, E., Bruno, R., & Conti, M. (2013). The role of communication systems in smart grids: Architectures, technical solutions and research challenges. Computer Communications. 36: 1665–1697

Azadeh, A., Ghaderi, S. F., & Sohrabkhani, S. (2008). A simulation-based neural network algorithm for forecasting electrical energy consumption in Iran. *Energy Policy*. 36: 2637–2644.

Azadeh, A., & Tarverdian, S. (2007). Integration of genetic algorithm, computer simulation and design of experiments for forecasting electrical energy consumption. *Energy Policy*. 35: 5229–5241,

Awais, M., Javaid, N., Shaheen, N., Iqbal, Z. Q., Rehman, G., Muhammad, K., & Ahmed, I. (2015). *An efficient genetic algorithm based demand side management scheme for smart grid*. 18th International Conference on Network Based Information Systems (NBIS-2015), Taipei, Taiwan.

Baimel, D., Tapuchi, S., & Baimel, N. (2016). Smart grid communication technologies. *Journal of Power and Energy Engineering*. 4: 1–8.

Bala, B. K. (1998). *Energy and Environment: Modelling and Simulation*. Nova Science Publishers, USA.

Bala, B. K. (2006). Computer modelling of energy and of environment for Bangladesh, *International Agricultural Engineering Journal*. 15: 151–160.

Bala, B. K., Alam, M. S., & Debnath, N. (2014). Energy perspective of climate change: The case of Bangladesh. *Strategic Planning for Energy and the Environment, Development*, 33(3): 6–22.

Bala, B. K., & Satter, M. A. (1986a). *Modelling of rural energy systems*. Presented at Second National Symposium on Agricultural Research, BARC, Dhaka 12 February 1986.

Bala, B. K, & Satter, M. A. (1986b). *Modelling of rural energy systems for food production in developing countries.* Energia and Agricoltura 2 Conferenza Internationale. N Sirmione/Brescia (Italia). 3, p. 306, 1986.

Bala, B. K., & Siddique, S. A. (2009). Optimal design of a PV-diesel hybrid system for electrification of an isolated island – Sandwip in Bangladesh using genetic algorithm. *Energy for Sustainable Development.* 13: 137–142.

Babu, N. R. (2019). *Smart Grid Systems: Modeling and Control.* Apple Academic Press, Canada.

Batty, M., Axhausen, K. W., & Giannotti, F. (2012). Smart cities of the future. European *Physics Journal Special Topics.* 214(1):481–518

Bharati, C., Rekha, D., & Vijayakumar, V. (2017). Genetic algorithm based demand side management of smart grid. *Wireless Personal Communications.* 93: 481–503.

Caamaño-Martín, E., Laukamp, H., Jantsch, M., Erge, T., Thornycroft, J., de Moor, H., Cobben, S., Suna, D., & Gaiddon, B. (2008). Interaction between photovoltaic distributed generation and electricity networks. *Progress in Photovoltaics Research and Applications*: 16(7), 629–643.

Carrascoo, J. M., Franquelo, L. G., Bialasiewicz, J. T., & Galvan, E. (2006). Power-electronic systems for the grid integration of renewable energy sources: A survey. *IEEE Transactions on Industrial Electronics.* 53: 1002–1016.

Chen, T. (2010). Smart grids, smart cities need better networks. IEEE Networks. 24: 2–3.

Coelho, V. N., Coelho, I. M., Coelho, B. N., de Oliveira., G. C., Barbosa, A. C., Pereira, L., de Freitas, A., Santos, H. G., Ochi, L. S., & Guimarães, F. G. (2017). A communitarian microgrid storage planning system inside the scope of a smart city. *Applied Energy.* 201:371–381.

Creery, A., & Byres, E. J. (2005). *Industrial cybersecurity for power system and SCADA networks.* Paper No. PCIC-2005-34.

Deb, K., Pratap, A., Agarwal, S., & Meyarivan, T. (2002). A fast and elitist multiobjective genetic algorithm: NSGA-II. *IEEE Transactions on Evolutionary Computation.* 6: 182–197.

Ekanayake, J. B., Liyanage, K., Wu, J., Yokoyama, A., & Nick Jenkins, N. (2012). *Smart Grid: Technology and Applications.* John Wiley and Sons, UK.

EPRI. (2011). EPRI smart grid demonstration initiative two - update. Figure 2 on page 7. www.epri.com/

Erinmez, I. A., Bickers, D. O., Wood, G. F., & Hung, W. W. (1999). *NGC experience with frequency control in England and Wales: Provision of frequency response by generators.* IEEE PES Winter Meeting, 31 January–4 February 1999, New York, USA.

Erol-Kantarci, M., Kantarci, B., & Mouftah, H. T. (2011). Reliable overlay topology design for the smart microgrid network. *Network IEEE.* 25: 38–43.

Eskandari, H., & Geiger, C. D. (2020). Solving expensive multiobjective optimization problems: A fast Pareto genetic algorithm approach. *Journal of Heuristics.* 14(3): 203–241.

European Commission (2006) *European Smart Grids Technology Platform: Vision and Strategy for Europe's Electricity Networks of the Future.* http://ec.europa.eu/resea rch/energy/pdf/smartgrids_en.pdf (accessed on 4 August 2011).

Farhangi, H. (2010). The path of the smart grid. *IEEE Power and Energy Magazine,* 8(1), January/February 2010, 18–28.

Farmanbar, M., Parham, K., Arild, O., & Rong, C. (2019). A widespread review of smart grids towards smart cities. *Energies.* 12(4484): 1–18.

Gao, J., Liu, J., Rajan, B., Nori, R., Fu, B., Xiao, Y., Liang, W., & Chen, C. L. P. (2014). SCADA communication and security issues. *Security and Communication Networks.* 7: 175–194.

Gaushell, D. J., & Block, W. R. (2002). SCADA communication techniques and standards. *IEEE Computer Application in Power.* 6: 45–50.

Grubb, M. J. (2003). The integration of renewable electricity sources. *Energy Policy.* 19. 670–688.

Gupta, R., Jha, D. K., Yadav, V. K., & Kumar, S. (2013). A multi-agent framework for operation of a smart grid. *Energy and Power Engineering.* 5: 1330–1336.

Hernández-Callejo, L., Zorita , C. B., Aguiar, J. M., Carro, B., Sanchez-Esguevillas, A. J. S., Lloret, J., & Massana, J. (2014). A survey on electric power demand forecasting: Future trends in smart grids, microgrids and smart buildings. *IEEE Communication Surveys & Tutorials* 16(3):1460–1495, 3rd Quart

Holland, J. H. (1975). *Adaptation in Natural and Artificial Systems.* University of Michigan Press, MIT, USA.

Hollands, R. G. (2008). Will the real smart city please stand up? *City.* 12:303–312.

Huq, A. Z. M. (1975). *Energy modelling for agriculture units in Bangladesh.* Paper presented at the National Seminar on Integrated Rural Development, Dhaka, 1975.

Ibrahim, H., Linca, A., & Perron, J. (2008). Energy storage systems – characteristics and comparisons. *Renewable and Sustainable Energy Reviews.* 12: 1221–1250.

IEEE PES Power & Energy Society. (2010). *Smart distribution systems tutorial.* IEEE PES General Meeting, Minneapolis, MN, July 2110.

Jacob, D., & Nithiyananthan, K. (2009). Smart and micro grid model for renewable energy based power system. *International Journal for Engineering Modelling.* 22 (1-4): 89–94.

Jin, X. (2015). *Analysis of micro grid comprehensive benefits and evaluation of its economy.* In Proceedings of 10th International Conference on Advances in Power System Control, Operation & Management (APSCOM 2015), Hong Kong, China, pp 1–4.

Kuzlu, M., Pipattanasomporn , M., & Rahman, S. (2014). Communication network requirements for major smart grid applications on HAN, NAN and WAN. *Computer Networks.* 67: 74–88.

Liang, C., & Shahidehpour, M. (2014). DC micro grids: Economic operation and enhancement of resilience by hierarchical control. *IEEE Transactions on Smart Grid.* 5(5):2517–2526

Liu, N., Chen, Q., Liu, J., Lu. X., Lo, P., Lei. J., & Zhang, J. (2015). A heuristic operation strategy for commercial building microgrids containing EVs and PV system. *IEEE Transactions on Industrial Electronics.* 62(4):2560–2570

López, G., Moura, P., Moreno, J. I., & Camacho, J. M. (2014). Multi-faceted assessment of a wireless communications infrastructure for the green neighborhoods of the smart grid. *Energies.* 7: 3453–3483.

Lund, H. (2007). Renewable energy strategies for sustainable development. *Energy.* 32: 912–919.

Ma, R., Chen, H. H., Huang, Y. R., & Meng, W. (2013). Smart grid communication: Its challenges and opportunities. *IEEE Transactions on Smart Grid.* 4(1):36–46.

Melhern, F. Y. (2018). *Optimization Methods and Energy Management in "Smart Grids".* PhD Thesis, Universite Bourgogne, Franche-Comte.

Mohamed, M. A., & Eltamaly, A. M. (2018). *Modeling and Simulation of Smart Grid Integrated with Hybrid Renewable Energy Systems.* Springer.

Momoh, J. A. (2012). *Smart Grid: Fundamentals of Design and Analysis*. IEEE Press.

Mondal M. A. H., Denich M., & Vlek P. L. G. (2010). The future choice of technologies and co-benefits of $CO_2$ emission reduction in Bangladesh power sector. *Energy*. 35, 4902–4909.

More, D., & McDonnell, D. (2007). Smart grid vision meets distribution utility reality. *Electric, Light & Power*, March 2007, 1–6.

Morris, G. Y., Abbey, C., Wong, S., & Joós, G. (2012). *Evaluation of the costs and benefits of microgrids with consideration of services beyond energy supply*. In Proceedings of IEEE Power and Energy Society General Meeting, San Diego, USA, pp 1–9

Pai, P., & Hong, W. C. (2005). Forecasting regional electricity load based on recurrent support vector machines with genetic algorithms. *Electric Power Systems Research*. 74: 417–425.

Parikh, J. (1997). *Energy Models for 2000 and Beyond*. Tata McGraw Hill, New Delhi, India.24.

Ranganathan, P. (2013). *Distributed Linear Programming Models in a Smart Grid*. PhD dissertation. North Dakota State University. USA.

Ranganathan, P., & Nygard, K. E. (2017). *Distributed Linear Programming Models in a Smart Grid*. Springer.

Sayed, K., & Gabbar, H. A. (2016). Chapter 18 SCADA and smart energy grid control automation in *Smart Energy Grid Engineering*. Edited by Hossam A. Gabbar. Academic Press/Elsevier. www.researchgate.net/publication/312000929.

Soetedio, A., Lomi, A., & Nakhoda, Y. I. (2017). Incorporating SCADA software and high level programming language for implementing the optimization technique in smart grid. *International Journal of Innovative Computing, Information and Control*. 13(3): 711–720.

Thomas, M. S., & McDonald, J. D. (2015). *Power System SCADA and Smart Grids*. CRC Press/Taylor and Francis.

U.S. Department of Energy. (2009). *Smart Grid System Report*, July 2009, www.oe.energy. gov/sites/prod/files/oeprod/DocumentsandMedia/SGSRMain_090707_lowres.pdf (accessed on 4 August 2011)

Wang, Y., Huang, Y., Wang, Y., Li, F., Zhang, Y., & Tian, C. (2018). Operation optimization in a smart micro-grid in the presence of distributed generation and demand response. *Sustainability*. 10(847): 1–25.

Warsi, N. A., Siddiqui, A. S., Kirmani, S., & Sarwar, M. (2019). Impact assessment of microgrid in smart cities: Indian perspective. *Technology and Economics of Smart Grids and Sustainable Energy*. 1–16.

Wood, A. J., Wollenberg, B. F., & Sheblé, G. B. (2014). *Power Generation, Operation and Control* (3rd ed.). IEEE John Wiley and Sons, UK.

Xu. G. (2016). *Optimization models and algorithms for demand response in smart grid*. University of Louisville.

Yu, F. R., Zhang, P., Xiao, W., & Choudhury, P. (2011). Communication systems for grid integration of renewable energy resources. *IEEE Networks*. 25: 22–29.

# Index

For Product Safety Concerns and Information please contact our EU
representative GPSR@taylorandfrancis.com
Taylor & Francis Verlag GmbH, Kaufingerstraße 24, 80331 München, Germany